Novel Materials
Design and Properties

Novel Materials Design and Properties

B.K. Rao and

S.N. Behera (Editors)

Nova Science Publishers, Inc.
Commack, New York

Assistant Vice President/Art Director: Maria Ester Hawrys
Graphics: Frank Grucci
Editorial Production: Susan Boriotti
Office Manager: Annette Hellinger
Acquisitions Editor: Tatiana Shohov
Book Production: Ladmila Kwartirof, Christine Mathosian,
                Joanne Metal and Tammy Sauter
Circulation: Cathy DeGregory and Maryanne Schmidt
Library of Congress Cataloging–in–Publication Data
available upon request

ISBN 1-56072-559-1
© *1998 Nova Science Publishers, Inc.*
  6080 Jericho Turnpike, Suite 207
  Commack, New York 11725
  Tele. 516-499-3103 Fax 516-499-3146
  E-Mail: Novascience@earthlink.net

Printed in the United States of America

CONTENTS

Contents   vii

PREFACE

This book presents the invited lectures given at the International Symposium on Novel Materials held at Puri during March 3-7, 1997. This was the second meeting jointly organized by the Institute of Physics at Bhubaneswar, India and the Department of Physics of Virginia Commonwealth University at Richmond, USA. The meeting was cosponsored by the Indo-US Science and Technology Collaboration Program, Naval Research Laboratory, Washington, D.C., USA, S. N. Bose National Center for Basic Sciences, Calcutta, India, and the Department of Science and Technology, Government of India, New Delhi, India.

The search for novel materials with tailor made properties is a continuous and ongoing process, which will keep the condensed matter and materials scientists occupied as long as there is a technological need to link up with the new materials discovered. At present, this search is spearheaded by developments in the fields of atomic clusters and cluster-assembled materials, nanostructured materials, multilayers, porous materials, materials following biomolecular paths and, of course, materials with a promise for room temperature super conductivity and novel magnetic properties. Understanding these materials requires an understanding of their growth, morphologies, electrical and magnetic properties etc. which depend on first principles electronic structure calculations, simulations, and modeling. All these aspects of novel materials were intensively discussed during the five days of the symposium. The gathering constituted of about ninety participants from ten countries. While the experts lectured on their subjects of specialization, the others benefited from these lectures and from mutual discussions. The symposium was designed with lectures in the mornings and the evenings while the afternoon was left free for the interaction of the participants. The mutual dependence of theoreticians and experimentalists needed for designing novel materials with unusual properties and the presentation of new techniques in both finally resulted in many collaborations. The beautiful surroundings of the sea beach at Puri provided an excellent back drop for the intellectual discussions that took place during the symposium.

The symposium featured ten plenary sessions which presented forty one invited talks. The topics on the designing of novel materials included:

1.   Atomic clusters and cluster reactions

2.   Surfaces and multilayers

FOREWORD

Novel Materials - Key to Future Technologies

Recent years have seen unprecedented advances in science and technology, leading to better understanding of the behavior of matter, and designing new functional and structural materials to meet a number of industrial needs. New material characterization tools, such as, the Scanning-Tunneling Microscopes, Atom-Force Microscope, AtomResolved TEM, Angle Resolved Auger and Synchrotron radiation; new synthesis and processing techniques, such as, MBE, MOCVD, PLD, Spray Pyrolysis and E-beam/X-ray Lithography; and high performance computing capabilities poised to predict the behavior of a billion atom aggregates, has brought rapid growth in various classes of materials including superconductors, ceramics, biomaterials, carbons, polymers, magnet, intermetalhcs and smart materials.

The theme of this International Symposium is to discuss the current state of our theoretical understanding and to elucidate voids in our knowledge base which needs addressing in order to predict and demonstrate novel and improve engineering material properties.

The symposium is one of a series supported by the Naval Research Laboratory (NRL) and the Office of Naval Research (ONR) funded under the Indo-US Science and Technology Initiative. The genesis of the Indo-US Science and Technology Initiative goes back to the utilization of U.S. Congress appropriated funds (PL-480) to build collaborations between the two countries through various intellectual collaborations. These funds, which were initially directed by the Agency for International Development to support the establishment of engineering education, National Center for Educational Research, and other collaborative activities in biomedicine, soon expanded to embrace other scientific and cultural disciplines.

During this period the NRL and its parent institution, the ONR, have been active participants in the program, in the areas of Physical and Materials Science, as well as in Ocean Sciences. Because of institutions'strong commitments to fundamental research, combined with the high intellectual vitality of India's premier research institutions, such as the Institute of Physics at Bhubaneswar, the NRUONR program has grown to be one of the largest funded programs of the U.S. Department of State, the principal sponsor of this initiative.

Over the last decade, NRIJONR have initiated over 60 research projects which have brought cooperation between most of the leading institutions of India and the United States. These projects have produced to date: (1) over 500 archival research papers published in international technical journals, such as, Nature, Phys. Rev. Let., Science, J. of Am. Chem. Soc., Proceedings of the National Academy of Sciences, and Materials Science and Engineering, (2) in excess of 50 Ph.D. theses, (3) four technical books and five book-chapters, and (4) over 20 topical workshops and conferences with their published proceedings on emerging areas in science and engineering. In recognition of research accomplishments, scientists in India have received several awards including the Royal Medal of the Royal Society (London), the Homi Bhabha Medal, C.V. Raman Centenary Medal, Mahendra Lal Sirkar Award, MRS of India Medal, Birla Prize, Bhatnagar Award, and a number of Fellowships to Societies and Academies. The program has supported more than 100 Indian scientists to visit U.S. research institutions to give seminars and conduct their collaborative efforts.

I am quite hopeful that the participation of leading scientists from India, U.S., and a number of other countries at this symposium will bring forward important insights and lead to many joint research projects.

I wish to thank Professor Bijan Rao of Virginia Commonwealth University, Dr. S. Behera and his associates of the Institute of Physics, and Ms. L. Kinger of the Science Office, U.S. Embassy for their dedication and commitments to make this symposium a success.

Bhakta B. Rath
Naval Research Laboratory
Washington, DC

THE DESIGN OF NOVEL MATERIALS:
AN OLD DREAM, NEW REALIZATIONS

Morrel H. Cohen
Exxon Research and Engineering Company, Retired
Annandale, NJ 08801, U. S. A.

ABSTRACT

In the early 50's, when I started my career in Physics, I would hold out the idea of designing materials to order as a defense of solid-state theory against experimentalists skeptical of its relevance. Now, over four decades later, that dream has become reality. In this paper, I shall summarize briefly several of the key advances in electronic structure theory which have made material design possible. First, however, I shall set the stage by reminding the audience of the confusion that existed about the foundations of condensed matter theory in the early fifties. The advances which lifted that confusion included pseudopotential theory and, especially, the Fermi-liquid theory, both of which will be included in the summary. Then I shall present recent advances my group has made in understanding surface chemistry and catalysis on transition metal surfaces. These include both new interpretive tools for electronic structure calculations, and new methods for the simulation of surface concentration profiles of alloys. They raise the possibility of the design of novel catalysts from first principles.

1. INTRODUCTION

I received my Ph.D. in Physics from the University of California at Berkeley in the Fall of 1952 under Charles Kittel in what was then called solid-state theory. I then became a member of the faculty of the University of Chicago with a joint appointment in the Physics Department and the Institute for the Study of Metals (ISM), later named the James Franck Institute. The ISM was one of the two prototypic interdisciplinary materials research institutions after which what became the NSF MRL's were modeled, the other being the Metallurgy Department of the General Electric Research Laboratory under Herbert Hollomon.

The founding director of the ISM was Cyril Stanley Smith, a major figure in metallurgy in this century. Cyril had grave reservations about the roles of theoretical physics and theoretical physicists in metallurgy and more generally in materials science, although that term was not in use then. While Cyril had a fruitful interaction with Philip Morrison which led to Cyril's introduction of topology into metallurgy and had a collegial relationship as well with Clarence Zener, my immediate predecessor in the ISM, his experiences with other theorists were not so happy. Accordingly, soon after my arrival, he challenged me to convince him that theoretical physics and physicists had a contribution to make to materials science.

Without giving the matter serious thought, my immediate response was that in time, advances in theoretical physics in general and electronic structure theory in particular would make possible the design of novel materials with desired properties. Given the immense barriers that would had to have been overcome, such material design was then no more than a dream. Now, 45 years later, that dream has become reality in some areas of materials science and is becoming so in others. Accordingly, it seemed fitting to me that in this opening lecture of the present International Conference on Novel Materials, I should review briefly the key developments that led to the translation of that dream into reality in some areas of materials science, section II. In that section I also cite a few existing examples of the design of novel materials. In section III, however, I shift from the past to the future and outline a strategy which could possibly lead to the design of novel catalysts. The system, the dissociation of H_2 on Pd(100), and the computational methods chosen by my collaborators and myself in our first attempt to implement that strategy are described in section IV. As discussed in section III, implementation of the strategy requires a deeper understanding of the physical concepts required to characterize the physical processes underlying the surface chemistry occurring during catalysis. Accordingly, in section V the concepts emerging from our study of H_2 on Pd(100) are discussed. We conclude in section VI with a few brief remarks about the ultimate possibility of the design of novel catalysts and of other types of novel materials.

2. THE KEY DEVELOPMENTS

There were two great puzzles confounding electronic structure theory in the 1950's. First, the independent-particle picture of the motion of valence and conduction electrons in crystals appeared to work at least qualitatively in a broad range of materials, suggesting that electron-electron correlations might not be important. This was puzzling because the electron-electron interaction was strong by any relevant measure and, even more so, because the Hartree-Fock approximation, the best one-electron approximation, yielded qualitatively incorrect results for the low-lying excitation spectrum of a metal [1]. Second, over the years, a substantial body of evidence had built up that in a broad range of the simpler elements and compounds, the energy bands were recognizable distortions of the free-electron energy bands folded back into the first Brillouin zone, suggesting that the electron-atom interaction was weak. This was puzzling because, once again, the interaction was strong by any relevant measure.

It became evident to me in 1952 that the resolution of the first puzzle lay in the notion that for normal metals, semiconductors, and insulators, the Coulomb interaction between electrons was not so strong as to destroy the convergence of adiabatic perturbation theory so that one could construct a one-to-one correspondence between the low-lying states of the independent-particle systems and of the interacting system. This insight remained unpublished, but, in 1956, Landau, following similar reasoning, introduced his Fermi-liquid theory and the quasi-particle concept, thereby providing the conceptual structure within which the first puzzle was fully resolved [2]. Concomitant and later development of many-body theory subsequently provided the basis for quantitative estimation of the Fermi-liquid parameters of the simplest Fermi liquids, conduction electrons in simple metals and He^3.

The resolution of the second puzzle emerged from the introduction of the Phillips-Kleinman pseudopotential in 1959 [3] together with the cancellation theorem of Heine and myself which explained how and when the pseudopotential could be weak [4].

The pseudopotential provided the enormous computational advantage of being compatible with the use of a plane-wave basis set for the calculation of the electronic structure of simple metals and semiconductors, and rapid progress in the study of the band structures of semiconductors and the Fermi surfaces of metals soon followed its introduction. In the work of Marvin Cohen and his colleagues, it has led to greatly improved understanding of a very broad range of materials and phenomena.

At that early stage, pseudopotential methods were unsuitable for application to transition metals, rare-earth metals, and other electronically complex materials. Two methods introduced earlier, the KKR method [5] and Slater's augmented plane wave method [6], were of general applicability but were computationally far more complex because they employed an energy-dependent basis set. The breakthrough which increased the utility of those and related methods was O. K. Anderson's introduction of the linearization of such energy dependences, as elaborated in his linear muffin-tin-orbital method [7].

Thus, by 1973, the foundations had been established for all of the one-electron computational methods currently in common use. What was lacking before 1964 was the basis for forming a clear understanding of what was being calculated. It was understood that the bare exchange of the Hartree-Fock equations should not be used, that the exchange interaction was in some way screened by electron correlation as it entered the determination of the quasiparticle spectrum. However, only heuristic local approximations such as Slater's $X\alpha$ approximation [6] were available, and just what the resulting band structure approximated was not clear. In particular, while many features of the optical spectra of semiconductors were well represented in the computed band structure, the optical band gaps were way off.

The missing elements were supplied by two landmark papers, one by Hohenberg and Kohn in 1964 [8] and one by Kohn and Sham in 1965 [9]. In the former, the general framework of density-functional theory was established. In the latter, it was shown that given the density functional entering the total energy, the ground-state electron density and thereby the total energy of an interacting N-electron system could be calculated from the solution to a particular independent-electron problem. The potential in the independent-electron problem, the Kohn-Sham potential, was expressed in terms of functional derivatives of the various contributions to the density functional for the total energy.

It immediately became clear that the electronic structure calculations carried out via the methods listed above yielded approximations to the exact Kohn-Sham eigenvalues and wavefunctions. Moreover, since the Kohn-Sham eigenvalues were distinct from the quasiparticle energies involved in the observable excitation spectrum of a material, it resolved in principle the optical band-gap problem for semiconductors. Quantitative resolution of that problem then followed in 1987 through direct computation via many body theory of the corrections to be added to the Kohn-Sham eigenvalues to obtain the quasiparticle energies [10].

Post 1965, there have been three parallel lines of continuing significant developments. First, there has been a steadily growing understanding of the various density functionals and a concomitant improvement in the accuracy of approximations to the true Kohn-Sham potential. At present, the so-called generalized gradient approximation [11] is the most accurate one widely used. Second, there has been a doubling of computer power per unit cost every eighteen months, Moore's law. Third, computer codes which take advantage of the steadily increasing speed and memory capacity of computers have continually grown in computational power and efficiency. This has led over the last eighteen years to a 50-fold increase in the computationally feasible size of the unit cell of the periodic structure to be analyzed, making possible, inter alia, the studies of the surface chemistry of transition metals I shall describe briefly in sections IV and V. More recently, powerful visualization tools have greatly aided the organization, presentation and especially the interpretation of the vast masses of data generated by the computations. Finally, there is emerging a deeper understanding of the minimum level of complexity required of the simple models used to interpret the results and, ultimately, to make predictions and thus to design novel materials.

All this is very well, but have novel materials yet been designed with the aid of electronic structure computations? The answer to this question is definitely yes. Marvin Cohen has used a simple model of the cohesive energy of a semiconductor well tested by pseudopotential computations to predict that carbon nitride, a new material, would be structurally stable and harder than diamond, which was subsequently observed [12]. Embedded atom methods, which emerge from much cruder versions of electronic structure theory, have been used in a wide variety of materials science contexts, leading in some cases to novel structures or even novel materials. Finally, I know of one instance in which band structure calculations play an essential and continuing role in the quality control of semiconductor laser manufacturing. While this last example does not concern the design of novel materials, it certainly provides a positive answer to Cryil Smith's implicit question of "What good is it?".

In the next section, section III, I shall lay out a strategy for achieving the goal of designing novel heterogeneous catalysts. In sections IV and V, I shall briefly describe work on the dissociation of H_2 by Steffen Wilke, Vincent Natoli, and myself [13] which constitutes our first steps towards implementing that strategy. The emphasis in that work is on teasing apart the elementary processes underlying a simple but representative catalytic process to develop the concepts to be embedded into models sufficiently simple computationally to be useful for prediction and design, an essential element of the strategy of section III.

3. TOWARDS THE DESIGN OF NOVEL CATALYSTS

A catalyst accelerates or makes possible reactions between reagents which otherwise would be impracticably slow or absent. In homogeneous catalysis, catalyst molecules are dispersed within a fluid phase. In heterogeneous catalysis, the catalyzed reactions take place on the external or internal surfaces of solid catalysts. It is the design of novel heterogeneous catalysts with which we are concerned here and, in particular, with the design of catalysts based on transition metals and their alloys. Such materials owe their effectiveness as catalysts, that is high yields, selectivity, and activity, to the presence of high densities of filled d-states below the Fermi level and empty ones above. The d-states, the central role of surfaces, and the presence of atoms and molecules reacting on the surfaces make detailed quantitative study of the resulting surface chemistry computationally complex. We are at present far from ready to make a direct attempt to design a novel material to catalyze a particular derived reaction. What is needed at this point is a strategy for getting from where we are now to that desired goal.

At present, there are several groups around the world with the capability of accurately computing the interaction of simple atoms and molecules with each other and with pure transition metal or ordered transition-metal compound surfaces. Steffen Wilke, Vincent Natoli, and I at Exxon together with our outside collaborators form one of those groups. What follows is a brief description of the multi-step strategy we have evolved for building that present capability into an ability to design novel catalysts.

1. Select a set of simple reactions which contains all of the elementary catalytic processes of scientific and/or technological interest.

2. Carry out first principles calculations of the potential energy surfaces (PES) of these simple reactions.

3. Use the PES to calculate reaction rates, yields, selectivity, etc. for the simple examples.

4. Develop interpretative tools and use them to identify, characterize and conceptualize the underlying physical processes active along the reaction pathway.

5. Construct models complex enough to embody the most important processes identified in step 4. but simple enough to be evaluated on a work station instead of the massively parallel super computers needed for the first-principle computations.

6. Evaluate the model parameters for the simple examples by fitting to the first-principles calculations.

7. Validate the models by achieving an acceptable goodness of fit.

8. Determine how the model parameters should vary with surface atomic and molecular species as well as with the transition metal.

9. At this point predictions can be made for pure transition metal catalysts, but only transition metal alloys can give the flexibility needed for catalyst design. The next stage of the strategy is to repeat steps 1.-8. for alloys, for which the computational complexity is far greater but not presently unfeasible.

4. THE MODEL SYSTEM: COMPUTATIONAL METHODS

The elementary steps in heterogeneous catalysis are very simple in outline and relatively few in number. For example, intramolecular bonds within a reactant are broken and replaced by bonds between the resulting fragments and the metal surface in a single continuous process. Chemisorbed fragments from different reactants can then combine and desorb, yielding the product. It all comes down to bond breaking and bond making. Turning now to the first element of the strategy outlined in the previous section, the dissociative adsorption of H_2 on a surface plane of a late transition metal provides the simplest possible example of intramolecular bond breaking with simultaneous formation of bonds between the fragments, H atoms, and the metal surface.

My colleagues, Steffen Wilke and Vincent Natoli, and I have carried out a detailed study of the changes of the electronic structure of the interacting molecule-metal system with the goal of carrying out step 4. of the strategy. The particular metal surface we have chosen to study is the Pd(100) surface for several different periodic coverages of H_2. We chose the dissociation of H_2 on Pd(100) [13] because steps 2. and 3. have already been carried out for that system. Steffen Wilke computed the six-dimensional potential energy surfaces of H_2 on Pd(100), fixing the 6 H_2 coordinates and completely relaxing all Pd positions within the slab geometry used in the computations [14]. Axel Gross and Wilke [15] then used these PES as input into a quantum dynamic simulation of the dissociative absorption probability of a normally incident beam of H_2 molecules of varying incident energy. Considering the compromises forced on the simulation by the computational complexity of the PES calculation, the agreement with experiment is excellent.

Those computations show that there are pathways down to the surface along which no barriers to dissociation exist. One such set of pathways lies in a plane with one coordinate, z, the distance of the center of the molecule from the (100) surface plane of atoms, the other coordinate, d, the proton-proton separation, with the molecular axis parallel to the surface, and with the molecular center above the four-fold hollow site of the surface. Within the d-z plane, there is a reaction pathway along which the total energy is a 5-dimensional minimum. We have calculated the electronic structure in detail at a sequence of points along this reaction pathway from z=3.2 Å and d=0.76 Å in the entrance channel, where the molecule approaches the surface but still interacts weakly with it so that the internuclear separation is unperturbed, out to z=1.06 Å and d=2.84 Å in the exit channel, where the H_2 molecule is dissociated and the H atoms chemisorbed into adjacent 4-fold hollow equilibrium sites.

We have of course obtained total energies and forces for these prescribed geometries, but to complete step 4. it is necessary to develop and apply a set of interpretive tools which assist in revealing the underlying physics. The interpretive or, somewhat more precisely, the diagnostic tools we have used include the electron density $\rho(\vec{r})$; the induced density change $\Delta\rho(\vec{r})$ defined as the difference between the actual electron density of the interacting metal and molecule for a given nuclear configuration and the superposed unperturbed electron densities of the isolated metal and molecule having the same nuclear configuration; the density of Kohn-Sham states g(E); the local density of Kohn-Sham states $g(\vec{r},E)$;

$$g(E) = \int d\vec{r}\ g(\vec{r},E); \tag{1}$$

and the isoelectronic softness kernel $S(\bar{r},\bar{r}')$ defined [16] as

$$S(\bar{r},\bar{r}')=\frac{\delta\rho(\bar{r})}{\delta v(\bar{r}')}\bigg|_N \quad, \qquad (2)$$

that is, the functional derivative of $\rho(\bar{r})$ with respect to the external potential $v(\bar{r}')$ at constant total electron number N.

Of all of these, $\Delta\rho(\bar{r})$ turned out to be the most revealing, particularly when augmented by the changes $\Delta E(\bar{r})$ in the local band energy density,

$$\Delta E(\bar{r}) = \int\limits^{E_F} dE\ \Delta g(\bar{r},E)\ E \quad, \qquad (3)$$

although all of the above tools were used in arriving at the conclusions outlined below.

The system actually studied was an infinite set of identical slabs of Pd metal each consisting of n layers of (100) atomic planes and separated from its neighbors by a vacuum layer. A periodic layer of H_2 molecules was placed symmetrically adjacent to each surface of each slab. The number of Pd atomic layers, the width of the vacuum layer, and the coverage of the hydrogen molecules were varied to find the most appropriate compromise between computational accuracy and cost as well as between cost and weakness of interaction between adjacent H_2 molecules. n=5 turned out to be satisfactory.

The computations were carried out with the first parallelized version of the WIEN 95 electronic structure code [13], an embodiment of the full-potential, linearized, augmented-plane-wave-method. The Kohn-Sham potential was evaluated via the generalized-gradient approximation [11].

Details of the computations will be presented elsewhere. Here I shall summarize in the next section the concepts which emerged from our studies of the isosurfaces of $\Delta\rho(\bar{r})$ in three dimensions.

5. PHYSICAL CONCEPTS IN THE SURFACE CHEMISTRY OF TRANSITION METALS

5.1. Orthogonality Repulsion

Transition metal elements and their alloys are understood to be more active catalytically than the simple or noble metals because their partially filled d-band provides high densities of filled and empty states near the Fermi energy E_F which are available for the formation of strong chemical bonds with separating molecular fragments which can compete with the intramolecular bonding of those fragments. In the case studied here, the dissociation of H_2 on Pd(100), the fragments are the individual H atoms.

The formation of such bonds is not a simple matter, however. The metallic d-states are tightly bound and do not project far from the surface. The metallic s-p states project further and dominate the tail of the metallic electron density first encountered by a molecule approaching the surface. Relatively far from the surface, i.e. at large z, the doubly occupied σ_g bonding orbital of H_2 is essentially unperturbed. Its energy lies below the bottom of the s-p band of H_2. The requirement that the states of the s-p tail be orthogonal to the σ_g orbital generates a repulsive contribution to the interaction between the molecule and the metal, the orthogonality repulsion or, equivalently, the Pauli repulsion. This orthogonality repulsion could, in principle, prevent the molecule from approaching closely enough to bond with the d-states.

This clearly does not happen, and Harris and Andersson [17] proposed in 1985 that electrons flow from the s-p tail away from the region occupied by the molecule, producing an orthogonality hole. This reduces the orthogonality repulsion and allows close approach to the surface. Screening of the orthogonality hole by the charge transferred into the d-states facilitates this.

Pseudopotential theory [3,4] tells us that the requirement of orthogonality of the metallic states to the molecular orbitals can be replaced by a repulsive pseudopotential inserted into the Kohn-Sham equation. Thus the Harris-Andersson proposal can be tested by comparing the electron density changes induced by the interaction of a H_2 molecule with the metallic surface with the changes induced by the perturbation of the metallic surface by a repulsive potential which mimics the pseudopotential.

In the left panel of Fig. 1, we show the isosurfaces of electron-density change, $\Delta\rho(\vec{r})$, (decrease - light shading and increase - dark shading) of magnitude 5×10^{-3} A^{-3} induced by interaction of a 1/4 monolayer of H_2 molecules with the Pd(100) surface. The centers of the molecules are above the four-fold hollow positions, the molecular axes are parallel to the cube axis, $z = 2.1$ Å, and the internuclear separation $d = 0.76$ A is that of the free molecule. The positions of all metallic atoms have been fully relaxed. The nuclear configuration corresponds to a point along the reaction pathway for molecular dissociation. The inset indicates the location of the protons by showing two spheres centered around the proton positions within the surface of induced electron deficiency surrounding the H_2 molecule. One sees clear evidence of an orthogonality hole. Moreover one sees clear evidence of electron transfer into d_{z^2} orbitals with axes rotated towards the orthogonality hole, screening it. Thus the three elements of the Harris-Andersson picture are present: 1. an orthogonality hole; 2. electron transfer into d-states; and 3. screening of the orthogonality hole by the transferred electrons.

Even given this confirmation of the Harris-Andersson picture, one can ask how closely these contours of constant density change resemble those induced by a repulsive potential perturbation. This question can be explored by computing the isoelectron softness kernel [16], Eq. (2). - $S(\vec{r}, \vec{r}')$ is the linear response of the electron density at $\vec{r}'$ to a delta-function potential perturbation at $\vec{r}'$. We calculated instead the response to a 1/4 monolayer of narrow Gaussian perturbations centered about $z' = 2.1$ Å above the 4-fold hollow sites, as indicated in the inset to the right-hand panel of Fig. 1. Comparing the contours in the right and left panels, we see remarkable similarities, with the principle differences occurring in the shape of what we have identified as the orthogonality hole.

To explore these differences, we have reduced the magnitude of $|\Delta\rho(\vec{r})|$ by a factor of 2, with the results shown in the left-hand panel of Fig. 2. We have also moved further out the exit channel along the reaction pathway to $z = 3.2$ Å and $d = 0.76$ Å. the resulting isosurfaces for $|\Delta\rho(\vec{r})| = 0.001$ Å^{-3} and 0.0005 Å^{-3} are shown in the left and right panels of

$$\Delta\rho(\mathbf{r}) \qquad\qquad -s(\mathbf{r},\mathbf{r'})$$

Z=2.1 A, d=0.76 A   Z'=2.1 A

$|\Delta\rho(\mathbf{r})| = 0.0050\ \text{Å}^{-3}$

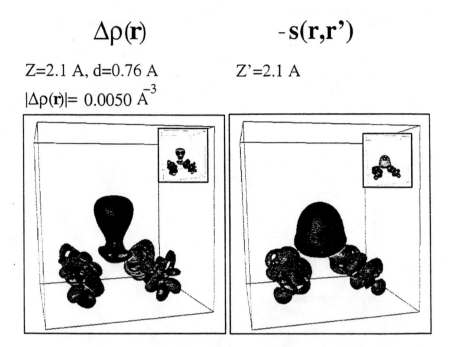

Figure 1. Left panel: isosurfaces of electron density change $\Delta\rho(r)$ induced by the interaction of a periodic array of H_2 molecules (1/4 monolayer) above the Pd (100) surface. The height of the molecular center z above the surface is 2.1 Å; the internuclear separation d is 0.76 Å; the molecular centers are above alternate 4-fold hollow sites; and the molecular axis is parallel to the cube edge. This configuration is a point on a reaction pathway along which no barrier to dissociation occurs. Isosurfaces of electron density increase are shaded darkly, those of density decrease lightly. The magnitude $|\Delta\rho(r)|$ is 0.0050 Å^{-3} in both cases. The isosurface surrounding the H_2 molecule is made transparent in the insert to show the location of the H atoms in the molecule at the centers of the two small dark spheres. Right panel: Similar isosurfaces for the negative of the isoelectronic softness kernel $S(r,r')$ with r' located at the same position as the molecular center in the left panel, so that $z'=2.1$ Å.

Fig. 3, respectively. We see clearly in Fig. 2 substantial changes in the shape of the orthogonality hole away from that expected from a repulsive potential perturbation as well as remarkable changes in the shape of the isosurfaces of electron density increase. In the left panel of Fig. 3, the shape of the orthogonality hole, while similar to that in Fig. 2, remains quite different from that expected from a potential perturbation, and there are small isolated

Morrel H. Cohen

dots of electron density increase outside the d-shells. These, as the resolution is increased in the right panel, grow into sizable regions, but the most unexpected result is the oscillation in sign of $\Delta\rho(\vec{r})$ with z below the center of the molecule.

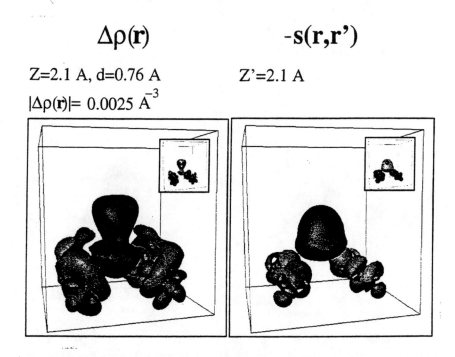

$$\Delta\rho(\mathbf{r}) \qquad\qquad -s(\mathbf{r},\mathbf{r}')$$

$$Z=2.1\ A,\ d=0.76\ A \qquad Z'=2.1\ A$$

$$|\Delta\rho(\mathbf{r})|=0.0025\ \overset{-3}{A}$$

Figure 2. The same as Figure 1 except that in the left panel $|\Delta\rho(r)|$ has been reduced by a factor of 2 to 0.0025 $\mathring{A}^{-3}$.

We can conclude from Figs. 1-3 that indeed the Harris-Andersson mechanism is present, but much more is going on, even at large z.

5.2. σ_g, σ_u^* Hybridization

The earliest stages of the formation of chemical bonds between the metallic surface and the individual hydrogen atoms is hybridization of both the occupied bonding σ_g orbital and the unoccupied antibonding σ_u^* orbital with the states of the metal. In the exit channel and when the dissociation is complete, the d-states play the dominant role. However, already in the entrance channel, hybridization of the metal s-p states in the tail of the electron density with both the σ_g and σ_u^* orbitals occurs.

$$\Delta\rho(\mathbf{r})$$

Z=3.2 A, d=0.76 A

$|\Delta\rho(\mathbf{r})|=0.0010\,\mathring{A}^{-3}$ \qquad\qquad $|\Delta\rho(\mathbf{r})|=0.0005\,\mathring{A}^{-3}$

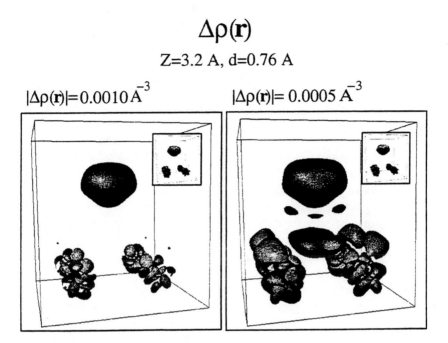

Figure 3. The same as the left panel of Figure 1 but with z=3.2 Å, d=0.76 Å and $|\Delta\rho(r)|$ = 0.0010 Å^{-3} in the left panel and 0.0005 Å^{-3} in the right.

The σ_g orbital lies below the bottom of the band, and hybridization raises states below E_F to energies above E_F, emptying them. Thus, an electron deficiency with a maximum below the center of the molecule indicates hybridization with σ_g. On the other hand, σ_u^* is above E_F so that hybridization pushes empty states above E_F down below E_F and fills them. Thus hybridization with σ_u^* tends to cut-off the lower outer regions, i.e. the lower corners, of the orthogonalization hole, while hybridization with σ_g tends to extend the lower middle of the orthogonalization hole. σ_u^* hybridization also raises the possibility of small regions of electron accumulation outside the lower outer of regions of the orthogonality hole.

All of these effects are clearly manifested in Figs. 1-3, in which the molecule is still in the entrance channel of the reaction pathway, that is, hybridization of the s-p tail with both σ_g and σ_u^* occurs even as far out as z = 3.2 Å, Fig. 3. However, more is going on than simple hybridization, as is clearly shown in Fig. 2, which we discuss in more detail in the next subsection.

5.3. Bridge Bonds

In the left panel of Fig. 1, we see a ring of electron density excess around each Pd atom perpendicular to the axis of each d_{z^2} orbital. That ring shows that electrons have been transferred into the Pd. $d_{x^2-y^2}$ and d_{xz} orbitals which themselves are already hybridized with s-p orbitals in the unperturbed metal. This could be interpreted as a part of the Harris-Andersson mechanism. However, when the sensitivity with which we probe electron-density changes is increased, as in the left panel of Fig. 2, we see a significant change in the geometry of the isosurface. In addition to the expected increase in the cross section of the ring, there emerge lobes or pillows of excess electron density between the rings and the hydrogen atoms.

Our interpretation of this phenomenon follows. The hybridization between the σ_u^* and the s-p states introduces an electron excess into the s-p tail, which is absorbed by the orthogonalization hole, leading to the change in shape of the orthogonalization hole described in the previous subsection. However, in addition, outside the region of the orthogonalization hole it generates an electron excess visible as the pillow pointed to above. Moreover, in the region where the Pd d-states contribute significantly to the electron density, they are already hybridized with the s-p states. Thus, the σ_u^* states are able to couple to the d-states even at values of z at which the direct overlap between the molecular orbitals and the d-states is negligible. The s-p states provide a *bridge* between the d-states and the molecular orbitals, the evidence for which is the pillow and the neck joining it to the ring. While change transfer may contribute to the ring, we expect that the dominant contribution comes from the *bridge bond* formation. We found that bridge bond formation emerges clearly in an analysis of simple models of the coupling of the molecular orbitals to the metal [13], which was based on the surface Green's function tight-binding LMTO method [18].

5.4. The Electrons Lead the Protons

Nakatsuji [19] has shown that the centroid of the change of electron density tends to lag the change in nuclear coordinate in movement away from a stable configuration and to lead it in movement away from an unstable one in an isolated molecule. We are examining changes in electron density along a reaction pathway leading from the free molecule to dissociated, chemisorbed H atoms. There is no barrier to dissociation along this pathway, and thus the movement is away from an unstable configuration. However, because we are dealing with a periodic extended system, there is no single centroid of the electron density change which is relevant. Instead, we focus on the pillows of excess electron density which, lying between the protons and the Pd atoms, are most relevant to the question of whether the electrons lead the protons, or vice versa.

Consider the sequence of six configurations displayed in Figs. 4-6. Starting at $z=1.59$ Å and $d=0.79$ Å, the configuration progresses along the reaction pathway until the molecule is completely dissociated at $z=1.06$ Å and $d=2.84$ Å, with the H atoms bonded to 2 Pd atoms in the bridge positions. The orthogonality hole shrinks and disappears, but the pillows grow and become the main concentration of electron-density excess, ultimately evolving into distorted and polarized H 1s orbitals.

$$\Delta\rho(\mathbf{r})$$

Z=1.59 A, d=0.79 A   Z=1.3 A, d=0.85 A

$|\Delta\rho(\mathbf{r})|= 0.01\,\text{Å}^{-3}$ $|\Delta\rho(\mathbf{r})|= 0.02\,\text{Å}^{-3}$

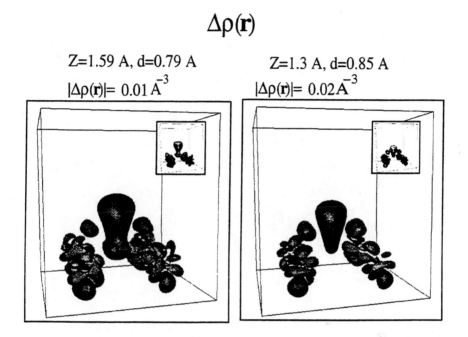

Figure 4. The same as the left hand panel of Figure 1 but with z=1.59 Å, d=0.79 Å and $|\Delta\rho(r)| = 0.01\,\text{Å}^{-3}$ for the left panel and z=1.3 Å, d=0.85 Å and $|\Delta\rho(r)| = 0.02\,\text{Å}^{-3}$ for the right.

We conclude that Nakatsuji's rule remains correct when applied to the bridge bonds. Within the bridge bonds the main concentration of electron-density excess leads the protons downward and outward into strong bonds with the d-states.

5.5. d-State Screening-Bonding Competition: Axis Rotation

The isosurfaces shown in Figs. 1-6 are very rich in important detail. It is beyond the scope of this paper to explore all significant features and show how they emerge from simple, quantitative models. However, certain of the most prominent phenomena can be appreciated just through inspection of the figures, as was already done in the above subsections, and we list several more of these here.

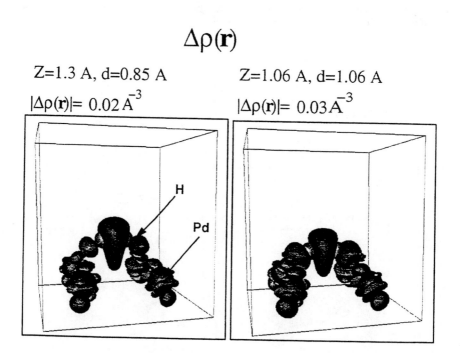

Figure 5. The same as the left panel of Figure 1 but with $z=1.3$ Å, $d=0.85$ Å, and $|\Delta\rho(r)|$ = 0.02 Å^{-3} for the left panel and $z=1.06$ Å, $d=1.06$ Å, and $|\Delta\rho(r)|$ = 0.03 Å^{-3} for the right.

Note that in Fig. 3, at $z=3.2$ Å, the d_{z^2} orbitals are rotated away form their unperturbed orientation parallel to the surface normal towards the H_2 molecule just as for $s(\vec{r}, \vec{r}')$ at $z'=2.1$ Å in Fig. 1. As pointed out in subsection 5.1, the electron excess in the screens the orthogonality hole. On the other hand, in Fig. 6 the d_{z^2} orbitals point towards the H atoms in the bridge positions, but there they are associated with an electron deficiency arising from hybridization with the H 1s orbitals now doubly occupied. This transition from excess to deficiency in the d_{z^2} orbital indicates that there is a competition between screening and hybridization in the metal's response to the presence of the molecules. More explicitly, between the two configurations at $z=2.1$, $d=0.76$ Å and $z=1.59$ Å, $d=0.79$ Å where the protons start to separate, the continually rotating d_{z^2} switches from screening to hybridization. This switching process is not simple. At $z=1.59$ Å, $d=0.76$ Å the d orbitals are distorted and cannot be simply labeled as d_{z^2}, $d_{x^2-y^2}$, etc., but screening is clear in one and hybridization clear in the other prominent orbital.

$$\Delta\rho(\mathbf{r})$$

Z=0.79 A, d=1.85 A   Z=1.06 A, d=2.84 A

$|\Delta\rho(\mathbf{r})| = 0.05\,\text{\AA}^{-3}$ $|\Delta\rho(\mathbf{r})| = 0.05\,\text{\AA}^{-3}$

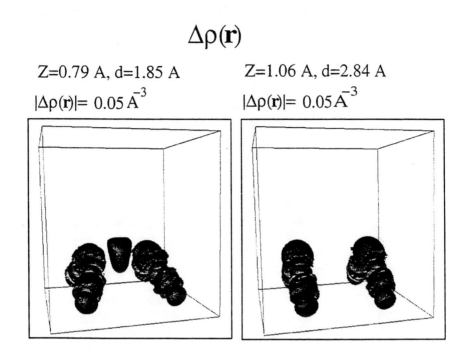

Figure 6. The same as the left panel Figure 1 but with z=0.79 Å, d=1.85 Å, and $|\Delta\rho(r)|$ =0.05 Å^{-3} for the left panel and with z=1.06 Å, d=2.84 Å, and $|\Delta\rho(r)|$ = 0.05 Å^{-3} for the right.

Note also that in Figs. 4-6 the d_{z^2} orbitals point towards the bridge-bond centers (the pillows), not the H-atoms. This is strong evidence for the bridge-bond concept.

5.6. Summary: H$_2$ Dissociation

Bond breaking on metallic surfaces has been extensively discussed in the literature. However, it is only in the last several years that the computational power of the available hardware and software has advanced to the point where first principle calculations of the accuracy and detail of those presented here have become possible. Accordingly, we confine our references to the extensive literature only to three contributions which we consider most relevant to the work presented here. These contributions contain three proposals for the key steps in H$_2$ dissociation on transition metal surfaces. The first, the formation of an orthogonalization hole by transfer of s-p electrons into empty d-states, by Harris and

Andersson [17] we have already discussed. Hammer and Scheffler [19] have emphasized the importance of d-σ_g^* hybridization, and Hammer and Nørskov [20] have emphasized the importance of d-σ_g^* hybridization. We find that all are present and important.

On the other hand, the concept of the bridge bond is new. While the idea that the molecular orbitals are renormalized by coupling to the s-p states and then hybridized with the d states is present in the literature and captures some aspects of the bridge-bond formation, it is incomplete. The bridge-bond plays a central role, indeed an essential role, in the hybridization of the molecular and metallic states. Direct hybridization of the d and the H 1s orbitals does not occur into well into the exit channel, where the pillow essentially evolves into the 1s orbital.

6. CONCLUDING REMARKS

The advances in our understanding of electronic structure theory and their applications to materials outlined in section II are rich with the possibility of the design of materials directly from first principles or from simple, efficient models based ultimately on first principles. Opportunities exist in many areas of condensed matter physics, materials sciences, device technology, etc. The dream of the beginning of my career has truly become reality towards the end of my career. But, there still remain challenges, problems of great computational and physical complexity, one of which I have discussed here, that of understanding catalysis on transition metal surfaces well enough to design novel catalysts. In section III I outlined a strategy through which that goal might be achieved. In sections IV and VI, I reported briefly on progress we have made in the first steps toward implementing that strategy. We conclude from what we have done thus far that the ultimate goal is feasible but that the effort required is considerable.

References

[1]   J. Bardeen, Phys. Rev. **50**, 1098 (1936).

[2]   L. Landau, J. Expt. Theor. Phys. (U.S.S.R.) **30**, 1058 (1956); **32**, 59 (1957); **35**, 97 (1958).

[3]   J. C. Phillips and L. Kleinman, Phys. Rev. **116**, 287,880 (1959); E. Antoncik, J. Phys. Chem. Solids **10**, 314 (1959).

[4]   M. H. Cohen and V. Heine, Phys. Rev. **122**, 1821 (1961).

[5]   J. Korringa, Physica **13**, 392 (1947); W. Kohn and J. Rostoker, Phys. Rev. **94**, 1111 (1954).

[6]   J. C. Slater, *Quantum Theory of Atomic Structure* (McGraw-Hill, N. Y., 1960); *Self-Consistent Field for Molecules and Solids* (McGraw-Hill, N. Y., 1974).

[7]   O. K. Anderson, Solid State Commun. **13**, 133 (1973).

[8]   P. Hohenberg and W. Kohn, Phys. Rev. **136**, B864 (1964).

[9]   W. Kohn and L. J. Sham, Phys. Rev. **140**, A1133 (1965).

[10] M. S. Hybertson and S. G. Louie, Comments Cond. Mat. Phys. **13**, 223 (1987).

[11] J. P. Perdew, *et al.*, Phys. Rev. B **46**, 6671 (1992).

[12] M. L. Cohen, Phys. Rev. B **32**, 7988 (1985); A. Y. Liu and M. L. Cohen, Science **245**, 841 (1989); K. M. Yu, *et al.*, Phys. Rev. B**49**, 5034 (1994).

[13] S. Wilke, V. Natoli and M. H. Cohen (to be published).

[14] S. Wilke and M. Scheffler, Phys. Rev. B **53**, 4926 (1996); Phys. Rev. Lett. **76**, 3380 (1996); Surface Sci. **329**, L605 (1995); S. Wilke, D. Hennig, and R. Löber, Phys. Rev. B **50**, 2548 (1994); S. Wilke, M. H. Cohen, and M. Scheffler, Phys. Rev. Lett. **77**, 1560 (1996); S. Wilke and M. H. Cohen, Surf. Sci. **380**, L446 (1997); S. Wilke, App. Phys. A. **63**, 583 (1996).

[15] A. Gross, S. Wilke and M. Scheffler, Phys. Rev. Lett. **75**, 2718 (1995).

[16] M. H. Cohen in *Density Functional Theory IV-Theory of Chemical Reactivity*, Topics in Current Chemistry, Vol. 183, Ed. R. Nalèwajski, (Springer, Berlin,1996), p. 143.

[17] J. Harris and S. Andersson, Phys. Rev. Lett. **55**, 1583 (1985); J. Harris, Faraday Discuss. **96**, 1 (1993).

[18] J. Kudrnovský, I. Turek and V. Drchal in *Lectures on Methods of Electronic Stucture Calculations*, Eds. V. Kumar, O. K. Anderson, and A. Mookerjee (World Scientific, Singapore, 1994), p. 231.

[19] B. Hammer and M. Scheffler, Phys. Rev. Lett. **74**, 3487 (1995).

[20] B. Hammer and J. K. Nørskov, Surf. Sci. **343**, 211 (1995).

A JOURNEY:
FROM CLUSTERS TO CRYSTALS

P. Jena, S.N. Khanna, S.K. Nayak, B.K. Rao, and B.V. Reddy
Department of Physics
Virginia Commonwealth University
Richmond, VA 23284-2000, U. S. A.

ABSTRACT

Atomic clusters provide an ideal link between atoms and the bulk and constitute a unique phase of matter with properties that depend strongly upon their size, composition, and structure. An understanding of the evolution of structural, electronic, optical, and magnetic properties of the clusters as a function of size is not only important from an academic point of view, but also has a strong technological merit as clusters can constitute the building block for novel materials. A combination of molecular dynamics and first principles molecular orbital theory based on spin density functional formalism have been used to study the evolution of structure, interatomic distance, ionization potential, and magnetism of Ni_n clusters with size and temperature. The results are compared with recent experimental data.

1. INTRODUCTION

The title of this paper is motivated by a poem written in 1911 by Constantine Cavafy, a Greek poet. The poem describes the journey of Ulysses as he was returning to his home Ithaca after fighting in the Trojan war. The poet wishes the journey to be long and full of adventure and knowledge, but advises the traveler that although arriving at Ithaca is the goal, he should not hurry the voyage. The journey from clusters to crystals can be viewed in a similar vein. As we add one atom after another, the cluster grows and eventually approaches its goal of being the crystal. During this voyage, the properties will change continually.

What one needs to do is to learn about the various properties of each of the clusters as they evolve. The knowledge one attains about the size-specific properties may be useful in designing new materials where clusters form the building blocks [1].

Like in the journey to Ithaca, the path of clusters to crystals is filled not only with beautiful and interesting science, but also numerous hurdles. One of the main hurdles is to determine the atomic structure of a cluster. There are no experimental techniques that would allow a direct determination of the atomic arrangements in a cluster. The clusters are too big for spectroscopic experiments and too small for diffraction experiments to be useful for structural determination. Since small clusters are characterized by a large surface to volume ratio, the atomic structure is not expected to mimic the crystal structure. Consequently, the electronic structure and properties of clusters that depend critically on the atomic arrangements cannot be bulk-like. The most interesting questions to ask are: (a) How do the various structural and electronic properties evolve ? (b) Do all properties evolve in a similar fashion ?

This paper is an attempt to answer these fundamental questions. We have chosen Ni clusters as an example. Ni is a transition metal element and its ferromagnetic behavior is attributed to the large density of d-electrons at the Fermi energy [2]. In addition, there is a wealth of experimental data that is available on this system. These include ionization potential [3], density of states [4], reactivity [5], and magnetic moment [6]. We have systematically studied the equilibrium geometry, electronic structure, binding energies, ionization potential, and magnetic moments in Ni clusters consisting of up to 23 atoms. In section II we outline briefly our theoretical procedure. Results on atomic structure and electronic properties are described in section III. A summary of our conclusions are given in section IV.

2. COMPUTATIONAL PROCEDURE

First principles theoretical calculations of the energetics and equilibrium geometries of atomic clusters containing tens of atoms are difficult due to the existence of many local minima in the potential energy surface. These calculations become prohibitively expensive if clusters consist of magnetic transition metal elements. Consequently, there are only a limited number of investigations [7-11] on the structure and properties of Ni_n clusters ($n \leq 6$) available to date.

In order to study the evolution of properties in clusters containing up to 30 atoms we have followed a dual theoretical procedure by combining classical molecular dynamics and first principles method. The latter is carried out using the self-consistent field-linear combination of atomic orbitals-molecular orbital (SCF-LCAO-MO) method [12] and local spin density approximation in the density functional theory [13]. The inner core-orbitals of Ni ($1s^2\ 2s^2\ 2p^6$) were frozen and the valence orbitals (3s 3p 3d 4s 4p) were represented by a double numerical basis. The electronic structure and properties were calculated using DMOL code [14]. For small Ni_n ($n \leq 6$) clusters, we have verified the accuracy of the DMOL results by using Gaussian basis sets and the Gaussian 94 software [15]. Calculations were carried out using all electrons as well as effective core potentials. The exchange-correlation contribution to the potential was incorporated at two levels: the local spin density approximation and generalized gradient approximation using the Becke-Perdew-Wang (BPW91) method [16].

Determination of equilibrium geometries of larger clusters (n > 6) at the first principles level is difficult. For these clusters, we used the constant energy molecular dynamics simulations and the semi-empirical interatomic potentials based on the tight binding method [17]. The velocity Verlet algorithm [18] was used to integrate the classical equations of motion with a time step of 5×10^{-15} s. The geometry corresponding to the global minimum was obtained by using simulated annealing. Once the geometries are obtained, we calculated the electronic structure and properties using the first principles approach described in the above. Since the interatomic potentials are obtained by fitting to bulk data, we compared the geometries for small clusters obtained from the molecular dynamics simulation with our first principles calculations [11]. The results were in excellent agreement with each other. In the following we discuss the evolution of atomic and electronic structure of Ni_n clusters.

3. RESULTS

We discuss the evolution of (a) the equilibrium geometry, coordination number, inter-atomic distance, binding energy; (b) electronic structure, density of states; (c) ionization potential; (d) magnetic moments; and (e) thermodynamic stability of Ni clusters as a function of size.

3.1. Atomic Structure and Stability

The equilibrium atomic structure of a cluster is governed by its underlying electronic structure. In the bulk phase Ni is fcc and the electronic behavior is metallic. In Ni_n clusters, the equilibrium geometries begin with symmetric and close-packed structures that resemble the geometries of rare-gas atom clusters. For clusters containing more than seven atoms, geometries with five-fold symmetries evolve. As more atoms are added, geometries with reduced symmetry (i.e. Jahn-Teller distorted) appear. This indicates that the bonding in clusters with n > 7 is characterized by delocalized (i.e. metallic) electrons while for n < 7 the bonding is mostly covalent. While it is difficult to visualize the geometries of larger clusters, meaningful insight into atomic structure can be obtained by calculating the average interatomic distance, <R>, and average coordination number, CN. These quantities are given by,

$$<R> = \frac{1}{N_b} \sum_{i<j} R_{ij} \quad and \quad CN = \frac{1}{N} \sum_{i} N_i \qquad (1)$$

Here R_{ij} is the bond distance between atom i and j and N_b is the total number of bonds. N_i is the number of nearest neighbors of the ith atom. Note that the nearest neighbor distance and the coordination number of bulk Ni are respectively 2.49 Å and 12. In Fig. 1 we plot the average interatomic distance and average coordination number for Ni_n clusters containing up to 23 atoms. Note that although both the quantities evolve fairly monotonically, they are far from being bulk-like. In particular, the coordination number, even in the largest cluster, is less than the coordination number of a surface atom on the <111> face. The low coordination also influences the stability of clusters. With increasing coordination the stability of the clusters increases but the binding energy of the largest cluster remains significantly below the bulk cohesive energy (see Fig. 2).

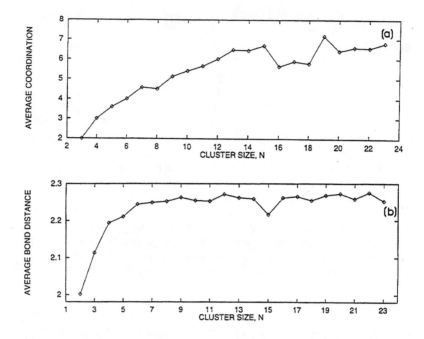

Figure 1. Average coordination and bond distance of Ni_n clusters as a function of size.

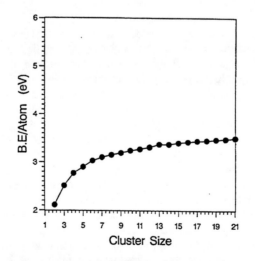

Figure 2. Binding energy/atom (eV) calculated by molecular dynamics as a function of cluster size. The experimental bulk cohesive energy is 4.4 eV.

As mentioned earlier, there is no direct experimental information on the geometries of clusters. However, attempts have been made to infer the atomic structure of clusters by analyzing their reaction data with molecules such as N_2. For example, Parks and co-workers [19] have measured the N_2 uptake by varying the temperature and pressure of N_2. Assuming certain characteristics of N_2 interaction with clusters, the authors predicted likely geometries of clusters. While our computed geometries agree with this prediction in some cases, discrepancies remain. We argue that the resolution of this discrepancy will remain until such time when direct experimental information on structure is available. In the meantime, we can calculate electronic properties of clusters using the present geometries and compare these with experiments. A good agreement between theory and experiment may serve as indirect evidence that theoretically determined geometries are indeed true.

3.2. Electronic Structure

We address the evolution of the electronic structure by concentrating on two different techniques: The photo-detachment spectroscopy [4] and near threshold photo ionization efficiency (PIE) [3]. In the former, a mass selected cluster anion is crossed by a fixed frequency laser, and the energy of the ejected electron is analyzed. This provides information on the electron density of states of the neutral cluster. In the PIE experiment, one measures the vertical ionization potential which corresponds to the energy difference between the neutral ground state geometry and the cationic cluster with the neutral geometry. Wang and Wu [4] have measured the photo-detachment spectra of Ni_n clusters containing up to 50 atoms. They observed that the spectra remained practically the same for clusters with $n > 14$. This means that the electronic structure of Ni_n clusters becomes metallic and bulk-like for clusters larger than Ni_{14}. We have analyzed the energy levels associated with the molecular orbitals (MO) of Ni_n ($n \leq 23$). We find that for clusters with $n > 14$, the MO levels essentially remain unchanged. This can also be seen more clearly by plotting the energy gap between highest occupied molecular orbital (HOMO) and lowest unoccupied molecular orbital (LUMO) (see Fig. 3). We note that the HOMO-LUMO gap nearly vanishes for $n > 14$ indicating that the clusters can be assumed to possess metallic character. On the other hand, the vertical ionization potential does not approach the work function of the $<111>$ surface of Ni even for very large clusters (see Fig. 4). Although the ionization potential varies strongly with size for $n \leq 7$, the variation is substantially reduced at large sizes. In this context the size-dependence of the ionization potential of Ni clusters is markedly different from that in alkali clusters where sharp drops [20] are seen for clusters corresponding to magic numbers at $n = 2, 8, 20...$. We also note that the theoretical results agree with experiment to better than 10% in all cases. Ironically, the disagreement between theory and experiment is worst in small ($n \leq 6$) clusters where all calculations are done at the first principles level of theory.

To understand the sources for the remaining discrepancy, we have repeated the calculations for $n \leq 6$ using all electrons and generalized gradient approximation (GGA) for the exchange-correlation potential. While inclusion of all electrons has negligible effect on the ionization potential, the GGA brought theory to better agreement with experiment. For example, the ionization potential of Ni_2 computed using local spin density approximation (LSDA) is 8.34 eV while that using GGA is 7.89 eV. The experimental value is 7.6 eV.

P. Jena, S.N. Khanna, S.K. Nayak, B.K. Rao and B.V. Reddy

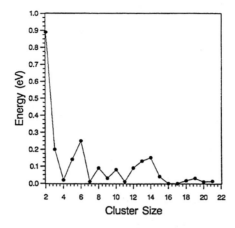

Figure 3. Dependence of HOMO-LUMO gap on cluster size.

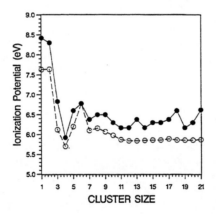

Figure 4. Calculated (solid circles) and experimental (open circles) [3] vertical ionization potential as a function of size.

Temperature could also influence the measured value of the ionization potential. This is particularly possible in transition metal clusters where electronic energy levels are closely spaced, and the spacings are comparable with the thermal energies. To study the temperature dependence, we have used the molecular dynamics simulation to probe the structures which a cluster would sample at elevated temperatures. We then calculated the vertical ionization potential of one of these structures assuming that at the experimental temperature, the structure of the cluster does not correspond to the ground state geometry. We found that [11] while for Ni_7 temperature did not influence the calculated ionization potential, it had significant impact on the ionization potential of Ni_{13} by lowering it by as much as 0.1 eV. Thus, it is possible that incorporation of temperature and GGA could bring quantitative agreement between calculated and experimental ionization potential. This suggests that the geometries calculated for Ni_n clusters ($n \leq 23$) from molecular dynamics simulation may be correct.

3.3. Magnetic Moments

The magnetic properties of clusters differ substantially from those in the bulk. This is due to low dimensionality, reduced size, symmetry and coordination that are hallmarks of clusters. It has long been established [21] that most of these characteristics lead to an enhancement of the magnetic moments of ferromagnetic elements. Since in clusters these quantities evolve differently, it is difficult to predict the size dependence of cluster magnetic moments without performing an *ab initio* calculation. For example, we see from Fig. 1 that the average coordination of an atom in a cluster increases with cluster size. Thus, one would expect that the magnetic moment/atom should decrease monotonically with size. However, experimentally one observes a strong non-monotonic dependence of magnetic moments with size [6]. This is brought about by special interplay between symmetry, coordination and interatomic separation in clusters.

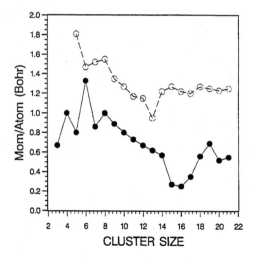

Figure 5. Calculated (solid circles) and experimental (open circles) [6] magnetic moments of Ni_n clusters vs. cluster size.

Using the geometrical structure obtained from molecular dynamics simulation, we have calculated the magnetic moment/atom for Ni clusters from first principles. The results are plotted in Fig. 5 and compared with the most recent experimental data [6]. Although there appears to be qualitative agreement between theory and experiment, the discrepancy is much worse than that in the ionization potential in Fig. 4. Interestingly, the disagreement between theory and experiment in Fig. 5 is much worse for small clusters than it is for larger clusters - similar to what we observed in Fig. 4. In particular, the experimental magnetic moment of Ni_5 is a factor of two larger than the calculated value. Since the geometry of Ni_5 was optimized using first principles theory, the overall disagreement in Fig. 5 cannot be blamed on poor choice of geometries that may have been brought about due to the use of molecular dynamics simulation based on empirical inter-atomic potential.

We have repeated the calculations of magnetic moments of Ni_n ($n \leq 6$, $n = 13$) clusters using all electrons as well as generalized gradient approximation. The calculated moments remained unaffected. We have also analyzed the effect of temperature using the same procedure as in the ionization potentials. This, too, had no effect on the calculated moments in Fig. 5. Recently, Castro *et al.* [10] have repeated the calculations of magnetic moments in small Ni clusters and their result agrees with what is given in Fig. 5.

To analyze why the same level of calculations yields nearly quantitative agreement between theory and experiment of ionization potentials and fails so badly for magnetic moments, we outline some of the experimental difficulties in measuring the magnetic moment. Conventionally, the average moments are measured by studying the deflection of a cluster in a Stern-Gerlach magnet [22]. The intrinsic moment of the cluster is calculated by relating it to the average magnetic moment and making use of the superparamagnetic model [23]. Here the cluster is viewed as a single ferromagnetic domain, but the giant moment of the cluster at a given temperature is allowed to explore the entire Boltzman distribution. Thus, the experimental value of the intrinsic moment depends not only on the validity of the super-paramagnetic model, but also on the reliability of the assumed temperature of the cluster. There is considerable debate in the literature about whether the temperature of a cluster can be determined uniquely [24]. Similarly, the validity of the super-paramagnetic model for very small clusters has also been questioned. It is worth pointing out that the measured magnetic moments of large Ni_n clusters (n ~ 200) obtained by two different groups [6,24] are in disagreement by as much as 50%. On the theoretical side, inclusion of spin-orbit effects, as well as orbital contribution to magnetic moments, needs to be explored. Thus, further experimental and theoretical work is needed to be done before we arrive at a satisfactory understanding of the magnetic moment of transition metal cluster systems.

3.4. Thermodynamic Stability

The study of the melting of clusters is a difficult problem for one does not know an unambiguous way of defining melting. In the solid phase, melting corresponds to a structural phase transition. However, in a cluster that has many local minima, it is difficult to identify melting from that when a cluster passes on from one local minimum to another. Experimental studies of thermodynamic stability of clusters are usually carried out by employing diffraction probes on supported clusters [25]. Here, too, melting is difficult to identify since a cluster structure can change when substrate temperature is raised as well as when the interaction between the substrate and cluster is strong. Nevertheless, certain qualitative aspects of cluster melting are independent of how one defines melting. For example, it is well known that the melting of a solid starts from its surface since the surface atoms have low coordination [26]. Since the coordination number of cluster atoms are less than that for bulk or surface atoms, one would expect a cluster to be thermodynamically less stable than the bulk. However, the dependence of this thermodynamic stability as a function of size and its relationship to energetics of the ground state cannot be easily inferred.

To analyze the effect of temperature on the stability of the clusters, we have followed the procedure described by Beck *et al.* [27]. They studied the melting of a cluster by monitoring the root mean square bond-length fluctuation as a function of temperature. A sharp change in this parameter at a fixed temperature indicates the thermal instability of the cluster, and this temperature can be interpreted as the cluster's "melting point." The abrupt changes in the bond-fluctuation can also happen when local minima in the potential energy surface of a cluster is separated by a small energy barrier and the cluster passes on from one structure to another. This phenomenon can be thought of as pre-melting.

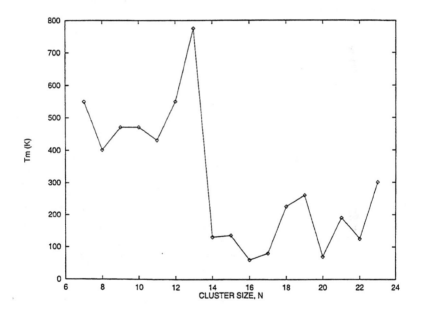

Figure 6. Melting point vs. size.

Using molecular dynamics simulation, we have studied the thermodynamic stability of Ni clusters ($7 \leq n \leq 23$). The computed melting points, as a function of size, are given in Fig. 6. We note that the thermodynamic stabilities of clusters falls into two categories. Clusters of less than 13 atoms, in general, have higher melting points than those consisting of more than 13 atoms. This feature is attributed to the geometry of clusters. Larger clusters ($n > 13$) have a more open shell structure with vacant "surface" sites. As the temperature is raised these atoms can move more freely than in clusters that have a compact or closed atomic shell structure. Note that the melting point of even the most thermodynamically stable cluster ($n = 13$) is almost half of the bulk value.

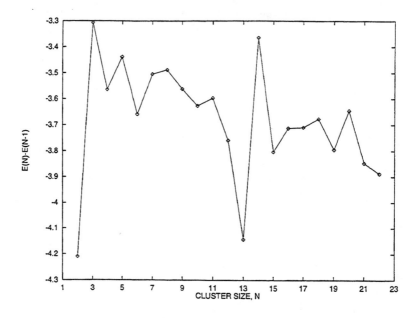

Figure 7. Plot of ΔE_N (= E(N) - E(N-1)) against cluster size.

To compare the thermodynamic stability of clusters with the energetic stability, we plot in Fig. 7 the energy gained, ΔE by adding an atom to an existing cluster as a function of size. The minima in Fig. 7 then correspond to energetically favored clusters. Note that there is a strong correlation between the minima in Fig. 7 with maxima in Fig. 6. This indicates that clusters that are energetically more stable at 0 K also tend to have a higher melting point.

4. CONCLUSIONS

A study of the evolution of the atomic structure, electronic structure, stability, ionization potential, magnetic moment and melting point of Ni_n clusters consisting of up to 23 atoms has been carried out using two complimentary theoretical approaches. The molecular dynamics simulations using empirical tight-binding inter-atomic potential were carried out to obtain the equilibrium geometries of clusters as well as the geometries of nearly degenerate isomers. Using these geometries, the electronic structure and properties were calculated from first principles.

The geometries for small clusters ($n \leq 6$) were found to be compact while the larger clusters have more open-shell structures. Although the coordination number, inter-atomic distance, and binding energy increase fairly monotonically with size, they are far from being bulk-like. The electron density of states, on the other hand, does not change when clusters contain more than 14 atoms.

The ionization potential shows a marked size dependence for $n < 7$, but varies slowly beyond this range. By including effect of temperature and gradient correction to the exchange-correlation potential, it is possible to achieve quantitative agreement with experiment. However, the situation appears to be much worse for the magnetic moment. A case is made for further work, both experimentally and theoretically, to sort out the remaining discrepancy in the magnetic properties of Ni_n clusters. The melting points of Ni_n clusters show a strong size dependence with small clusters ($n < 13$) exhibiting higher melting points than larger ($n > 13$) ones. This is attributed to open-atomic shell structures of larger clusters which show pre-melting behavior. The thermodynamic stabilities of clusters are also found to be correlated with their energetic stability at 0 K. We have shown that a combination of molecular dynamics simulation and first principles electronic structure theories can be used to study the properties of transition metal clusters.

ACKNOWLEDGMENTS

This work is supported in part by grants from the Army Research Office (DAAH04-95-0158) and the Department of Energy (DEFG05-87E61316)

References

[1] S.N. Khanna and P. Jena, Phys. Rev. Lett. **69**, 1664 (1992); ibid. 71, 208 (1993).

[2] V.L. Moruzzi, J.F. Janak, and A.R. Williams, *Calculated Electronic Properties of Metals* (Pergamon, N.Y., 1978).

[3] M.B. Knickelbein, S. Yang, and S.J. Riley, J. Chem. Phys. **93**, 94 (1990).

[4] L.S. Wang and H. Wu, J. Phys. Chem. (to be published); G. Gantefor, W. Eberhardt, H. Weidele, D. Kreisle, and E. Recknagel, Phys. Rev. Lett. **77**, 4524 (1996).

[5] Y. Hamrick, S. Taylor, G.W. Lemire, Z.W. Fu, J.C. Shui, and M.D. Morse, J. Chem. Phys. **88**, 4095 (1988); J.L. Elkind, F.D. Weiss, J.M. Alford, R.T. Laaksonen, and R.E. Smalley, J. Chem. Phys. **88**, 5215 (1988).

[6] S.E. Apsel, J.W. Emmert, J. Deng, and L.A. Bloomfield, Phys. Rev. Lett. **76**, 1441 (1996).

[7] M. Tomonari, H. Tatewaki, and T. Nakamura, J. Chem. Phys. **85**, 2975 (1986).

[8] P. Mlynarski and D.R. Salahub, J. Chem. Phys. **95**, 6050 (1991).

[9] F. A. Reuse and S.N. Khanna, Chem. Phys. Lett. **234**, 77 (1995).

[10] M. Castro, C. Jamorski and D. R. Salahub, Chem. Phys. Lett. **271**, 133 (1997).

[11] B.V. Reddy, S.K. Nayak, S.N. Khanna, B.K. Rao, and P. Jena (submitted).

[12] W.J. Hehre, J.L. Radom, P.V.R. Schleyer, and J.A. Pople, *Ab-initio Molecular Orbital Theory* (Wiley, New York, 1986).

[13] W. Kohn and L.J. Sham, Phys. Rev. **140**, A1133 (1965); P. Hohenberg and W. Kohn, Phys. Rev. **136**, B864 (1964).

[14] *DMOL code*, Biosym Technologies, Inc., San Diego.

[15] *Gaussian 94*, Revision B.1, M.J. Frisch, G.W. Trucks, H.B. Schegel, P.M.W. Gill, B.G. Johnson, M.A. Robb, J.R. Cheeseman, T. Keith, G.A. Petersson, J.A. Montgomery, K. Raghavachari, M.A. Al-Laham, V.G. Zakrzewski, J.V. Ortiz, J.B. Foresman, J. Closlowski, B.B. Stefanov, A. Nanayakkara, M. Challacombe, C.Y. Peng, P.Y. Ayala, W. Chen, M.W. Wong, J.L. Andres, E.S. Replogle, R. Gomperts, R.L. Martin, D.J. Fox, J.S. Binkley, D.J. Defrees, J. Baker, J.P. Stewart, M. Heads-Gordon, C. Gonzalez, and J.A. Pople, Gaussian, Inc., Pittsburgh, PA (1995).

[16] A.D. Becke, J. Chem. Phys. **88**, 2547 (1988); C. Lee, W. Yang, R.G. Parr, Phys. Rev. B **37**, 786 (1988).

[17] A.P. Sutton and J. Chen, Philos. Mag. Letts. **50**, 45 (1984).

[18] M.P. Allen and D.J. Tildesley, *Computer Simulation of Liquids* (Oxford University Press, 1987).

[19] E.K. Parks, L. Zhu, J. Ho and S.J. Riley, J. Chem. Phys. **100**, 7206 (1994); J. Chem. Phys. **102**, 7377 (1995); E.K. Parks and S.J. Riley, Z. Phys. D **33**, 59 (1995).

[20] W.D. Knight, K. Clementer, W.A. de Heer, W.A. Saunders, M.Y. Chou, and M.L. Cohen, Phys. Rev. Lett. **52**, 2141 (1984); S. Saito and S. Ohnishi, *ibid*, **59**, 190 (1980).

[21] F. Liu, M.R. Press, S.N. Khanna, and P. Jena, Phys. Rev. B **39**, 6914 (1989).

[22] J.G. Louderback, A.J. Cox, L.J. Lising, D.C. Douglass, and L.A. Bloomfield, Z. Phys. D **26**, 301 (1993).

[23] S.N. Khanna and S. Linderoth, Phys. Rev. Lett. **67**, 742 (1991).

[24] W.A. de Heer, P. Milani, and A. Chatelein, Phys. Rev. Lett. **65**, 488 (1990).

[25] D.J. Smith, A. Petford-Long, L. Wallenberg, and J. Boven, Science **233**, 872 (1986).

[26] A. Bouten, B. Rousseau, and A.H. Fuchs, Europhys. Lett. **18**, 245 (1992).

[27] T.L. Beck, J. Jellinek and R.S. Berry, J. Chem. Phys. **87**, 545 (1987).

STABILITY OF CLUSTER MATERIALS: CLUSTER-CLUSTER AND CLUSTER-SURFACE INTERACTION

Matti Manninen and Hannu Häkkinen
Department of Physics
University of Jyväskylä
P.O. Box 35, FIN-40351 Jyväskylä, Finland

ABSTRACT

Ab initio molecular dynamics is used to study cluster-cluster and cluster-surface interactions. The electronic structure and the total energy is calculated using the Kohn-Sham density functional formalism, *ab initio* pseudopotentials and a plane wave basis. In the dynamical calculation the electronic structure is recomputed at every time step so that the ions move on the Born-Oppenheimer energy surface. The interaction between Na clusters and NaCl surface is weak physisorption, the adsorption energy being about 0.5 eV per atom. Sodium clusters on Na surface loose immediately their identity and the atoms take their epitaxial positions on the adsorption layer. Dynamical simulation of two magic Na_8 clusters indicate that the clusters, although having a marked HOMO-LUMO gap, are not bonded by weak van der Waals bonds, but melt together forming a Na_{16} cluster. Simulations with two CLi_4 clusters suggest that these clusters form a metallic bond, not a van der Waals bond like two CH_4 molecules.

1. INTRODUCTION

The cluster-cluster interaction and the internal stability of clusters are the key ingredients in determining the possibility to form cluster-based materials. In the extreme case, the intracluster bonding between atoms is much larger than the intercluster bonding. In this case the border between a cluster material and a molecular solid becomes hazy and for example the fcc crystal of fullerene molecules is sometimes called as an example of a cluster material whereas crystalline methane, consisting of CH_4 molecules is considered as a molecular solid. In both examples the clusters are bound via van der Waals interactions whereas the atoms in a cluster/molecule are covalently bound.

Only clusters with a full electronic shell, i.e a large energy gap between the highest occupied orbital (HOMO) and the lowest unoccupied orbital (LUMO), can form cluster materials with weak intercluster interactions. Metal clusters usually have a small HOMO-LUMO gap and are thus expected to have strong cluster-cluster bond which can drastically change the properties of the clusters in a cluster material as compared to those of free clusters. However, metal clusters with closed electronic shells, often referred to as magic, have been considered as candidates of interesting cluster materials, especially if they have also a closed packed atomic arrangement [1]. Indeed, model calculations [2] have shown that cluster materials composed of magic metal clusters could have a band gap, like semiconductors, if the inter-cluster bonds do not change the cluster geometries.

The key question in forming cluster materials from magic metal clusters is then whether the cluster geometry is rigid enough to maintain the magic features also when the clusters are brought together. This question also arises when studying the interaction of clusters with surfaces, i.e. is the surface-cluster interaction weaker than the interatomic interactions inside the cluster. We have studied these questions using *ab initio* molecular dynamics [3] and in this paper we will review results of three different systems.

First we will discuss a possibility of a new class of "double magic" metal clusters, i.e. cluster which have both closed electronic shell and high-symmetry closed packed geometry [4]. These are the tetrahedrons in 3D and triangles in 2D. It turns out that these clusters have as strong electronic magic numbers as spherical clusters and they coincide with the number of atoms in complete tetrahedra and triangles.

As a second problem we study the interaction of sodium clusters with surfaces [5, 6]. We show that sodium clusters interact only weakly with an insulating NaCl surface, maintaining the properties of free clusters. In contact with the metallic sodium surface, however, even the magic sodium cluster immediately "melts" to an epitaxial layer.

The third problem is the interaction of two magic clusters in vacuum. We show that the bond between two magic Na_8 clusters is as strong as the intracluster bonds, indicating that metallic clusters cannot be used as basis of those kind of cluster materials as fulleride. We also studied interactions between two CLi_4 clusters. Being closely related to methane they could be more likely to form weak van der Waals bonds. However, our simulations showed that this is not the case, rather the clusters favor formation of a strong C-C bond.

2. METHOD

The calculations reported here have been done using BO-LSD-MD (Born-Oppenheimer Local-Spin-Density Molecular Dynamics) method developed by Barnett and Landman [3]. Briefly, one solves for the Kohn-Sham one-electron equations for the valence electrons of the system corresponding to a fixed position of ions. The converged solution gives the Helmann-Feynman forces which together with the classical Coulomb repulsion between the ions are the basis for classical molecular dynamics. The fully converged solution for the electronic structure is obtained for each successive ionic positions during the molecular dynamics. Consequently the system follows dynamics on the Born-Oppenheimer energy surface. The method can be used for finite systems or, using periodic boundary conditions, to infinite systems.

The electronic structure in each ionic configuration is computed using *ab initio* pseudopotentials [7] and plane-wave technique. For the systems studied we had from 8 to 300 valence electrons. A satisfactory accuracy was obtained with a kinetic energy cutoff from 10 to 40 Ry, meaning from 10000 to 216000 plane waves depending on energy cutoff and on the system size.

The time step of the molecular dynamics was 1 to 5 fs which ensures an excellent conservation of total energy [8]. Naturally, the total simulation times were short, typically of the order of 10 ps.

3. TETRAHEDRAL METAL CLUSTERS

The shell structure of valence electrons in metal clusters was observed by Knight *et al.* [9, 10] and was explained with a spherical mean field potential. The electronic shells corresponded to degenerate angular momentum eigen values in a simple spherical potential well. While the clusters with open shell of electrons are deformed [11, 12, 13] it was generally assumed that the magic clusters are nearly spherical in shape. Since metallic bonding favors closed packed structures in bulk it is expected that geometries which maximize the coordination will be most strongly bound also in clusters. A cluster with a nearly spherical closed packed geometry with a magic number of electrons would then be a candidate of extreme internal binding.

Icosahedral shape is the most spherical shape of closed packed geometries, and indeed it has been shown that an icosahedral potential has the same magic numbers as the spherical potential up to at least several hundreds of particles [14, 15, 16]. The numbers of atoms corresponding to complete icosahedra are 13, 55, 147, 309 etc. However, it is easy to notice that none of these numbers correspond to the electronic magic numbers of metal clusters, 8, 20, 40, 58, 92, etc., not even if one considers metals with valence two or three. Using combination of two elements it is possible to find icosahedral metal clusters with magic electron numbers. The simplest of such clusters consists of a four-valent atom at the center and 12 trivalent atoms at the surface of a 13 atom icosahedron, e.g. $SiAl_{12}$ or CAl_{12} as suggested by Khanna and Jena [1]. Indeed, calculations show that these clusters have reduced reactivity with gas atoms [17], but when brought together to form a solid they seem to melt together with a metallic bond [18, 19].

Recently, Reimann *et al.* [4] have shown that a strong electronic shell structure exists also in tetrahedral clusters. The first few magic numbers in a tetrahedron shape cavity correspond to those of 3D harmonic oscillator: 8, 20, 40, 70, 112. The first two of these are the same as in spherical cavity (8, 20), but surprisingly, in tetrahedral cavity the magic numbers follow those of harmonic oscillator at least to the magic number 112. Already the magic number 40 is much stronger in the tetrahedral cavity than in spherical cavity and indeed the magic number 40 observed in alkali metal clusters may be due to tetrahedral deformation [4].

Tetrahedral clusters are especially interesting since they naturally correspond to closed packed structures of atoms, the number of atoms being 4, 10, 20, 35, 56 etc. When multiplied with two these numbers are exactly the same as the electronic magic numbers in tetrahedral cavity (or finite potential well). This means that any small tetrahedral cluster of divalent metal is both electronically and geometrically magic and expected to have enhanced stability.

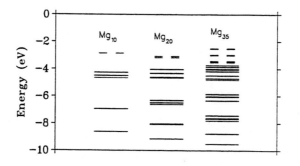

Figure 1. Single-particle energy levels of tetrahedral Mg_{10}, Mg_{20} and Mg_{35} clusters. The empty levels are shown as a dashed lines.

We have calculated the electronic structure and geometry of small tetrahedral magnesium clusters [4] using the BO-LSD-MD method. Fig. 1 shows the electronic level structure of 10, 20 and 35 atom tetrahedra. The electronic shell structure is clearly seen in all cases. We have also tested the stability of 10 atom cluster by heating it up to 80 K. Within the simulation time of 10 ps it remained in the tetrahedral geometry.

The situation is similar in two-dimensional clusters. Triangular clusters have the same electronic magic numbers than 2D harmonic oscillator and again these numbers coincide with the numbers of electrons in complete closed packed triangles made of divalent atoms [4].

Khanna and Jena [1] have studied the interaction between two Mg_4 triangles. Indeed they found that the bond between the two clusters is relatively weak and only slightly distorts the geometries of the individual tetrahedra. However, only a limited set of geometries were considered.

4. STABILITY OF MAGIC Na_8

Alkali metals are the best examples of nearly free electron metals. The Fermi surface is nearly spherical, the pseudopotential is weak and cohesion and bulk modulus are mainly determined by the properties of a homogeneous electron gas. The electronic shell structure is most clearly seen in alkali metal clusters. It is then interesting to study a magic alkali metal cluster to see how "magic" is a metal cluster when its reactivity with other cluster is considered. Does it behave like a rare gas atom with only weak van der Waals interactions with other systems or does it tend to form metallic bonds?

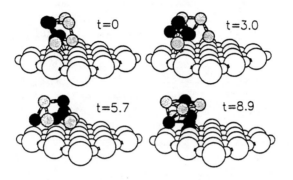

Figure 2. Selected snapshots of Na_8 on NaCl(100) surface. Half of the cluster atoms are shown in black and half in grey in order to visualize the diffusion process. The unit of time is ps.

We have studied these questions using Na_8 as a test cluster. Fig. 2 shows snapshots of BO-LSD-MD calculations of Na_8 on top of a NaCl(100) surface. The temperature of the cluster is 550 K, i.e. well above the bulk melting temperature of sodium. The small sodium cluster is also liquid as seen in the figure. Nevertheless, the cluster floats on top of the surface, keeping its magic electronic structure intact, and the adsorption energy to the surface is rather weak, about 0.5 eV per atom. The formation of strong chemical bond is prevented by the fact the electron levels of the cluster lie in the large energy gap of the NaCl. Also nonmagic sodium clusters on NaCl(100) have nearly same geometries as in the vacuum[6, 20].

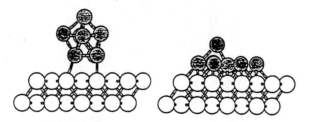

Figure 3. The initial (left) and the final (right) configuration of the collapse of Na_8 on Na(110) surface. At the final configuration the atoms are at their epitaxial positions.

Fig. 3 illustrates the reaction of Na_8 cluster with $Na(110)$ surface [6, 20]. The cluster was initially placed just above the surface and then let to relax with molecular dynamics. The resulting structure is an epitaxially grown cluster on the surface. The closed shell electronic structure does not prevent the cluster to break on the reactive surface of the metal. We expect similar result also for other metallic surfaces.

As a third example for Na_8 we have studied the interaction between two clusters. The clusters were initially placed so far from each other that the electron densities slightly overlapped. The initial temperature was zero and the clusters were at rest. The molecular dynamics was done in the constant energy mode. The clusters were pulled together and the temperature increased due to the binding energy. The resulting bond between the cluster was a strong chemical bond and the two clusters lost their identity, forming together a strongly deformed Na_{16} cluster.

In conclusion, the magic sodium cluster has an increased stability when compared to free sodium clusters of other sizes, but when in contact with a metallic surface or an other metal cluster it looses its identity. Magic sodium clusters cannot form cluster assembled materials in the same way as for example the fullerenes.

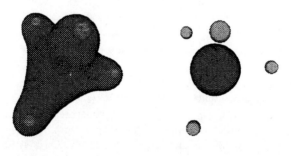

Figure 4. The constant density surface of valence electrons in CH_4 (left) and CLi_4 (right). The bonding in CH_4 is covalent while it is ionic in CLi_4. The density surfaces are plotted at values of 0.17 a.u. and 0.13 a.u. for CH_4 and CLi_4, respectively, and correspond to about 50 % of the maximum density.

5. INTERACTION BETWEEN TWO CLi_4 CLUSTERS

The total number of valence electrons in CLi_4 is eight and consequently it has a full electronic shell if we consider it as a metallic cluster. It has also a compact geometry. The carbon atom is at the center of a tetrahedron formed by the lithium atoms. A related molecule is methane, CH_4. The bonding in CH_4 is covalent while in CLi_4 it is more ionic, Fig. 4. Methane molecules do not bind chemically and C_2H_8 molecule does not exist. It is then interesting to study the interaction of the CLi_4 clusters to see if they behave as methane molecules or if they form a metallic bond like two Na_8 molecules.

Two CLi_4 clusters were first aligned along a common C_{3v} axis and let relax dynamically together. During a short (0.4 ps) simulation at about 200 K they kept their original geometry, but were strongly bond together, see Fig. 5 (a). After that the symmetry of the system was artificially broken by small random components in the velocities of atoms, and the system was heated up to 500 K. The clusters "melted" and at the end of the simulation (1.7 ps) the initial geometry had drastically changed, Fig. 5 (b), and the binding energy was 1.4 eV larger than in the initial geometry (a).

Finally, we wanted to test if the carbon atoms prefer to form a C-C bond. To this end we started the simulation from a geometry where a C_2 molecule was surrounded by lithium atoms. The atom positions were relaxed to the closest local minimum shown in Fig. 5 (c). The binding energy of this geometry is still 1.1 eV larger than in (b). It should be stressed that the geometry (c) is not necessarily the ground state of C_2Li_8 cluster. However, it seems clear that the ground state will have a C-C bond.

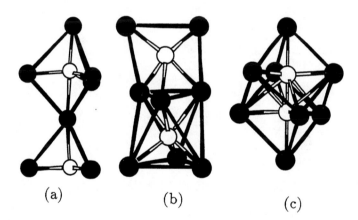

(a)   (b)   (c)

Figure 5. Interaction of two CLi_4 clusters. Left (a): A dynamical configuration where the clusters share a common C_{3v} axis. Middle: (b) The local minimum after the symmetry of (a) is broken and relaxed. Right: (c) A configuration with a C-C bond. Configuration (c) has the largest binding energy.

The binding in CLi_4 molecules is different than in CH_4 and they do not prefer to form weakly bound cluster materials. Two CLi_4 molecules can form a very strongly bound C_2Li_8 whereas C_2H_8 does not exist.

6. CONCLUSIONS

We have studied possibilities to use magic metal clusters as basis of cluster materials with weak intercluster bonding. We have shown that several electronically magic clusters with compact geometry exist. Especially, tetrahedrons of divalent metals form a whole set of different size "double magic" clusters.

Molecular dynamics simulations seem to indicate that metal clusters, although magic, react strongly with each other and with metallic surfaces, preventing formation of such cluster materials where the intercluster bonding is weaker than the intracluster bonding.

ACKNOWLEDGMENTS

We would like to thank U. Landman, R.N. Barnett, S. Reimann, M. Koskinen, and P. Jena for many valuable discussions during this research. This work was supported by the Academy of Finland.

References

[1] S.N. Khanna and P. Jena, Phys. Rev. Lett. **69**, 1664 (1992).

[2] M. Manninen, J. Mansikka-aho, S. H. Khanna, and P. Jena, Solid State Commun. **85**, 11 (1993).

[3] R. N. Barnett and U. Landman, Phys. Rev. B **48**, 2081 (1993).

[4] S.M. Reimann, M. Koskinen, H. Häkkinen, P.E. Lindelof, and M. Manninen, to be published.

[5] H. Häkkinen and M. Manninen, Phys. Rev. Lett. **76**, 1599 (1996).

[6] H. Häkkinen and M. Manninen, Europhys. Lett. **34**, 177 (1996).

[7] N. Troullier and J.L. Martins, Phys. Rev. B **43**, 2081 (1991).

[8] H. Häkkinen, R.N. Barnett, and U. Landman, Europhys. Lett. **28**, 263 (1994).

[9] W. D. Knight, K. Clemenger, W. A. de Heer, W. A. Saunders, M. Y. Chou, and M. Cohen, Phys. Rev. Lett. **52**, 2141 (1984).

[10] W. de Heer, Rev. Mod. Phys. **65**, 611 (1993).

[11] K. Clemenger, Phys. Rev. B **32**, 1359 (1985).

[12] W. Ekardt and Z. Penzar, Phys. Rev. B **38**, 4273 (1988).

[13] H. Häkkinen, J. Kolehmainen, M. Koskinen, P.O. Lipas, and M. Manninen, Phys. Rev. Lett. **78**, 1034 (1997).

[14] J. Mansikka-aho, M. Manninen, and E. Hammarén, Z. Phys. D **21**, 271 (1991).

[15] J. Mansikka-aho, E. Hammarén, and M. Manninen, Phys. Rev. B **46**, 12649 (1992).

[16] N. Pavloff and S. Creagh, Phys. Rev. B **48**, 18164 (1993).

[17] S.N. Khanna and P. Jena, Chem. Phys. Lett. **218**, 383 (1994).

[18] A.P. Seitsonen, M.J. Puska, M. Alatalo, R.M. Nieminen, V. Milman, and M.C. Payne, Phys. Rev. B **48**, 1981 (1993).

[19] A.P. Seitsonen and K. Laasonen, J. Chem. Phys. **103**, 8075 (1995).

[20] H. Häkkinen and M. Manninen, J. Chem. Phys. **105**, 10565 (1996).

CLUSTER SURFACE INTERACTION:
SHAPES AND MORPHOLOGIES

C. Bréchignac,[1] Ph. Cahuzac,[1] F. Carlier,[1] C. Colliex,[1,2] M. de Frutos,[1]
A. Masson,[1] C. Mory,[2] and B. Yoon[1,2]
[1] Laboratoire Aimé Cotton, CNRS, Bât 505
[2] Laboratoire de Physique des Solides, Bât 510
Campus d'Orsay, 91405 Orsay cedex, France

ABSTRACT

This paper presents island formation from preformed antimony clusters soft landed on thin films of amorphous carbon and cleavage surfaces (0001) of graphite. By exploiting the decrease of cluster mobility as cluster size increases as well as the cluster surface interaction we are able to modify the kinetic of nucleation, growth and coalescence processes. Compact islands have been observed on amorphous carbon, whereas an evolution from compact to dendritic shapes with a fractal dimension occurs on graphite substrate as the mean size of the deposited clusters increases. The supported island shapes are discussed in terms of kinetic of nucleation and growth processes and size effects in coalescence between two adjacent clusters.

1. INTRODUCTION

Up to now thin films have been obtained either by classical metallurgical approaches or by more sophisticated molecular beam techniques. However these techniques present some limitations such as the obtention of a high island density with a narrow range of size, which constitutes a real challenge in order to explore specific properties of small particles. The development of cluster sources, delivering a high particle flux over an adjustable size distribution, offers a new method for the realization of these nanostructured thin films.

The aim of this paper is to discuss nucleation and growth of islands [1] on different substrates, amorphous or crystalline, from preformed clusters. Striking differences are observed concerning the island growth and the evolution of island morphology with the size of the deposited clusters and also with the nature of interacting partners. Which parameters are relevant for compact or fractal structures, how the growth and the structure of the final object can be controlled are among the fundamental questions that we want to answer.

2. EXPERIMENTAL PROCEDURE

In our standard metal cluster source [2], the metal vapor produced by thermal evaporative process is quenched in a flow of helium gas cooled by liquid nitrogen. The helium carrier gas cools the metal which then condenses into clusters. Our cluster source produces a distribution of cluster masses which follow a pseudoGaussian distribution. Both the maximum of the cluster distribution and its width are determined by the partial pressures of the metallic vapor and of the carrier gas, and by the nozzle diameter. The cluster distribution is monitored by a time of flight mass spectrometer.

The neutral cluster beam is deposited at low kinetic energy, less than 0.05 eV per atom two meter downstream from the source onto a prepared surface maintained at room temperature in a vessel with a residual pressure of 10^{-8} torr. Under such conditions we know from previous unimolecular dissociation experiments that fragmentation does not occur [2]. The deposition time is varied from 15 seconds to few minutes. The deposition rate is measured by a microbalance previously calibrated from Rutherford Back Scattering (RBS).

The specimen obtained from deposition of antimony clusters on amorphous carbon or on graphite are transferred into an electron microscope. We use the high angle annular dark field mode in the VGHB 501 scanning transmission electron microscope (STEM) to visualize the shape, the size and the density distributions of the deposited antimony islands. With this detector configuration, one collects all electrons of 100 keV incident energy scattered at a relatively large angle, between 30 and 150 mrad. At those angles scattering processes are mostly elastic processes depending strongly on the Z number of the concerned atoms. Consequently they can be adequately used to discriminate heavy antimony atoms from light carbon ones. Quantitatively the relevant cross-section ratio between those two atomic species is of the order of 15, which means that a double intensity level between adjacent pixels corresponds in one case to a given number of C atoms and in the other one to the same number of C atoms on which are sitting a number of Sb atoms reduced by a factor of 15 with respect to the number of carbon atoms. One thus understands how quantitative maps of the intensity distribution in such images can be used to plot the number of atoms in different islands [3].

The chemistry of the observed material has also been checked by XPS or electron energy loss spectroscopy (EELS). It has been shown that a noticeable oxidation process only occurs after a long preservation of the specimens in the air (more than one week) and that a rapidly transferred specimen maintains its antimony character. Moreover we have checked that the oxidation does not affect the shape of the deposited islands.

3. RESULTS AND DISCUSSION

3.1. Antimony deposition on amorphous carbon

It has been already shown that molecular antimony growth on carbon substrates occurs in the 3D island or Volmer-Weber mode, where small particles are nucleated on the substrate and then grow into islands of the condensed phase [4]. In this mode, the concentration of adsorbed molecules is found to be very low at room temperature, due to fast re-evaporation from the weakly binding substrate. The condensation coefficient is defined as the ratio of the number of atoms adhering to the substrate to the number of atoms arriving. In our case the density of adhering atoms is deduced from the intensity distribution of micrographs and the density of arriving atoms from the microbalance. We found at room temperature an increase of the condensation coefficient from 0.3 to 1 as increasing the size of the incident clusters from Sb_4 to Sb_{150}. For incident preformed clusters having more than 150 atoms per cluster, all the arriving material is condensed onto the substrate.

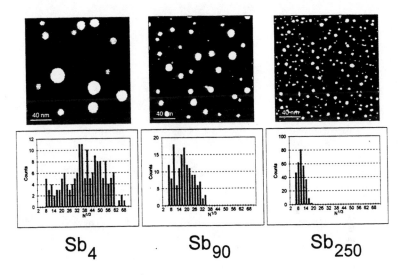

Figure 1. Annular dark field STEM micrographs of Sb islands grown from the slow deposition of Sb_n clusters of average size respectively n = 4, n = 90, n = 250, together with the histograms of their size distribution.

Fig. 1 shows a sequence of images for increased mean size $\bar{n}$ of incident clusters Sb_n, while keeping constant a coverage of about 0.5 ML. For each micrograph an histogram of the number of particles with a given radius is shown and constitutes a first approach to the evaluation of the characteristics of the deposited material for a given $\bar{n}$. All antimony islands exhibit compact spherical shapes. However it is clearly seen that at room temperature, for the same coverage, the islands grown from molecular Sb_4 deposition are larger than those grown from clusters containing about 90 atoms which are themselves larger than those obtained from an incident cluster distribution peaking at 250 atoms. Moreover in addition to a decrease of the mean value of the supported island size distribution, the increase of the incident cluster size gives rise to a narrowing of the supported island distribution. Such a narrow width of the supported island distribution from cluster deposition constitutes an interesting approach to cluster-controlled granular film formation.

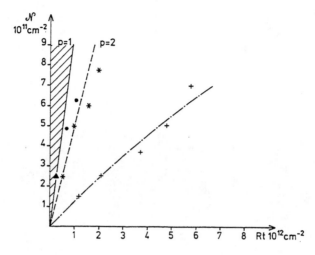

Figure 2. Plot of the average density $\mathscr{N}$ of Sb islands as a function of deposition rate Rt. The symbols correspond to different mean sizes $\bar{n}$ of the incident clusters : (+) n = 150 ; (*) n = 350 ; (·) n = 600; (Δ) n = 2200. The curves of slope p = 1, p = 2 ... identify behaviors corresponding to different numbers p-1 of collision processes involved in the production of the observed islands.

This behavior is interpreted in terms of kinetic processes. Assuming that antimony clusters are mobile and that their mobility decreases as cluster size increases as shown already for Ag [5], the diffusion length of Sb_n decreases as n increases. The diffusion of a cluster stops when it hits a critical nucleus or a stable island. With increasing cluster size the probability for a mobile cluster to find a second mobile cluster to create a stable nucleus is higher than the probability to attach an existing island and both the mean size of the supported island distribution and its width decrease. Values of island densities $\mathscr{N}$ versus incident cluster dose Rt, where R is the cluster flux and t the deposit time, are shown in Fig. 2 for various incident cluster size. For a mean value $\bar{n}$ of the incident cluster size distribution larger than 150 we have measured that all the arriving material is condensed onto the substrate. The conservation of mass leads to $\bar{n}$ Rt = $\mathscr{N}\bar{N}$ ($\bar{N}$ stands for the average number of atoms constituting the supported islands). Hence $\mathscr{N}$ / Rt = $\bar{n}/\bar{N}$ = 1 / p where p-1 is the mean number of collisions to form an island from preformed clusters of mean size $\bar{n}$. We have found that the average size of the islands is approximately twice the average size of the incident clusters for $\bar{n}$ = 350. The classical atomic nucleation transposed here to cluster gives Sb_{350} and Sb_{700} as critical and stable nucleus respectively. For larger incident cluster size distribution the deposited clusters are very weakly mobile and the supported island distribution is identical to the incident one. Table 1 provides an estimate of the average number $\bar{N}$ of atoms constituting the observed islands as a function of the average value $\bar{n}$ of the incident cluster size distribution for the same coverage $\theta = 0.7$ ML.

$\bar{n}$	4	90	150	250	350	600	2200
$\bar{N}$		7×10^4	4×10^3	1300	750	700	1000	2200

The shape of the supported islands formed from preformed cluster deposition informs about the coalescence processes. The micrographs of Fig. 1 recorded with the STEM show islands with compact hemispherical shapes in agreement with a liquid like coalescence process. In the small cluster size range, the decrease of the melting temperature with decreasing cluster size [6,7] and the energy release produced by the reduction in the total surface area during coalescence are in favor of a liquid-like behavior for the small cluster size deposition. For larger cluster sizes, such a behavior is no longer valid as shown in Fig. 3 in which a bunch of incident clusters is observed. At that stage it should be very interesting to compare the atomic-scale structures of the granular films obtained from Sb_{120} leading to spherical islands of 2200 atoms and those obtained from deposition of Sb_{2200}, which may reflect the structure of free clusters.

20 nm

Figure 3. Annular dark field image of a family of Sb_{2200} clusters deposited on amorphous carbon. Even when they collide to form an island of bigger size, the primary clusters maintain their individuality.

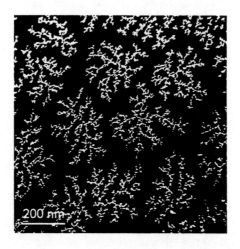

Figure 4. Annular dark field STEM micrograph of the fractal-type topography of islands issued from the deposition of Sb_{250} clusters on a defect-free surface of graphite for a total coverage of 0.7 ML.

3.2. Antimony deposition on graphite

As previously shown the Sb-films obtained by cluster deposition on graphite substrate exhibit dendritic shapes [8] as shown in Fig. 4. A dimensional analysis has been performed following the procedure described in ref. [9,10]. Subsets of the islands were defined by concentric circles about the island center. The mass M, deduced from STEM measurements, inside a circle is plotted versus the radius R of the corresponding circle in a log-log plot. Fractal behavior would imply a power-law dependence of the mass M with its radius R of the form $M \sim R^D$. Despite a reduced range for the power-law behavior, a fractal dimension of D $= 1.72 \pm 0.008$ is obtained for deposition of Sb_n with n > 150. This value is in agreement with that determined from simulation of 2D DLA growth [11]. In the original DLA model, Witten and Sander [11] have simulated fractal growth from atom deposition assuming that atoms migrate via random walks with small jump lengths over the surface until two of them meet. Then the two atoms stick and stop to form the center of the island which in turn grows by the same process. Such a process of growth implies that the edge diffusion of atom along an island is sufficiently low while free adatom diffusion is still appreciable [12]. Pronounced dendritic growth requires large defect free regions and low densities of nuclei as compared to the fractal sizes. In our case the mobility of clusters on graphite is larger than on amorphous carbon due to the free defect cleavage graphite surface. The transposition of DLA growth from atom to cluster may imply that the coalescence of clusters constituting the island is sufficiently low while cluster diffusion is still fast.

Fig. 5 shows a sequence of morphologies of islands for different incident cluster size. The evolution from compact equilibrium shape to fractal non equilibrium or growth shape as the deposited cluster size increases should be determined by the kinetic of the growth mechanism. From Fig. 5 it is seen that the thickness of the dendritic branches decreases with increased the incident cluster size until it reaches a width of about 60 Å. This smallest observed width has been obtained from deposition of a cluster distribution peaking at 2000 atoms of 60 Å diameter. For Sb_4 deposition the edge diffusion is very rapid and the shape of the islands relaxes to their equilibrium spherical shape. As increasing cluster size, the time to coalesce and the energy release decrease whereas the cluster mobilities are still appreciable leading to branches with decreasing width.

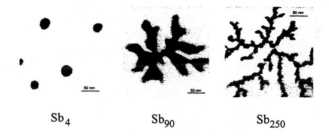

$$Sb_4 \qquad\qquad Sb_{90} \qquad\qquad Sb_{250}$$

Figure 5. Sb island shapes on graphite (0001) surface after deposition of preformed cluster Sb_n at room temperature, for a total coverage of about 1 ML.

4. CONCLUSION

In conclusion the manometer scale structures obtained from controlled cluster deposition open new opportunities for creating metastable states. By exploiting the mobility of adcluster on surfaces we are able to grow densely nanostructures of various morphologies.

References

[1]   J.A. Venables, G.D.T. Spriller, M. Hanbiicken, Rep. Prog. Phys. 47, 399 (1984).

[2]   C. Bréchignac, Ph. Cahuzac, F. Carlier, M. de Frutos, J.Ph. Roux, J. Chem. Phys. 102, 763 (1995).

[3]   C. Colliex and C. Mory in *Quantum electron microscopy*, Edited by J. N. Chapman and A.J. Craven, The Scottish Universities Summers Schools in Physics, 25, 149 (1984).

[4]   M. Hashimoto, M. Itoh, H. Kuramochi, and J. Takei, Thin Solid Films 161, 123 (1989).

[5]   J.M. Wen, S.L. Chang, J.W. Burnett, J.W. Evans, and P.A. Thiel, Phys. Rev. Lett. 73, 2591 (1994).

[6]   J.P. Borel, Surface Science 106, 1 (1981).

[7]   F. Ercolessi, W. Andreoni, and E. Tosatti, Phys. Rev. Lett. 66, 911 (1991).

[8]   L. Bardotti, P. Jensen, A. Hoareau, M. Treilleux, and B. Cabaud, Phys. Rev. Lett. 74, 4694 (1995).

[9]   P. Meakin, Phys. Rev. A, 27, 1495 (1983).

[10] M. Tence, J.P. Chevalier, and R. Jullien, J. Physique 47, 1989 (1986).

[11] T.A. Witten and L.M. Sander, Phys. Rev. Lett. 47, 1400 (1981).

[12] G.S. Bales and D.C. Chrzan, Phys. Rev. B 50, 6057 (1994).

FULLERENE MOLECULES COATED WITH TRANSITION METALS

F. Tast, N. Malinowski[1], S. Frank, M. Heinebrodt, I.M.L. Billas,
and T.P. Martin

Max-Planck-Institut für Festkörperforschung
Heisenbergstrasse 1, 70569 Stuttgart, Germany

ABSTRACT

Compound clusters of fullerene molecules and transition metal atoms having the composition $C_{60}M_x$ and $C_{70}M_x$ with x = 0..150 and M $\in$ {Ti, Zr, V, Y, Ta, Nb} were produced using laser vaporization in a low-pressure inert gas aggregation cell. Intensity anomalies in the mass spectra correlate with the atomic radii of the different metals indicating the formation of complete metal layers around the central fullerene molecule. Using high laser intensities the metal-fullerene clusters can be transformed into met-cars and metal-carbides. Photofragmentation spectra of preselected $C_{60}Ta_x$ indicate that the fullerene cage is destroyed for x $\geq$ 3.

1. INTRODUCTION

An early experiment immediately following the discovery of C_{60} [1] was the endohedral doping with a lanthanum atom [2, 3]. The fullerene can then be shrink-wrapped by photofragmentation until a critical size is reached and the lanthanum atom has to be released because the carbon cage has become too small to fit around the dopant. Later a series of different atoms and molecules were successfully shrink-wrapped in this manner, yielding information about the available volume inside fullerene molecules [4, 5, 6, 7]. Here we report on a complementary experiment: fullerene molecules are 'wrapped' with metal atoms.

---

[1]*Permanent address:* Central Laboratory of Photoprocesses, Bulgarian Academy of Science, 1040 Sofia, Bulgaria

Earlier experiments on alkali metal coated C_{60} indicated that the stability of these clusters seems to be determined predominantly by electronic effects [8]. Similar experiments for alkaline earth metals led to a different result. Here the stability is of geometric origin determined by the formation of multiple icosahedral metal layers around a central fullerene molecule [9].

In this paper we report on transition metal coated fullerene clusters. Geometric magic numbers in the mass spectra are observed and interpreted as the result of the formation of a first complete metal layer covering the fullerene molecule. Surprisingly, we do not find just one magic number for each metal, but a set of distinct intensity anomalies of comparable intensity, indicating that there are different ways of building a complete metal layer on a fullerene molecule surface.

2. EXPERIMENTAL

As cluster source we used a modified low-pressure, inert gas condensation cell [10]. This standard technique is, however, limited to materials with sufficiently high vapor pressures at the temperatures attainable in resistively heated ovens. This limitation could be overcome using laser desorption. The transition metals with their very low vapor pressure were evaporated using the second harmonic output of a pulsed Nd:YAG laser focused onto a metal target inside the condensation cell. The laser vaporized material mixes with fullerene vapor from a resistively heated oven. This vapor mixture is quenched in a low pressure helium atmosphere (ca. 1 mbar) cooled by liquid nitrogen. The composition of the resulting clusters can be controlled by changing the temperature of the fullerene oven and the power and focusing conditions of the vaporization laser. At moderate fullerene oven temperatures it is possible to find operating conditions where almost exclusively clusters of the composition $C_{60}M_x$ or $C_{70}M_x$ containing only one fullerene molecule and a varying number of metal atoms are formed. The clusters are transported from the condensation cell through a nozzle and a differential pumping stage into a high vacuum chamber where they are photoionized and mass analyzed in a time-of-flight mass spectrometer (TOF).

For the ionization we used excimer laser pulses at three different laser wavelengths (308nm, 248nm, and 193nm). In order to study the relative stability of the clusters it is useful to heat them until they evaporate atoms. The rate of the evaporation is determined by the stability of the different species. Clusters of low stability will quickly lose atoms until a more stable configuration is reached. Thus the intensity of mass peaks corresponding to especially stable clusters is enhanced. In our experiments heating and ionization were achieved simultaneously using a sufficiently high excimer laser fluence ($\approx 10 \, mJ / cm^2$).

3. RESULTS AND DISCUSSION

A mass spectrum of $C_{60}Ti_x$ is shown in Fig. 1. The spectrum contains peaks corresponding to both singly and doubly ionized clusters. The peaks of each charge state series are connected with a line to emphasize the intensity fluctuations. Especially strong peaks are observed for x = 62, 72 and 80. Note also the weak edge for x=50. The same magic numbers are repeated in the doubly charged series indicating that the special stability of $C_{60}Ti_{62}$, $C_{60}Ti_{72}$, and $C_{60}Ti_{80}$ is independent of the charge state of the clusters and therefore presumably of geometric origin.

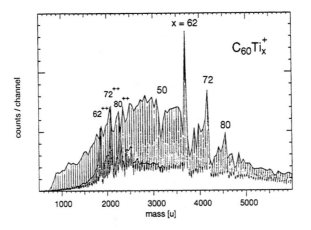

Figure 1. Mass spectrum of singly and doubly photoionized $C_{60}Ti_x$ clusters. Magic numbers are observed for x=62,72, and 80 in both the singly and doubly charged series, indicating that the enhanced stability of the corresponding clusters is independent of the number of electrons and therefore presumably of geometric origin. Note also the weak step for x=50.

Using zirconium instead of titanium, we get similar results. Clearly discernible steps in the mass spectrum of $C_{60}Zr_x$ (not shown) indicate that again clusters with x=50,62,72, and 80 are exceptionally stable. In the following, we will present further results indicating that this enhanced stability is presumably due to the completion of a first layer of metal atoms on the fullerene molecule surface.

First, let us compare the observed magic numbers with earlier results for alkaline earth metal coated fullerenes, where the proposed layer growth mechanism could be established [9]. For $C_{60}Ca_x$, $C_{60}Sr_x$, and $C_{60}Ba_x$ the first metal layer is completed at x=32. Additional weaker magic numbers at x=35,38, and 43 were observed. Considering the significantly smaller size of Ti and Zr, a larger number of atoms is expected to fit on the surface of one C_{60} molecule. Using the atomic radius as an estimate for the size of the metal atoms (r_{Ca} = 3.731a.u., r_{Sr} = 4.065a.u., r_{Ba} = 4.109a.u., r_{Ti} = 2.737a.u., and r_{Zr} = 3.005a.u.) the surface area occupied by a Ca, Sr or Ba atom is about $2^2/1.5^2 \approx 1.8$ times larger than the area covered by a Ti or Zr atom. Thus about 1.8 times more Ti or Zr atoms are necessary for the formation of a complete metal layer on a C_{60} molecule. This agrees with the ratio of the observed magic numbers.

The fact that we observe several different magic numbers is probably due to different possibilities of arranging the metal atoms to form a closed layer. In the case of the alkaline earth metals, not only one but up to four successive layer closings could be observed. This additional information allowed the determination of a growth pattern and the probable structure of the clusters. For the transition metals, the second layer closing could not be determined because of the experimentally limited metal content. Therefore the interesting question of the actual structure of the layer responsible for the observed magic numbers cannot be answered from the available experimental data. However, we can check the expected size dependence by looking at other materials.

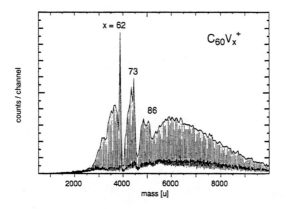

Figure 2. Mass spectrum of photoionized $C_{60}V_x$ clusters. The peak with x=62 is especially strong. Weaker steps are visible for x=73 and 86.

Therefore, let us now turn to the results for vanadium having an even slightly smaller size ($r_V = 2.478$a.u.). The mass spectrum of $C_{60}V_x$ in Fig. 2 shows intensity anomalies similar to the features observed for $C_{60}Ti_x$ and $C_{60}Zr_x$. Again x=62 is a very strong magic number. Two weaker steps are observed at x=73 and 86. Compared to the Ti and Zr, these two magic numbers are slightly shifted to larger values. In the case of the smaller vanadium atom the corresponding packing patterns are apparently slightly modified accommodating one or respectively four additional metal atoms in a complete layer. Note also that there is no intensity anomaly at x=50 in contrast to the two previous spectra. Comparing the mass spectra in decreasing order of the size of the three different metal atoms, the magic number x=50, clearly visible in the $C_{60}Zr_x$ spectrum, gets weaker for $C_{60}Ti_x$ and finally disappears in the case of $C_{60}V_x$. Obviously, 50 vanadium atoms are not sufficient to form a complete coverage of a C_{60} molecule.

Instead of varying the size of the metal atoms it is also possible to change the central molecule from C_{60} to C_{70}. In the simplest picture this amounts to changing the available surface area by approximately a factor of $70/60 \approx 1.2$. This is about equal to the ratio r_{Ti}^2/r_V^2 between the areas occupied by Titanium and Vanadium atoms. Thus the relation between metal atom size and available fullerene surface area in the case of $C_{60}V_x$ is comparable to that of $C_{70}Ti_x$. And in fact the mass spectrum of $C_{70}Ti_x$ shown in Fig. 3 reveals the same magic number pattern x=62, 73, and 86 as observed in the $C_{60}V_x$ system plus an additional very weak step at x=95 which can be seen more clearly in the doubly charged series. This indicates that the structure of the metal layer is not determined by the detailed structure the fullerene molecule. It is rather the available fullerene surface area compared to the size of the metal atoms that predominantly influences the magic numbers.

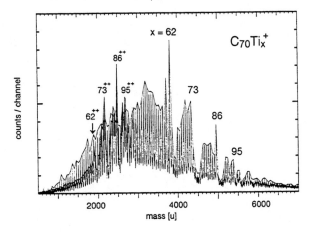

Figure 3. Mass spectrum of photoionized $C_{70}Ti_x$ clusters showing especially strong peaks for x=62, 73, and 86. The weak edge at x=95 can be seen more clearly in the peak series of the doubly charged clusters.

This leads to the following simple geometrical model. The number N of metal atoms that can be accommodated on the surface of a fullerene molecule is given by the ratio between the available fullerene surface area $A_{fullerene}$ and the area A_{metal} occupied by one metal atom

$$N = f \cdot \frac{A_{fullerene}}{A_{metal}} .$$

The filling factor f takes care of the fact that spheres cannot be packed together without leaving interstitial voids. Introducing the fullerene radius $R_{fullerene}$ and metal atom radius r_{metal} and substituting $A_{fullerene} = 4\pi(R_{fullerene} + r_{metal})^2$ and $A_{metal} = \pi r_{metal}^2$ we get

$$\sqrt{N} = 2\sqrt{f}(\frac{R_{fullerene}}{r_{metal}} + 1) . \tag{1}$$

In Fig. 4 this is compared with the experimental data. The square root of the observed magic numbers is plotted over $R_{fullerene}/r_{metal}$, where we used $R_{C_{60}} = 9.62 a.u.$ and an effective $R_{C_{70}} = \sqrt{70/60}\, R_{C_{60}} = 10.395 a.u.$ which ensures the correct surface area ratio between C_{60} and C_{70} in our model. This allows us to show the results for C_{60} (triangles) and C_{70} (diamonds) in a single plot. The data points for the alkaline earth metals are taken from reference [9]. We also include the magic numbers found for $C_{70}V_x$ (x=62, 75, 88, and 104) and $C_{60}Y_x$ (x=41) in the diagram without showing the measured mass spectra. The solid lines represent equation (1) for three different packing densities corresponding to the indicated patterns. These planar packing schemes are of course not strictly applicable for a spherical surface. They only serve as an estimate for reasonable packing densities. Altogether the size dependence of the presented magic numbers agrees well with this simple picture, apparently there are certain numbers which allow a geometrically favorable arrangement of metal atoms on a fullerene molecule provided that the atoms have the right size to cover the fullerene molecule. This size criterion has to be fulfilled only approximately, allowing several different magic numbers for the same metal corresponding to different packing schemes with different packing densities.

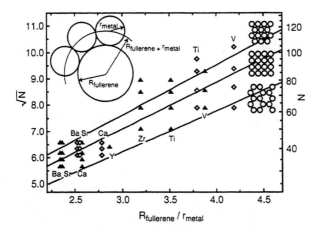

Figure 4.Observed magic numbers of metal fullerene clusters. Results for C_{60} are represented by triangles and for C_{70} by diamonds. Details see text. The data for the alkaline earth metals is taken from reference [9].

4. PHOTODISSOCIATION EXPERIMENTS

To study the photodissociation of these transition metal coated fullerenes we used a modified experimental set-up. An additional preselection TOF, followed by controlled photofragmentation with a second heating laser pulse and subsequent mass analysis in the main TOF allows a detailed exploration of the fragmentation process. Details of these tandem-TOF experiments are described elsewhere [11].

Starting with $C_{60}V_x$ clusters and preselecting a single peak, e.g. $C_{60}V_{20}$, we observe at low fragmentation laser intensities the successive loss of single metal atoms. When the fragmentation laser fluence is increased, we start to see carbon loss also, indicating a disruption of the fullerene cage. At sufficiently high laser intensities the most abundant product of this fragmentation process is V_8C_{12}, a so called met-car cluster. The special stability of this species was observed for the first time by Castleman and coworkers [12]. They created this cluster in a plasma reaction from hydrocarbon gas and metal vapor and coined the name metallo-carbohedrene (met-car). Analogous clusters M_8C_{12} containing other transition metals were also observed in similar experiments [13, 14]. Actually, the titanium met-car Ti_8C_{12} was the first member of the met-car family to be discovered [15]. This prototype met-car could also be produced in our experiments as photofragmentation product, starting from metal coated fullerenes $C_{60}Ti_x$ and $C_{70}Ti_x$.

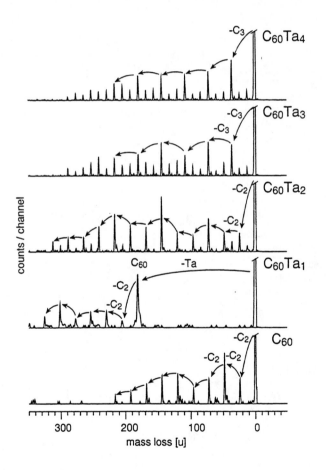

Figure 5. Fragmentation spectra of preselected $C_{60}Ta_x$ with x = 0.4.

We also investigated fullerene molecules doped with tantalum and niobium. Here the dissociation behavior is different. The final products of the fragmentation process are metal carbide clusters with a metal carbon ratio of about 1:1 in agreement with the stoichiometry of bulk NbC and TaC. These carbide clusters show the typical magic number pattern of cubic nano-crystals in accordance with the bulk lattice structure.

Tandem-TOF experiments on these clusters gave surprising results. For preselected clusters $C_{60}Ta_x$ with $x \geq 3$ the first observable decay channel is C_3 loss. This is in contrast to the results for $C_{60}V_x$, for which we always observe metal loss at low laser intensities before carbon loss sets in at higher heating laser fluences. This evaporation of neutral C_3 molecules observed here is well-known in small carbon cluster C_n with $n \leq 20$ which are too small to form fullerene cages [16]. Regarding this C_3 loss as a fingerprint of non-fullerene clusters, the observed fragmentation behavior indicates that in the case of $C_{60}Ta_x$ clusters, the fullerene cage is already broken when the first fragmentation step occurs. This is true for C_{60} doped with more than two tantalum atoms. For $x = 1$ and $x = 2$ however, the fullerene cage seems to be still intact, even when the cluster is heated until fragmentation sets in. This can be concluded from Fig. 5 showing the first fragmentation steps of preselected $C_{60}Ta_x$ for $x = 0..4$. $C_{60}Ta_1$ first loses its tantalum atom. The remaining C_{60} then displays the usual C_2 loss sequence. For $C_{60}Ta_2$ this C_2 loss is observed immediately, signaling that the cage structure of C_{60} is still intact, although the two metal atoms are firmly bonded to the cluster. The loss of C_3 is also observed as a second minor decay channel. This C_3 loss becomes predominant for clusters containing three or more tantalum atoms, indicating a non-fullerene structure.

5. CONCLUDING REMARKS

We have presented a series of mass spectra of transition metal coated fullerenes showing strong intensity anomalies. They are interpreted as the result of an enhanced stability of metal-fullerene clusters with a complete metallic layer. Exposed to high laser intensities these metal coated fullerene molecules transform into met-car or metal-carbide clusters. Photodissociation measurements on preselected $C_{60}Ta_x$ clusters reveal a severe destabilization of the fullerene cage in clusters with more than two metal atoms.

References

[1] Kroto, H.W., Health, J.R., O'Brien, S.C., Curl, R.F., Smalley, R.E.: Nature **138**, 165 (1985).

[2] Health, J.R., O'Brien, S.C., Zhang, Q., Liu, Y., Curl, R.F., Kroto, H.W., Tittel, F.K., Smalley, R.E.: J. Am. Chem. Soc. **107**, 7779 (1985).

[3] Cox, D.M., Trevor, D.J., Reichmann, K.C., Kaldor, A.: J. Am. Chem. Soc. **108**, 2457 (1986).

[4] Weiss, F.D., Elkind, J.L., O'Brien, S.C., Curl, R.F., Smalley, R.E.: J. Am. Chem. Soc. **110**, 4464 (1988).

[5] Chai, Yan, Guo, Ting, Jin, Changuning, Haufler, R.E., Chibante, L.P.F., Fure, J., Wang, Lihong, Alford, J.M., Smalley, R.E.: J. Phys. Chem. **95**, 7564 (1991).

[6] Weiske, T., Bohme, D.K., Hrusak, J., Krätschmer, W., Schwarz, H.: Angew. Chem. Int. (Ed. Engl.) **30**, 884 (1991).

[7] Martin, T.P., Heinebrodt, M., Näher, U., Göhlich, H., Lange, T., Schaber, H.: Int. J. mod. Phys. B **6**, 3871 (1992).

[8] Martin, T.P., Malinowski, N., Zimmermann, U., Näher, U., Schaber H.: J. Chem. Phys. **99**, 4210 (1993).

[9] Zimmermann, U., Malinowski, N., Näher, U., Frank, S., Martin, T.P.: Phys. Rev. Lett. **72**, 3542 (1994).

[10] Zimmermann, U., Malinowski, N., Näher, U., Frank, S., Martin, T.P.: Z. Phys. D **31**, 85 (1994).

[11] Tast, F., Malinowski, N., Frank., S., Heinebrodt, M., Billas, I.M.L., Martin, T.P.: Phys. Rev. Lett. **77**, 3529 (1996).

[12] Guo, B.C., Wei, S., Purnell, J., Buzza, S., Castleman, Jr., A.W.: Science **256**, 515 (1992).

[13] Wei, S., Guo, B.C., Purnell, J., Buzza, S. Castleman, Jr., A.W.: Science **256**, 818 (1992).

[14] Pilgrim, J.S., Duncan, M.A.: J. Am. Chem. Soc. **115**, 9724 (1993).

[15] Guo, B.C., Kerns, K.P., Castleman, Jr., A.W.: Science **255**, 1411 (1992).

[16] Geusic, M.E., McIlrath, T.J., Jarrold, M.F., Bloomfield, L.A., Freeman, R.R., Brown, W.L.: J. Chem. Phys. **84**, 2421 (1986).

A MOLECULAR FP-LMTO STUDY OF COPPER CLUSTERS

A. Mookerjee,[1] R.P. Datta,[1] A. Banerjea,[1] and A.K. Bhattacharyya[2]
[1] S.N. Bose National Centre for Basic Sciences
JD Block, Sector 3, Salt Lake City, Calcutta 700091, INDIA.
[2] Department of Engineering
University of Warwick, Coventry CV47AL, U.K.

1. INTRODUCTION

Simple alkali metals clusters are fairly well described by the spherical jellium model. The quasi-free valence electrons occupy single-particle states in an effective spherically symmetric box potential. This is rather insensitive to the geometry of the atomic arrangement inside the cluster. Consequently one obtains a pronounced shell closing effect [1]. Although the noble metals Cu, Ag, and Au have closed d-shells and singly occupied outermost s-shell structures and several authors have suggested that there should be a close similarity to the shell closing effect in simple alkali metals, cohesive studies in the bulk metal and a series of EXAFS studies of Cu clusters supported on carbon [2] - [4] indicate that the d-electrons, through their hybridization with the s-electrons, play an important role in the electronic structure and binding energy of these systems.

The most powerful first-principles molecular dynamics method, the Car-Parinello technique [5] has been eminently successful in dealing with s-p valence clusters : e.g. the alkali and alkaline earth metals, Al, and Si. The basic pseudopotential approach has not been that successful with noble and transition metals. There have been attempts at combining other techniques with simulated annealing. The first one in this regard is the empirical tight-binding (TB) or the Linear Combination of Atomic Orbitals (LCAO) methods [6] - [7], which are at best qualitative, since the assumption of transferability of the hamiltonian parameters is definitely of questionable validity. The next one is the TB-Linearized Muffin-Tin Orbitals (TB-LMTO) method [8] coupled to simulated annealing, which faces several problems when applied to clusters. The treatment of the interstitial region outside the muffin-tin spheres centered at the atomic positions is difficult. Unlike the bulk, where the interstitial region is small and inflating the muffin-tins to slightly overlapping atomic spheres can do away with the interstitial altogether, for clusters this is certainly not so. As the atoms move about, the atomic spheres may not overlap and the interstitial contribution is significant. One may try to overcome this by enclosing the cluster with layers of empty spheres carrying charge but not atoms. This complicates the actual calculations enormously and it is entirely unclear if this has been done in the work referenced.

In the work reported here, we turn to the molecular version of the Full-Potential LMTO (FP-LMTO) suggested by Methfessel and van Schilfgaarde [9] - [10] and combine it with the Harris-functional technique to carry out a molecular dynamical study of Cu clusters ranging in size between 10 and 20.

2. THE FP-LMTO MOLECULAR DYNAMICS

The molecular version of the Full Potential-LMTO utilizes the basic philosophy to Muffin-Tin Orbitals methods. It is based on the Density Functional Theory in the Local Density Approximation. The electron-electron interaction is treated approximately. In practice :

$$\left[-\nabla^2 + V_{eff}(\underline{r})\right]\psi_i(\underline{r}) = \varepsilon_i\psi_i(\underline{r})$$
$$\rho(\underline{r}) = \sum_i f_i|\psi_i(\underline{r})|^2$$
$$where \ \ V_{eff} = V_N(\underline{r}) + 2\int\frac{\rho(\underline{r'})}{|r-r'|}d^3\underline{r'} + V_{xc}\left(\rho(\underline{r})\right). \quad (1)$$

The first step is the solution of the Schrödinger equation in a very unpleasant potential with Coulomb singularities. As in most approaches we use the variational approach. We choose a basis of representation $\{\phi_m(\underline{r})\}$ such that

$$\Psi(\underline{r}) = \sum_m c_m\psi_m(\underline{r}).$$

The problem reduces to a matrix eigenvalue problem :

$$\mathcal{H}\underline{c} = \varepsilon S\underline{c}.$$

Computational effort scales as $\sim$ (matrix dimension)^3. For free clusters, we cannot reduce this through Bloch's theorem and we do not wish to follow a supercell technique. Our approach tries to use a minimal basis set at the expense of a rather complicated formulation. The basis is built up of Hänkel functions $H_{iL}(\underline{r})$ diverging at $\underline{r} = \underline{R}_i$, augmented inside the muffin-tin spheres by solutions $u(r)Y_L(\hat{r})$ of the Schrödinger equation :

$$u''_L(r) = \left[\frac{\ell(\ell+1)}{r} + V(r) - \varepsilon\right]u_L(r)$$

with boundary conditions such that its logarithmic derivative matches that of the Hänkel function. Any matrix element of an arbitrary operator $\mathcal{O}$ in this basis then can be written as :

$$\langle \phi_{iL} | \mathcal{O} | \phi_{jL'} \rangle = \left[\sum_k \int_{S_k} + \int_I \right] \phi_{iL}^*(\underline{r}) \mathcal{O} \phi_{jL'}(\underline{r}) d^3\underline{r}. \tag{2}$$

Here S_k is the muffin-tin sphere centered at $\underline{R}_k$. The Hänkel functions associated with a muffin-tin at $\underline{R}_i$ can be written in terms of a Bessel function at $\underline{R}_j$ as $\sum_{L''} S_{iL,kL''} J_{kL''}$. The structure matrix S depends entirely on the geometric arrangement of the muffin-tins. The first integral becomes

$$= \sum_{k \neq i,j} \sum_{L''L'''} S_{iL,kL''}^* S_{jL',kL'''} \langle J_{L''} | \mathcal{O} | J_{L'''} \rangle_{S_k} + \sum_{k=i,\neq j} \sum_{L''} S_{jL',iL''} \langle H_L | \mathcal{O} | J_{L''} \rangle_{S_i} \cdots$$

$$+ \sum_{k=j,\neq i} \sum_{L''} S_{iL,jL''}^* \langle J_{L''} | \mathcal{O} | H_{L'} \rangle_{S_j} + \langle H_L | \mathcal{O} | H_{L'} \rangle_{S_i}$$

$$= \mathcal{O}^{HH} + S^\dagger \mathcal{O}^{JH} + \mathcal{O}^{HJ} S + S^\dagger \mathcal{O}^{JJ} S \tag{3}$$

There is a *separation* of atomic and structural information. This simplifies the actual calculations enormously. When the atoms move, the structural part changes because of change in geometry while the atomic part changes because of change in charge density (hence the potential).

Most of the interstitial integral can be obtained from the muffin-tin spheres because, in the interstitial, the basis are solutions of the Helmholtz equation, and using the Green theorem :

$$\int_I \phi_1^* \phi_2 d^3\underline{r} = \frac{1}{\kappa_1^2 - \kappa_2^2} \sum_k \int_{S_k} [\phi_1^* \nabla \phi_2 - \phi_2 \nabla \phi_1^*] d^2\underline{r}$$

$$\int_I \phi_1^* \left(-\nabla^2\right) \phi_2 d^3\underline{r} = \kappa_2^2 \int_I \phi_1^* \phi_2 d^3\underline{r} \tag{4}$$

If the potential here is a constant we can get by with the above. But for clusters this is definitely not so. In the molecular FP-LMTO we use a tabulation technique. We expand the product :

$$\phi_i^*(\underline{r}) \phi_j(\underline{r}) = \sum_m C_m^{ij} \chi_m(\underline{m})$$

where $\chi_m(\underline{r})$ is another set of muffin-tin centred Hänkel functions. In practice we put two atoms along the z-axis and make accurate numerical expansion by least squares fit for different distances and tabulate $C_m^{ij}(d)$:

$$A_{mn} = \int_I \chi_m^*(\underline{r}) \chi_n(\underline{r}) d^3 r$$

$$B_m = \int_I \phi_i^*(\underline{r}) \phi_j(\underline{r}) \chi_m(\underline{r}) d^3\underline{r}$$

$$C = A^{-1} B$$

This is the two-centre fit table (TCF). For arbitrary geometry then we may easily calculate the necessary matrix elements by a fitting procedure to the table. The procedure is *fast*.

For molecular dynamics, problem arises from the fact that the Pulay terms in the force are impossibly difficult to calculate directly as the basis set changes in a complicated manner when the atoms move. To do the molecular dynamics, we use the Harris functional procedure as follows : At a time step τ, we obtain the self-consistent charge density $\rho(\underline{r},\tau)$ using the FP-LMTO procedure. At a neighboring time $\tau+\delta\tau$, we hazard a guess $\rho_g(\underline{r},\tau+\delta\tau)$ and obtain

$$\tilde{E}(\tau) = E_H[\rho_g(\underline{r},\tau+\delta\tau)] = \sum_i \varepsilon_i \left[V_{eff}\right] - \int \rho_g(\underline{r})V_{eff}\left[\rho_g(\underline{r})\right] + U\left[\rho_g(\underline{r})\right] + E_{xc}\left[\rho_g(\underline{r})\right]$$

We build up the guess charge density by bodily moving the muffin-tin spheres containing the atoms and charge densities and simply overlapping the resultant charges. Then the force is given by

$$\left.\frac{\partial \tilde{E}}{\partial \tau}\right|_{\tau\to 0}.$$

For *dynamics* we use the Verlet algorithm :

$$\underline{r}_{n+1} = 2\underline{r}_n - \underline{r}_{n-1} + \frac{F}{m}(\Delta t)^2$$

where n denotes the time step of length Δt. We can now do straightforward molecular dynamics. But this often leads to unphysical heating/cooling of the system if our time steps are too large. For small time steps the procedure is inordinately slow. We use simulated annealing by adding an extra friction term carefully $\underline{F} \Rightarrow \underline{F} - \gamma m \underline{\dot{r}}$. Methfessel and Schilfgaarde [9] have also used a free dynamics with feedback to overcome the above difficulty.

3. RESULTS

We have chosen the various parameters for the FP-LMTO based on optimizing results for the bulk Cu and the dimer. The values of κ^2 were chosen from optimum bulk calculations. The muffin tin radii were chosen as 1.9 A° to produce the bond length and binding energies of the Cu dimer correctly. For augmentation within the sphere we have used 4s,4p,3d,4f, and 5g functions (ℓ_{max}=3). For representation of interstitial functions we have used five κ^2 values with angular momentum cutoffs ℓ_{max} = 4,4 6,2, and 1.

We may now examine Fig. 1. For N=13, we obtain the lowest energy structure to be an icosahedron with two five-fold rings, an atom in the center, and two atoms capping the rings. We have started from both a cubo-octahedral and a random structure. Both flow into the icosahedral structure, which is the stable form. For N = 12, one of the capping atoms is missing and a distortion of the uncapped pentagonal ring occurs. The structures are qualitatively similar to 12 and 13 atom clusters in Al [11]. As expected, an extra atom added to the 12 atom cluster preferentially caps the uncapped pentagon. This should be true for heteroatomic additions.

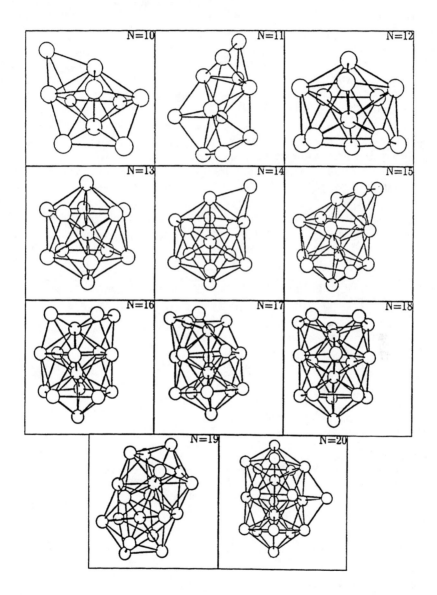

Figure 1. Energetically stable clusters of Cu of sizes 10-20.

For N = 11, the icosahedral basic makeup begins to break up. We still see a pentagonal ring capped on both the sides, but the remaining four atoms cap triangular faces of the basic seven-vertices polyhedron. For N = 10, we still see the underlying seven vertex polyhedron. The three remaining atoms do seem to cap triangular faces, but two of them (lower ones in the figure) cap symmetrically, while the third capping is quite asymmetric. As further work, it should be instructive to add a different atom to these clusters and see in which positions they bond preferentially.

As we go on adding atoms to the icosahedral structure at N = 13, atoms begin to cap one of the five faces formed by a pentagonal ring with its capping atom in the icosahedron. Atoms preferentially go on capping faces on one end of the icosahedron. For N = 15 and 16, we see some near degenerate structures. The lower energy structure has atoms on neighboring faces of the icosahedron. Other structures, differing in energy by about 1 per cent, has extra atoms on non-neighboring faces of the icosahedron. These isomers cause difficulties during the molecular dynamical evolution and care has to be taken about the time steps taken suitably small and we had to reheat and cool the structures repeatedly in order to escape from a metastable minimum.

This capping goes on till N = 18 when all the five faces are capped, forming a third pentagonal ring. At N = 19, the extra atom caps this ring to form a stable double icosahedron.

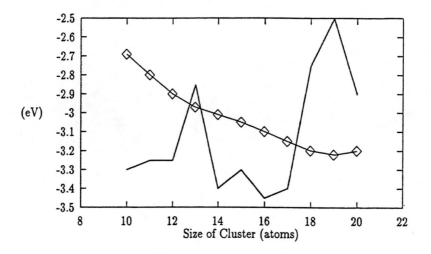

Figure 2. Binding energies (eV) and its second derivative as a function of cluster size.

Finally, at N = 20, the additional atom in stead of capping the top faces, now begins to cap one of the equatorial long side-faces of the double icosahedron. This is expected, as otherwise one would have the formation of elongated clusters, which we expect to be unstable against the formation of more geometrically compact structures. This is very different from the TB-LMTO simulated annealing which finds N = 20 to be a magic number and displays a structure with a central atom virtually surrounded by nineteen atoms [8]. This result is highly suspect as the structure is very open with almost all the atoms on the surface.

Fig. 2 shows the binding energy. There are very weak minima at N = 13 and N = 19, but unless one is determined to seek it, there are no evident signs of magic numbers. We do not find any evidence of odd-even alternation nor traditional free atom-like shell closure. We rather see evidence of stability at "icosahedral closures" in this range. Our study shows that in the range $10 \leq N \leq 20$, the most stable configurations are those that form specific tight-knit geometric configurations. This has been conjectured earlier and there is evidence of this for Ni clusters as well [12].

ACKNOWLEDGMENTS

We should like to thank Profs. M. Methfessel and M. van Schilfgaarde for making the entire mechanism of the FP-LMTO available to us and enthusing us to make use of this powerful technique.

References

[1] Christensen O.B., Jacobsen K.V., Norskov J.K. and Manninen M., Phys. Rev. Lett. **66**, 2219 (1991).

[2] Apai G., Hamilton J.F., Stohr J. and Thompson A., Phys. Rev. Lett. **43**, 165 (1979).

[3] Balerna A., Bernicri E., Piccozi P., Reale A., Santucci S., Burrattini E. and Mobilio S., Surf. Sci. **156**, 206 (1985).

[4] Montano P.A., Purdum H., Shenoy G.K., Morrison T.I. and Schultze W., Surf. Sci. **156**, 216 (1985).

[5] Car R. and Parinello M., Phys. Rev. Lett. **55**, 2471 (1985).

[6] Datta A., Mookerjee A. and Bhattacharyya A.K., Mod. Phys. Lett. **B8**, 883 (1994).

[7] Datta R.P., Banerjea A., Mookerjee A. and Bhattacharyya A.K., Mod. Phys. Lett. **B10**, 211 (1996).

[8] Lammers U. and Borstel G., Phys. Rev. **B 49**, 17360 (1994).

[9] Methfessel M.S. and van Schilfgaarde M., Phys. Rev. **B48**, 4937 (1993); Int. J. Mod. Phys. **B7**, 262 (1993).

[10] Methfessel M.S., Schilfgaarde M. and Scheffler M., Phys. Rev. Lett. **70**, 29 (1993).

[11] Das G.P., Kanhere D.G., Panat V., Shah Vaishali, J. Phys. Condensed matter (1996).

[12] Jena P., *see article in this volume* (1997).

CLUSTER INTERACTIONS

B. K. Rao, S. E. Weber, and P. Jena
Department of Physics
Virginia Commonwealth University
Richmond, VA 23284-2000, U. S. A.

ABSTRACT

Self-consistent first principles calculations have been performed to study the interaction of small atomic clusters with H_2, N_2, NO, and CO molecules as well as with substrates. The large surface-to-volume ratio and finite size of clusters give rise to many unusual features which can lead to production of novel materials. While the reactivity of clusters with these molecules provides a fundamental understanding of heterogeneous catalysis, the study of relative stability and magnetic properties of supported clusters enable us to examine the critical parameters for synthesizing cluster-assembled materials. The interdependence of reactivity and magnetism of clusters is discussed.

1. INTRODUCTION

Even though atomic clusters have been known to chemists for quite a while, they have attracted researchers only since 1984 when Knight *et al.* [1] observed that small clusters of sodium atoms showed unusual stability at specific sizes. Following this, intense research activities [2] both in theory and experiment have showed that small atomic clusters show nonmonotonous variation in their properties as a function of size. The properties can be changed by adding/removing an atom, adding a "foreign" atom, adding/removing an electron, or just simply by changing the geometry of the cluster. The possibility of such manipulations of properties have led to a potential of producing materials by assembling suitably chosen clusters [3]. This search for novel materials with tailor made properties requires the simultaneous efforts of theoreticians as well as experimentalists.

Most of the experiments have been dealing with gas phase clusters till now. These serve well to provide fundamental information on the properties of clusters. However, to be able to produce useful materials with unusual combinations of properties, one must deposit the free flying clusters on substrates [4] or isolate them in cage-like structures [5]. In each case, one must have a complete understanding of the microscopic interactions of the clusters among themselves as well as with the isolating material or substrate. Even though some "cluster materials" have already been produced (e. g. nanowires, cluster-assembled electronic circuits), the interplay among various properties of clusters and surfaces needs to be thoroughly investigated.

Small clusters contain mostly surface atoms. However, they are still not expected to behave completely like bulk surfaces because bulk surfaces follow specific long range order. Large surface-to-volume ratio is expected to make small clusters very reactive. On the other hand, clusters of specific composition, charge state, and size have been known to be nonreactive. The case of Al_7 is a typical example. The mass spectra of small Al clusters do not show any large peak at a specific size. However, after reacting the cluster beam with a stream of oxygen, Castleman *et al.* [6] have found that the peak for Al_7 stayed very large while the others almost vanished from the spectrum. As the clusters were ionized for the analysis through mass spectra, Al_7 was actually observed as Al_7^+ which has 20 valence electrons. Following the criterion of Knight [1], this corresponds to a "magic number" cluster and shows lack of reactivity with oxygen while the other clusters combine with oxygen readily and are removed from the spectrum. This cluster depletion spectroscopy shows the lack of reactivity for Al_7^+ clusters and proves that cluster interactions are intimately related to the electronic structure.

In this paper, we shall examine the interaction of Ni atoms and cations with clusters of H_2, N_2, NO, and CO to understand the effect of electronic structure on cluster reactivity. Increasing the size to include small clusters, we shall then investigate the interaction of H_2 and N_2 with Ni_n and Rh_n clusters. The results for these will be presented in section 3 after the theoretical methods used for the calculations are provided in section 2. The interaction of clusters with surfaces will be discussed in section 4. A summary will be presented in section 5.

2. THEORETICAL METHOD

Our theoretical procedure is based on two different approaches. In the quantum chemical approach we use the self-consistent field linear combination of atomic orbitals - molecular orbital (SCF-LCAO-MO) method [7]. In this case we use the Hartree-Fock (HF) theory to calculate the electrostatic and exchange contributions to energy. The self-consistent energy calculated by this process still needs to be corrected for the correlation effects among electrons which had been approximated by the HF theory. We have chosen to use the Möller-Plesset perturbation theory up to the fourth order (MP4) [7,8] for this correction. The quantum chemical method, as discussed, is computationally very demanding. The other method used by us is based on density functional theory (DFT) [9,10]. This method groups together the spin exchange and the effect of correlation as exchange-correlation (XC). In spite of the simplification for the XC energy, density functional theory used with the local density approximation (LDA) [11] has been very successful in calculating properties of atomic clusters although there are cases where the LDA is inadequate in explaining experimental results. Efforts have been made to go beyond LDA by including gradient corrections (GGA) [12]. We include nonlocal corrections in our studies whenever necessary. Computationally, the LDA and the nonlocal versions of DFT are less demanding than the HF-MP4 formalism. Both of these methods are used by us through the GAUSSIAN-94 software [13]. In this program, linear combinations of gaussian functions are used to describe the atomic functions. In each case considered, we have checked the accuracy of the gaussian basis sets through calculation of properties of small clusters and comparison with experimental values. Details of the basis functions are provided at suitable places in the following sections.

The equilibrium geometry of a cluster can be obtained, in principle, by minimizing the total energy of the cluster with respect to all possible bond lengths, bond angles, and spin multiplicities. Interestingly enough, spin multiplicity is not a variable in the case of bulk solids or surfaces. However, interactions of small clusters are strongly dependent on the spin configuration. This links the geometry, reactivity, and the magnetism of atomic clusters intimately. Therefore, a procedure for global optimization becomes prohibitively difficult if clusters contain more than a few (typically about 8) atoms per cluster. Our geometry optimizations, therefore, utilize different procedures for different size clusters. For very small clusters ($n < 8$), we begin with the atoms at arbitrary positions. The total energy of the cluster is then calculated self-consistently using one of the procedures described earlier. The force at every atom site is then calculated by the gradient technique. The atoms are moved along the path of steepest descent to a new adjacent location and the above two steps are repeated until the force at every atomic site vanishes. Since it is possible to reach a local minimum in the energy surface by this procedure, one must be careful to search for the global minimum in the total energy by starting from different initial configurations. For larger clusters, one follows optimizations within specific symmetries depending upon the nature of the problem.

3. INTERACTION OF ATOMS, IONS, AND SMALL CLUSTERS WITH CHEMICAL SPECIES

The unusual properties of small clusters suggest that catalytic activities of these entities might lead to efficient hydrogen storage devices, cleavage of N_2 bonds followed by nitrogen fixation, and possibly help in pollution control through removal of CO molecules. Our previous studies have shown that when a H_2 molecule interacts with an atom of Al, Co, Ni, or Cu [14], the molecule dissociates before binding with the metal atom. However, when the molecule is allowed to interact with the corresponding cations, the molecule does not dissociate. It binds with the cations associatively. The dissociative binding of the molecules with atoms can be explained in a simple manner from the energetics of such interactions. Let us examine the energies gained in the following exothermic reactions involving a metal atom M and hydrogen atoms.

$$H + H => H_2 + \Delta E_1$$

$$M + H => MH + \Delta E_2$$

If ΔE_1 is larger than $2\Delta E_2$, the dissociation of the molecule will not be preferred in comparison to the retention of the H-H bond. For example, Table 1 shows that $2\Delta E_2$ for H-Ni interaction is 6.14 eV which is greater than the binding energy ($\Delta E_1 = 4.48$ eV) for the H_2 molecule. Therefore, the dissociation is favored. For the H-Ni$^+$ interaction, on the other hand, the energy is 3.36 eV. This prevents the molecule from dissociation and the binding is associative. Our previous studies with cations have showed that cations of Ni [14,17] and V [18] bind with a large number of H_2 molecules in this manner. It is now established [14,17] that the cations polarize the molecules and the long-range electrostatic interaction is mostly responsible for the attachment of the molecules in successive solvation shells surrounding the cation. The mechanism has been observed to be active in the case of ions of even simple metals like Li [19] and one may consider using this effect for hydrogen storage devices. An examination of the binding energies shows that cations of most transition metals (except Ti and Sc) can be used for this purpose.

Table 1. Binding energies of various atoms and ions [15,16].

Interaction	Binding Energy (eV)
H + H	4.48
C + O	11.09
N + N	9.76
N + O	6.51
H + Ni	3.07
N + Ni	2.28
O + Ni	3.70
H + Cu$^+$	0.92
H + Ni$^+$	1.68
H + Co$^+$	1.98
H + Fe$^+$	2.12
H + Mn$^+$	2.06
H + Cr$^+$	1.36
H + V$^+$	2.05
H + Ti$^+$	2.39
H + Sc$^+$	2.40

Ni   **Ni⁺**

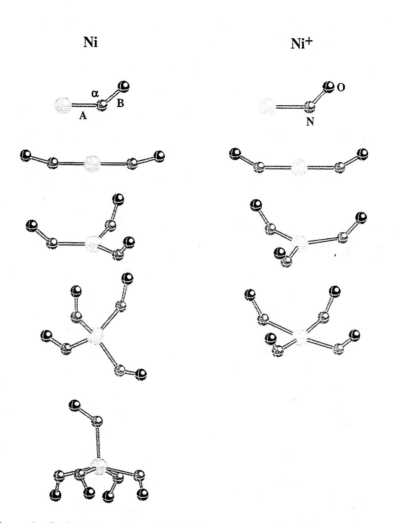

Figure 1. Optimized structures for $Ni(NO)_n$ and $Ni^+(NO)_n$ clusters.

A similar analysis with the binding energies of NO, CO, and N_2 (which are quite large) shows that one may not expect any dissociative binding of these molecules with metal atoms and ions. Starting with a few molecules of these species we have performed global geometry optimizations at the HF-MP4 level with Ni atoms and cations. All the calculations have been done with triple zeta quality basis functions including polarization functions. The results are presented in Table 2. The corresponding geometries are given in Figs. 1 through 3. As expected, these molecules do not dissociate before binding with the atoms or cations. The nature of binding of NO, N_2, and CO with Ni^+ ion is similar to that of H_2. The cation causes polarization which leads to the binding of a large number of molecules without dissociation. The binding energy of each successive molecule with the cation decreases as usual. The bond lengths of the molecules increase slightly when a small number of molecules are bound to the atom or ion. With the increase of the number of molecules, their bond lengths approach the equilibrium value.

Ni   Ni+

Figure 2. Equilibrium geometries of $Ni(N_2)_n$ and $Ni^+(N_2)$ clusters.

When the size of the metal cluster ion increases, the polarization effect decreases because the slight amount of charge transferred from the molecules gets distributed among more number of metal atoms. This leads to a rapid decrease of the number of molecules bound to the metal cluster cation with the increase of the cluster size. This has been observed in experiment [20] and explained by theory [21]. For example, our calculations show that the binding energy of a H_2 molecule with Al^+, Al_2^+, and Al_3^+ goes from 1.09 eV to 0.58 eV, and finally to 0.016 eV. The reduction of the binding is not as dramatic in the case of neutral clusters interacting with H_2 molecules. Our calculations show that the binding energy of the H_2 molecule is 1.99 eV, 2.14 eV, and 1.21 eV for AlH_2, Al_2H_2, and Al_3H_2 respectively. In each case, the H_2 molecule is dissociated, as expected.

Table 2. Binding energy/molecule (eV) for interactions of various chemical species with Ni and Ni^+.

n	Binding Energy/molecule (eV)		
	$Ni(NO)_n$	$Ni(N_2)_n$	$Ni(CO)_n$
1	2.45	1.33	2.43
2	2.86	1.77	2.55
3	2.28	1.47	2.30
4	1.94		2.03
5	1.69		
n	$Ni^+(NO)_n$	$Ni^+(N_2)_n$	$Ni^+(CO)_n$
1	3.69	1.79	2.64
2	2.95		2.39
3	2.54		
4	2.04		

It is observed from Figs. 1-3 that the binding of CO and N_2 occurs with a metal atom or ion in a linear chain with the metal atom/ion staying at the end of the chain. This is similar to the case observed with metal surfaces. This characteristic of interaction of N_2 with small metal clusters has been used by Riley et al. [22] in conjunction with plateaus observed at saturation coverage of the clusters with these molecules, to determine the structure of small Ni clusters. Using LSD, Reuse et al. [23] have calculated the binding energies of N_2 molecules with small Ni clusters. They have verified that for small cluster sizes, each Ni atom binds with two N_2 molecules at saturation as observed by Riley et al. They have also observed that in each case the binding of the N_2 molecules with a metal atom occurs as a linear chain as seen in our Fig. 2. However, upon detailed examinations of larger clusters we have found that the empirical rules suggested by Riley et al. [22] need to be applied with caution because these rules depend heavily upon knowledge obtained from observations with bulk materials while small clusters usually do not resemble the bulk. For example, our calculations on Ni_7 clusters [24] with double numeric basis set and DFT show that there are two isomers for this cluster. In comparison to the capped octahedron predicted by Riley et al. [22] the pentagonal bipyramid has a slightly lower energy (0.1 eV lower). The experiment shows two plateaus, one for N_2 uptake of 8 molecules, and the other for 7 molecules. While the former is explained by the capped octahedral structure, the latter fits the pentagonal bipyramid. Moreover, magnetic moment measurements [25] show that there are two possible isomers for Ni_7. This shows how reactivity and magnetic moment can both be used together to obtain information on cluster geometries which cannot be obtained by any other means at present.

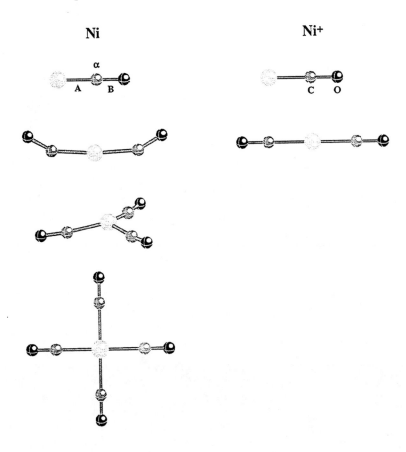

Figure 3. Optimized geometries of $Ni(CO)_n$ and $Ni^+(CO)_n$ clusters.

From the above discussions it is evident that the reactivity and magnetism of a cluster must be related. We have performed calculations on the interaction of H_2 molecules with Rh clusters [26] using double zeta basis functions and DFT-GGA technique. The results are given in Table 3. In each case the H_2 molecule dissociates before binding with the cluster. The equilibrium structure of Rh_4 is a tetrahedron with a spin multiplicity of 1 which changes to 5 after the hydrogen atoms are bound to the cluster. The interaction with the square shaped isomer (with multiplicity of 5) produces a binding which is almost half of the binding produced by the tetrahedral Rh_4. This indicates that the reactivity of H_2 may depend on the underlying magnetic structure of the cluster. The process can be explained in terms of the relative positions of the electronic energy levels of the constituents before and after the interaction and the occupancy of these levels by the electrons of the system. Electronic structure calculations can, therefore, suggest methods for activation/passivation of clusters for reactivity with different molecules through a change in geometry and a related change in the magnetism of the cluster. Detailed calculations have been performed on transition metal clusters interacting with N_2 molecules [27]. These calculations have used triple zeta basis sets and DFT-GGA with Becke-Perdew-Wang functional [12]. The results show that while single transition metal atoms do not dissociate N_2 molecules, dimers of most of the 3d and 4d transition metal series dissociate a nitrogen molecule before binding it. This can find important application in the field of agriculture and drugs. Moreover, it is also noticed that the interaction often changes the spin multiplicity of the metal atoms and clusters as seen with the case of Rh clusters discussed above. For example, nitrogen changes the coupling between the spins of two Cr atoms in Cr_2 from antiferromagnetic to ferromagnetic [27]. Such phenomena can be important in designing nanoscale magnets.

Table 3. Binding energies of H_2 and spin multiplicities of Rh_nH_2 clusters. For comparison, the spin multiplicities of the Rh_n clusters is also given.

n	Multiplicity of Rh_n	Multiplicity of Rh_nH_2	H_2 binding energy (eV)
1	4	2	1.62
2	5	3	1.08
4 (square)	5	3	0.56
4 (tetrahedron)	1	5	1.04

4. CLUSTER-SURFACE INTERACTIONS

Going from gas phase clusters to clusters deposited on surfaces the first question that arises in the mind is whether the clusters change in any manner due to the surface. Any such change will depend upon the nature of the interaction of the cluster atoms with the atoms of the surface. If the interaction is strong, the clusters will be expected to change their original properties completely. For example, when gold clusters are deposited on gold surface [28], local relaxation results in the gold clusters completely losing their identity and forming just another layer of gold on the surface. Theoretical calculations have also demonstrated the same behavior by Li clusters interacting with Li surface [29] and Na clusters interacting with Na surface [30]. However, in the case of epitaxial growth of Ag clusters on Ag surface [31], Pt clusters on Pt surface [32] and Rh clusters on Rh surface [33] show preponderance of specific sized clusters. The question is if this is related to the well known magic numbers in these clusters. Experiment with Pt clusters [32] shows that the epitaxial growth results mostly in Pt_7 clusters and this is not the usual magic number and gas phase Pt_7 is not expected to be two-dimensional in geometry. Clearly, cluster-surface interaction is responsible for Pt clusters to grow only in two dimensions and finally form closely packed 7-atom islands.

B.K. Rao, S.E. Weber and P. Jena

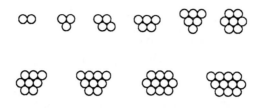

Figure 4. Equilibrium structures of Al clusters growing on the fcc (111) surface of Al.

To study the effect of the substrate we have investigated [34] the two dimensional growth of Al clusters on fcc (111) and (100) planes of bulk Al. Similar study has also been done with Li clusters on the (111) plane of Li lattice [34] and Ag clusters on Ag (001) substrate [35]. For the Al and Li systems, we have used HF-MP4 method with triple zeta basis and for the Ag system, self-consistent Korringa-Kohn-Rostoker Green's function method [36] has been used. In each case it is observed that the underlying geometry of the substrate influences the growth of the clusters and leads to stabilities of clusters which have no correlation with the magic numbers in the gas phase. To demonstrate this we present the equilibrium geometries of the first system in Fig. 4. The corresponding binding energies of these clusters are plotted in Fig. 5. The distinct peaks at n=4, 7, and 10 indicate that close packed islands are the preferred geometries in this case. Similar results have been obtained for the other two systems also. Experiment with Pt clusters on Pt surface [32] show that Pt_4, Pt_7, and Pt_{10} show relatively less mobility (more binding energy). In particular, the preponderance of Pt_7 is now understood. As the calculation of the Ag clusters on Ag surface has incorporated the effect of the bulk surface and has yielded identical results [35], one is confident that the nature of cluster-surface interaction is well explained by theory. The effect of such interactions on the magnetism of the deposited clusters is discussed elsewhere [37] in this book.

5. SUMMARY

Using *ab initio* calculations, we have investigated the interactions of atoms, ions, and small atomic clusters with various small molecules with a view to developing new materials with unusual properties that may provide means for cheap source of energy, increase of global food production, and improvement of the environment. It has been observed that small cluster ions can bind unusually large amounts of hydrogen in the molecular form. The nature of binding of N_2, NO, and CO molecules with the cations is similar to that of the H_2 molecules. However, it is noticed that some neutral clusters can break the bonds of these species. The interaction of small clusters with surfaces leads to a different growth pattern of the clusters and a corresponding change of the electronic and other properties of the clusters. Reactivity and magnetism of small clusters are intimately related and one can be used to drive the other in desired directions.

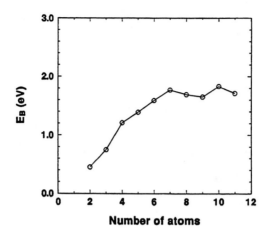

Figure 5. Plot of binding energy of Al$_n$ clusters on the (111) plane of the Al fcc lattice.

ACKNOWLEDGMENTS

The authors wish to acknowledge support from the Department of Energy (DEFG05-87E61316) and the Army Research Office (DAAH04-95-0158).

References

[1] W.D. Knight, K. Clementer, W.A. de Heer, W.A. Saunders, M.Y. Chou, and M.L. Cohen, Phys. Rev. Lett. **52**, 2141 (1984).

[2] *Physics and Chemistry of Finite Systems: From Clusters to Crystals*, Eds. P. Jena, S. N. Khanna, and B. K. Rao (Kluwer, Dordrecht, 1992); W. A. de Heer, Rev. Mod. Phys. **65**, 611 (1993); M. Brack, Rev. Mod. Phys. **65**, 677 (1993).

[3] S. N. Khanna and P. Jena, Phys. Rev. Lett. **69**, 1664 (1992); S. N. Khanna and P. Jena, Phys. Rev. Lett. **71**, 208 (1993); S. N. Khanna and P. Jena, Chem. Phys. Lett. **219**, 479 (1994); S. N. Khanna and P. Jena, Phys. Rev. B **51**, 13705 (1995).

[4] I. Yamada in *Physics and Chemistry of Finite Systems: From Clusters to Crystals*, Eds. P. Jena, S. N. Khanna, and B. K. Rao (Kluwer, Dordrecht, 1992), p. 1193; See also A. F. Hebard, *ibid*, p. 1213; M. Krohn, Thin Solid Films **163**, 291 (1988); K. Sattler in *Physics and Chemistry of Small Clusters*, Eds. P. Jena, B. K. Rao, and S. N. Khanna (Plenum, New York, 1987), p. 713.

[5]   P. A. Jacobs in *Metal Clusters in Catalysis*, Eds. B. C. Gates, L. Guczi, and H. Knözinger (Elsevier, Amsterdam, 1986); M. Ichikawa, Adv. Catal. **38**, 283 (1992); Y. Nozue, T. Kodaira, and T. Goto, Phys. Rev. Lett. **68**, 3789 (1992).

[6]   B. C. Guo, K. P. Kerns, and A. W. Castleman, Jr., Science **255**, 1411 (1992); B. C. Guo, S. Wei, J. Purnell, S. Buzza, and A. W. Castleman, Jr., Science **256**, 515 (1992). See also references in these papers.

[7]   W.J. Hehre, J.L. Radom, P.V.R. Schleyer, and J.A. Pople, *Ab-initio Molecular Orbital Theory* (Wiley, New York, 1986).

[8]   C. Möller and M. S. Plesset, Phys. Rev. **46**, 618 (1934).

[9]   W. Kohn and L.J. Sham, Phys. Rev. **140**, A1133 (1965); P. Hohenberg and W. Kohn, Phys. Rev. **136**, B864 (1964).

[10] R. G. Parr and W. Yang, *Density Functional Theory of Atoms and Molecules* (Oxford UniversityPress, London, 1989).

[11] J. P. Perdew and A. Zunger, Phys. Rev. B **23**, 5048 (1981).

[12] A.D. Becke, J. Chem. Phys. **88**, 2547 (1988); A. D. Becke, J. Chem. Phys. **84**, 4524 (1988); C. Lee, W. Yang, and R.G. Parr, Phys. Rev. B **37**, 786 (1988); J. P. Perdew and Y. Wang, Phys. Rev. B **45**, 13244 (1992).

[13] *Gaussian 94*, Revision B.1, M.J. Frisch, G.W. Trucks, H.B. Schegel, P.M.W. Gill, B.G. Johnson, M.A. Robb, J.R. Cheeseman, T. Keith, G.A. Petersson, J.A. Montgomery, K. Raghavachari, M.A. Al-Laham, V.G. Zakrzewski, J.V. Ortiz, J.B. Foresman, J. Closlowski, B.B. Stefanov, A. Nanayakkara, M. Challacombe, C.Y. Peng, P.Y. Ayala, W. Chen, M.W. Wong, J.L. Andres, E.S. Replogle, R. Gomperts, R.L. Martin, D.J. Fox, J.S. Binkley, D.J. Defrees, J. Baker, J.P. Stewart, M. Heads-Gordon, C. Gonzalez, and J.A. Pople, Gaussian, Inc., Pittsburgh, PA (1995).

[14] J. Niu, B. K. Rao, P. Jena, and M. Manninen, Phys. Rev. B **51**, 4475 (1995).

[15] K. P. Huber and G. Herzberg, *Constants of Diatomic Molecules* (Van Nostrand Reinhold, New York, 1979).

[16] J. L. Elkind and P. B. Armentrout, J. Phys. Chem. **91**, 2037 (1987); J. Am. Chem. Soc. **108**, 2765 (1987); J. Phys. Chem. **90**, 6576 (1986); J. Phys. Chem. **90**, 5736 (1986).

[17] J. Niu, B. K. Rao, and P. Jena, Phys. Rev. Lett. **68**, 2277 (1992).

[18] J. Niu, B. K. Rao, S. N. Khanna, and P. Jena, Chem. Phys. Lett.230, 299 (1994).

[19] B. K. Rao and P. Jena, Europhys. Lett. 20, 307 (1992).

[20] D. Cox, P. Fayet, R. Brickman, M. Y. Hahn, and A. Kaldor, Catal. Lett. 4, 271 (1990).

[21] B. K. Rao, S. N. Khanna, and P. Jena in *Proceedings of SSPFA '94*, Ed. V. Lozovski (NASU Institute of Semiconductor Physics, Kiev, 1994), p. L27.

[22] S. J. Riley in *Clusters and Nanostructured Materials*, Eds. P. Jena, and S. N. Behera (Nova Science, New York, 1996), p. 77; E.K. Parks, L. Zhu, J. Ho and S.J. Riley, J. Chem. Phys. 100, 7206 (1994); J. Chem. Phys. 102, 7377 (1995); E.K. Parks and S.J. Riley, Z. Phys. D 33, 59 (1995).

[23] F. A Reuse, S. N. Khanna, B. V. Reddy, and J. Butter, Chem. Phys. Lett. 267, 258 (1997).

[24] S. K. Nayak, B. Reddy, B. K. Rao, S. N. Khanna, and P. Jena, Chem. Phys. Lett. 253, 390 (1996).

[25] S.E. Apsel, J.W. Emmert, J. Deng, and L.A. Bloomfield, Phys. Rev. Lett. 76, 1441 (1996); W.A. de Heer, P. Milani, and A. Chatelein, Phys. Rev. Lett. 65, 488 (1990); I. M. L. Billas, A. Châtelain, and W. A. de Heer, Science 265, 1682 (1994).

[26] S. K. Nayak, S. E. Weber, P. Jena, K. Wildberger, R. Zeller, P. H. Dederichs, V. S. Stepanyuk, and W. Hergert, Phys. Rev. B 56, 8849 (1997).

[27] S. E. Weber, B. V. Reddy, B. K. Rao, and P. Jena, Phys. Rev. Lett., *submitted*, S. E. Weber and P. Jena, Science, *submitted*.

[28] C. S. Chang, W. B. Su, and T. T. Tsong, Surf. Sci. Lett. 304, L456 (1994).

[29] B. F. Constance, B., K. Rao and P. Jena in *Physics and Chemistry of Finite Systems: From Clusters to Crystals*, Eds. P. Jena, S. N. Khanna, and B. K. Rao (Kluwer, Dordrecht, 1992), p. 1065.

[30] H. Häkkinen and M. Manninen, Europhys. Lett. 34, 177 (1996); J. Chem. Phys. 105, 10565 (1996); Phys. Rev. Lett. 76, 1599 (1996).

[31] B. D. Todd and R. M. Lynden-Bell, Surf. Sci. 281, 191 (1993).

[32] G. Rosenfeld, A. F. Becker, B. Poelsema, L. K. Verheij, and G. Comsa, Phys. Rev. Lett. 69, 917 (1992).

[33] G. L. Kellogg in *Science and Technology of Atomically Engineered Materials*, Eds. P. Jena, S. N. Khanna, and B. K. Rao (World Scientific, Singapore, 1996), p. 91; G. L. Kellogg, Phys. Rev. Lett. **73**, 1833 (1994).

[34] A. K. Ray, B. K. Rao, and P. Jena, Phys. Rev. B **48**, 14702 (1993).

[35] S. K. Nayak, P. Jena, V. S. Stepanyuk, W. Hergert, and K. Wildberger, Phys. Rev. B **56**, 6952 (1997).

[36] R. Zeller, P. Lang, B. Drittler, and P. H. Dederichs in *Application of Multiple Scattering Theory to Materials Science*, Eds. W. H. Butler, P. H. Dederichs, A. Gonis, and R. Weaver (Materials Research Society, Pittsburgh, 1992), p. 357; K. Wildberger, V. S. Stepanyuk, P. Lang, R. Zeller, and P. H. Dederichs, Phys. Rev. Lett. **75**, 509 (1995).

[37] See the paper of W. Hergert, V. S. Stepanyuk, P. Rennert, K. Wildberger, R. Zeller, and P. H. Dederichs in this book.

NICKEL-ALUMINUM ALLOY CLUSTERS: STRUCTURES, ENERGETICS, AND DYNAMICS

J. Jellinek and E. B. Krissinel¶
Chemistry Division, Argonne National Laboratory, Argonne
Illinois 60439, USA

ABSTRACT

Results of computational studies of Ni_nAl_m alloy clusters, for all possible compositions (n, m) such that n+m=13, are presented and discussed. These include structural and energy characteristics, an analysis of mixing of the two components and of its role in defining the hierarchy of the structural forms, and a description of the peculiarities of the dynamical features as defined by the cluster composition. New definitions of mixing energy and mixing coefficient, applicable to systems described by either many-body (as is the case for metals) or pairwise-additive potentials, are introduced and used to quantify the degree of mixing. It is shown that the clusters undergo a stage-wise solid-to-liquid-like transition as their energy is increased. The nature of the stages of the transition depends on the composition of the cluster.

1. INTRODUCTION

The field of atomic clusters has undergone a spectacularly rapid development and evolved into a sovereign discipline in its own right [1]. Historically it emerged as a "laboratory" for studies of surface and bulk properties of materials on the microscopic (atomic) level. The initial goal determined the nature of the early theoretical studies, most

---

¶ Permanent address: Institute for Water and Environmental Problems, Siberian Branch, Russian Academy of Sciences, Barnaul 656099, Russia.

of which considered clusters of atoms arranged into slabs of lattices. Relatively soon it became clear that clusters are physical systems with very interesting and diverse properties in their own right, and that more often than not, especially in the small and intermediate size range, their properties are different from the properties of the corresponding materials in bulk amounts. For example, the equilibrium geometric structures of clusters, especially those of higher stability, are usually, except for clusters of very large sizes, different from lattice structures of bulk crystalline materials. Size-dependence characterizes virtually every physico-chemical property of clusters, and different properties converge to their bulk values, as the clusters grow in size, with different rates and not necessarily monotonically.

One of the distinguishing characteristics of the cluster field is its inherently interdisciplinary nature. Concepts and techniques of diverse areas of physics, chemistry, and materials science and technology are invoked and used to study cluster systems and cluster-related phenomena. An indication of the degree of maturity of the cluster field is its "inverse" impact on the bordering "traditional" disciplines. For example, the new area of nanophase [2] and cluster-assembled materials [3] is ultimately rooted in the cluster field. Using clusters of different material(s) and size as building blocks opens principally new possibilities in rational design and fabrication of novel materials. These materials are expected to have structural, mechanical, electronic, optical, magnetic, and other properties that are not only superior to those of the traditional (atom- or molecule-based) materials, but are also fine-tuned to particular needs and specifications.

Metal clusters are of especial interest because of their direct relevance to many modern technologies, which include microelectronics, thin films, catalysis, etc. Theoretical studies of one-component metal clusters have been performed using first-principles-based approaches and semiempirical potentials. One of the distinguishing features of the interatomic interactions in metals is their many-body character [4], which is a consequence of the characteristic electron delocalization, and which makes an adequate theoretical description of metallic cohesion a challenging task. Mixing two types of metals only complicates this task. Although mixes of metals, alloys, are known to mankind and used for millennia (at least since the bronze age), a systematic microscopic understanding of their properties is lacking. Studies of two-component metal clusters should aid in gaining such an understanding. Bimetallic clusters are of great interest also in their own right. Selection of the component metals and their relative concentrations (stoichiometries), as well as size, are those additional "knobs" that can be used to "tune" various physicochemical properties to desired specifications. For such tuning, however, an atomic-level understanding and characterization of the properties of alloy clusters is needed. As heterogeneous systems, bimetallic or, more generally, two-component clusters are considerably more diverse and complex than their one-component counterparts, and analyses of their properties require new concepts and techniques. Such concepts and techniques are only beginning to emerge.

In this communication we give a brief review of our recent studies of nickel-aluminum alloy clusters [5-8]. Results and their analyses are presented for mixed Ni-Al 13-mers considered as a paradigm. The peculiarities of the structural and dynamical properties of the clusters are examined as a function of their composition. New general definitions of mixing energy and mixing coefficient are introduced and used to quantify the degree of mixing of two components in a cluster. A classification scheme that allows one to understand the systematics and hierarchy of the structural forms of a mixed cluster with a specified size and composition is formulated. The effect of the composition on the mechanism(s) of the solid-to-liquid-like transition in mixed Ni-Al 13-mers, induced by increase of their internal energy, is examined. The theoretical and methodological background is outlined in the next section. Results on the structural forms and their energetics are presented and discussed in section 3. An analysis of the dynamical properties is given in section 4. We conclude with a summary in section 5.

2. THEORETICAL AND METHODOLOGICAL BACKGROUND

The interatomic interactions in the model Ni_nAl_m clusters are described by the many-body Gupta-like potential [9]

$$V = \sum_{i=1}^{N} \left\{ \sum_{j=1(j\neq i)}^{N} A_{ij} \exp\left(-p_{ij}\left(\frac{r_{ij}}{r_{ij}^o} - 1 \right) \right) - \left[\sum_{j=1(j\neq i)}^{N} \xi_{ij}^2 \exp\left(-2q_{ij}\left(\frac{r_{ij}}{r_{ij}^o} - 1 \right) \right) \right]^{1/2} \right\}, \quad (1)$$

where N is the total number of atoms in the cluster; i and j label the individual atoms; A_{ij}, ξ_{ij}, p_{ij}, q_{ij}, and r_{ij}^o are adjustable parameters, the values of which depend on the type of the i-th and the j-th atoms; and r_{ij} is the distance between the i-th and the j-th atoms. The values of the parameters for the Ni-Ni, Al-Al, and Ni-Al pairs of atoms are adopted from Ref. [10] and are listed in Table 1. For details of the fitting procedure we refer the reader to Ref. [10]. Here we mention only that the values of the parameters for the Ni-Ni and Al-Al interactions are fitted to bulk properties of the pure nickel and aluminum, respectively. We use the potential, Eq. (1), as a model. Our emphasis is on introducing new tools and concepts appropriate for the analyses of two-component systems and on applying them to study structural and dynamical properties of bimetallic clusters.

Table 1. Parameters of the Gupta-like potential [10].

Parameters	Ni-Ni	Al-Al	Ni-Al
A (eV)	0.0376	0.1221	0.0563
ξ (eV)	1.070	1.316	1.2349
p	16.999	8.612	14.997
q	1.189	2.516	1.2823
r^o (Å)	2.4911	2.8638	2.5222

The stable (globally or locally) structural forms of the Ni_nAl_m 13-mers correspond to the minima of the potential energy surfaces generated by Eq. (1). A different surface corresponds to each fixed composition (n, m). As in the case of homogeneous clusters, two-component clusters form many different geometric forms. We refer to these as *isomers*. A two-component cluster, however, even when its composition and isomeric (geometric) form are fixed, may possess structures that differ by the distribution of the two types of atoms between the sites of the chosen isomer. We call these structures *homotops* ("the same topography or geometry"). For example, each isomer of Ni_7Al_6 (or Ni_6Al_7) forms $13!/(6!7!)=1716$ homotops. Because of the possible symmetry of an isomer, not all these homotops are necessarily different. But, overall, the number of different structural forms (isomeric *and* homotopic) of two-component clusters is considerably larger than those of their one-component counterparts. It is not our goal here to perform an exhaustive search for the structural forms of the Ni-Al 13-mers. Instead, we have chosen the first six lowest energy isomers of pure Al_{13}, Fig. 1, and used them as geometric templates for structures of the mixed Ni-Al 13-mers. The search for isomers of Al_{13} (overall, 129 stable structures were found) was performed using simulated thermal quenching from many configurations generated along a high-energy trajectory (cf. below); the details of the procedure are described in Refs. [5,6]. For each composition (n, m) and each template isomer we obtained all the corresponding homotops. This was accomplished by considering all the possible replacements of n Al atoms by n Ni atoms and allowing for complete relaxation of all the interatomic distances (for technical details see Refs. [5,6]). Normal mode analysis was performed for each resulting equilibrium (stationary) structure to filter out those configurations that correspond to saddle points, rather than to minima, of the potential energy surfaces. The characterization of the structures includes an analysis of the energy spectra of homotops grouped into classes specified by the isomer and the composition (stoichiometry) of the cluster (see next section).

An important structural aspect of two-component clusters is the degree of mixing of the component elements. The tendency of two types of atoms to mix or to segregate in a cluster is ultimately defined by energy considerations. It has been suggested [11] to use the number of pairs of unlike atoms that are first neighbors of each other as a measure of the degree of mixing. It is clear that this definition translates into an energy measure ("mixing energy") only when the total configurational (potential) energy of a two-component system is pairwise-additive *and* the ranges of the pair interactions do not extend beyond the first neighbors. For many systems neither of these two conditions is valid. As mentioned above, the interatomic interactions in metals are of inherently many-body character. An exact definition of the mixing energy V_{mix}, which is free from the above-mentioned shortcomings (it is valid for systems characterized by either many-body or pairwise-additive potentials with arbitrary ranges), is given by [5]

$$V_{mix} = V_{A_nB_m} - \left[V_{A_n}^{(A_nA_m)} + V_{B_m}^{(B_nB_m)} \right], \qquad (2)$$

where A and B are the types of atoms, $V_{A_nB_m}$ is the configurational energy of the A_nB_m cluster, $V_{A_n}^{(A_nA_m)}$ is the configurational energy of the A_n subcluster in the A_nA_m cluster, and $V_{B_m}^{(B_nB_m)}$ is the configurational energy of the B_m subcluster in the B_nB_m cluster; the configurations of the pure A_nA_m and B_nB_m clusters are identical to that of the mixed A_nB_m cluster, and the configurations of the n-atom and of the m-atom subclusters are the same in A_nB_m, A_nA_m, and B_nB_m. Taking into account that

$$V_{A_nB_m} = V_{A_n}^{(A_nB_m)} + V_{B_m}^{(A_nB_m)}, \tag{3}$$

where $V_{A_n}^{(A_nB_m)}$ and $V_{B_m}^{(A_nB_m)}$ are the configurational energies of, respectively, the A_n and B_m subclusters in the A_nB_m cluster, Eq. (2) can be rewritten as

$$V_{mix} = \left(V_{A_n}^{(A_nB_m)} - V_{A_n}^{(A_nA_m)} \right) + \left(V_{B_m}^{(A_nB_m)} - V_{B_m}^{(B_nB_m)} \right). \tag{4}$$

The physical meaning of V_{mix} is, thus, the total change in the configurational energy of the A_n and B_m subclusters caused by removing them from the A_nA_m and B_nB_m clusters, respectively, and bringing together to form the A_nB_m mixed cluster. It is clear that in the case of pairwise-additive potentials with interaction ranges that do not extend beyond the first neighbors V_{mix} is proportional to the number of pairs of first-neighbor atoms of unlike type. (For the proportionality to hold rigorously one has to neglect the possible minor differences in the distances between first-neighbor atoms of unlike types.) The mixing coefficient K_{mix} is defined as

$$K_{mix} = \frac{V_{mix}}{V}. \tag{5}$$

In the next section we show that K_{mix} plays an important role in defining the hierarchy of the homotopic forms of two-component clusters.

The dynamical properties are derived from long (up to 16 ns) molecular dynamics simulation runs. The forces acting on the atoms are computed from the potential, Eq. (1). The dynamics of the atoms are obtained by solving Newton's equations of motion. The velocity version [12] of the Verlet algorithm is used to time-propagate the trajectories. These are run on a fine grid of cluster energies covering a broad range. The initial conditions are chosen so as to supply zero total linear and angular momenta to the clusters. The trajectories are propagated with a step size of 2 fs, which assures conservation of the total energy within 0.01% even in the longest runs. The quantitative characterization of the cluster dynamics is performed in terms of:

1. Caloric curve, i.e. time-averaged kinetic energy (per atom) as a function of the total energy (per atom);
2. Relative root-mean-square (rms) bond length fluctuation δ,

$$\delta = \frac{2}{N(N-1)} \sum_{i<j} \frac{\left(\left\langle r_{ij}^2 \right\rangle_t - \left\langle r_{ij} \right\rangle_t^2 \right)^{\frac{1}{2}}}{\left\langle r_{ij} \right\rangle_t}, \tag{6}$$

as a function of total energy (per atom). $\langle \ \rangle_t$ denotes time-averaging over the entire trajectory;

J. Jellinek and E.B. Krissinel

3. Specific heat (per atom) C [13],

$$C = \left[N - N\left(1 - \frac{2}{3N-6}\right)\langle E_k \rangle_t \langle E_k^{-1} \rangle_t \right]^{-1} , \qquad (7)$$

as a function of total energy (per atom). E_k is the (internal) kinetic energy of the cluster;
4. Cluster temperature T,

$$T = \frac{2\langle E_k \rangle_t}{(3N-6)k} , \qquad (8)$$

where k is the Boltzmann constant.

3. STRUCTURES AND ENERGETICS

The first six lowest energy structures of Al_{13}, which we use as template isomers for the mixed Ni-Al 13-mers, are shown in Fig. 1. The icosahedron, the most stable form of Al_{13}, is also the most stable isomer of Ni_{13} (cf. Refs. [14,15]) and of all the mixed $Ni_n Al_m$,

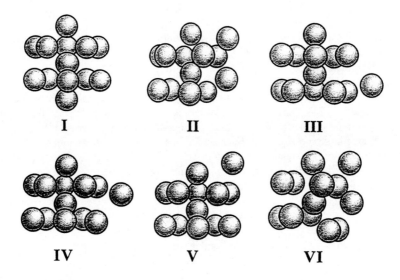

Figure 1. The first six isomers of Al_{13} as defined by the Gupta-like potential with parameters listed in Table 1. Their energies are: (I) -33.812 eV; (II) -33.085 eV; (III) -33.066 eV; (IV) -33.060 eV; (V) -33.037 eV; and (VI) -33.013 eV.

n+m=13, clusters. The ordering of isomers of a mixed cluster with a fixed composition is defined by the energies of the most stable homotops of these isomers. The ordering of the isomers depends, in general, on the composition (stoichiometry). Not all the stable

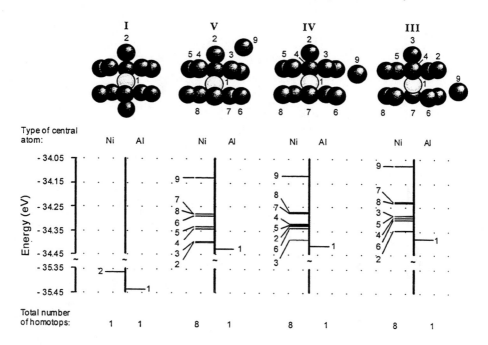

Figure 2. Four isomers of $Ni_{12}Al$ and the energy levels of the homotops corresponding to them (the darker spheres represent Ni, the lighter ones Al). The homotops are separated into classes defined by the type of the central atom. The pictures show the lowest energy homotop of each isomer. The numbers labeling the energy levels indicate the position of the Al atom in the corresponding homotop. The Roman numerals identify the template isomer of Al_{13} with the same geometric structure.

geometric structures of one-component counterparts of mixed clusters necessarily survive as stable (i.e., corresponding to minima of the potential energy surfaces) isomers of mixed clusters, and vice versa. For example, isomers II and VI of Al_{13} are not stable geometries of the $Ni_{12}Al$ cluster. The ordering of the four surviving isomers of $Ni_{12}Al$ is shown in Fig. 2, which also displays the energy spectra of the homotops corresponding to each isomer. The characteristics to notice are the total number of homotops associated with the individual isomers and the patterns of the homotop energy spectra. As is clear from the figure, and also from general considerations, isomers of lower symmetry give rise to a larger number of homotops . Three features are to be noted with regard to the homotop energy spectra. First, the lowest energy homotop of each isomer is the one with the Al atom in the center (or, more generally, in the site of the highest coordination). Second, within each isomer, the energy gap between the (only) homotop with Al in the center and the closest to it homotop with Ni in the center is small. Third, the gap between the energies of the homotops corresponding to the icosahedral isomer, on the one hand, and the energies of the homotops corresponding to the higher energy isomers, on the other, is large (about an order of magnitude or more larger than the energy separation between the two most stable homotops within the individual isomers).

J. Jellinek and E.B. Krissinel

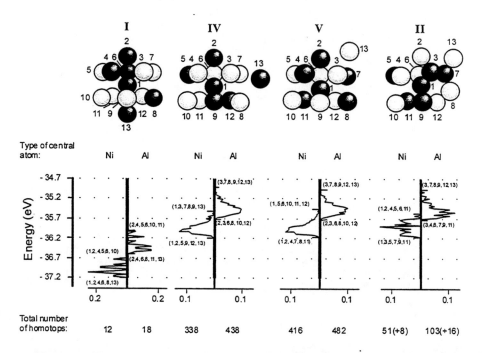

Figure 3. The same as Fig. 2, but for the Ni₆Al₇ cluster. Because of the large number of homotops associated with each class (normalized) distributions of their energies, rather than the energy levels themselves, are shown. The distributions (histograms) are obtained with an energy box size of 0.03 eV. The homotops of the lowest and the highest energy in each class are labeled. The numbers forming a label indicate the positions of the Ni atoms in the corresponding homotop. The numbers in parenthesis with the "+" sign show the number of additional stationary configurations in the individual classes, which represent saddle points, rather than minima, of the corresponding potential energy surface (the energies of these unstable configurations are not included in the distributions).

One of the relevant questions is: How do the mentioned characteristics depend on the composition of the mixed Ni-Al 13-mer? Figure 3 displays the same type of data as Fig. 2, only for the Ni₆Al₇ cluster. Characteristics of the first four isomers of Ni₆Al₇ are shown, although all six template geometries survive as isomers for this cluster. Inspection of Figs. 2 and 3 confirms that, indeed, the energy ordering of the isomers changes with the cluster composition. Other effects of changing the stoichiometry from (n=12, m=1) to (n=6, m=7) are: 1) The number of homotops associated with each isomer increases dramatically. As a consequence, distributions of homotop energies, rather than the energies themselves, are plotted; 2) Ni is the central atom in the lowest energy homotop of each isomer; 3) The distributions of the homotop energies within the individual isomers become bimodal. The lower-energy branch of the distributions represents homotops with Ni in the center, whereas the higher-energy branch corresponds to homotops with Al in the center. The two branches overlap only a little: the energy gap between the most stable homotop with Ni in the center and that with Al in the center becomes large; 4) In contrast, the energy gap between the icosahedral homotops, on the one hand, and those corresponding to the higher energy isomers, on the other, vanishes.

A more complete picture of the evolution of the structural and energy characteristics of the Ni_nAl_m 13-mer, as its composition is gradually changed from (n=12, m=1) to (n=1, m=12), is presented in Fig. 4. The figure displays data for the first two isomers corresponding to six different stoichiometries (n, m); for each stoichiometry, the pattern of the homotop energy spectrum of the second isomer is typical also for the other high-energy isomers (cf. Figs. 2 and 3). The major features of the mentioned evolution can be summarized as follows: 1) $Ni_{12}Al$ is the only composition of the mixed Ni-Al 13-mer with a preference for Al as the central (or the most coordinated) atom. All other compositions prefer Ni as the central atom; 2) The closer the composition of the cluster to 50/50%, the larger the number of homotops associated with each isomer; 3) As the number of the Al atoms in the cluster increases, the bimodal character of the distribution of the homotop energies associated with each individual isomer becomes more pronounced: the energy range of homotops with Ni in the center becomes more separated from the energy range of homotops with Al in the center; 4) As the number of the Al atoms increases, the energy gap between the icosahedral homotops, on the one hand, and the homotops corresponding to the higher energy isomers (of the same composition), on the other, decreases and eventually vanishes.

Implicit in the above analysis of the systematic changes in the structural and energy properties caused by changes in the stoichiometry is a subdivision of the manifold of structural forms of the mixed Ni-Al 13-mer (overall 13603 stable structures were found) into classes defined by the cluster composition, its geometry (isomer), and the type of the central atom. In fact, performing the analysis in terms of these classes is what makes the systematic nature of the changes in the properties transparent. The remaining question is: What defines the energy ordering of the homotops within each individual class? The answer to this question is given in Fig. 5, which displays graphs of the mixing coefficient as a function of the equilibrium configurational energy of the homotops. Data are presented for four classes corresponding to the first two isomers of Ni_6Al_7. As is clear from the graphs, within each class the mixing coefficient changes monotonically (exactly or globally) with the homotop energy: the more stable the homotop (i.e., the lower its equilibrium energy) the higher the degree of its mixing. Nickel and aluminum exhibit a propensity to mix. The monotonicity of the graphs holds exactly when the energy gaps between neighboring homotops in a class are larger than the energy changes associated with the individual equilibrium relaxations of the homotops. Otherwise, the monotonicity holds globally. The reason for the oscillations and the local violation of the monotonicity in the graphs corresponding to the two classes of the second isomer of Ni_6Al_7 is that the energy spacing between neighboring homotops in these classes is very small: 338 homotops with Ni in the center and 438 homotops with Al in the center have their energies within ranges with widths of 0.557 eV and 0.722 eV, respectively (cf. Fig. 3). The validity of the mixing coefficient as a parameter that defines the energy ordering of the homotops within the individual structural classes (these latter are specified by the cluster composition, isomer, and type of the central atom) holds for all compositions and all isomers.

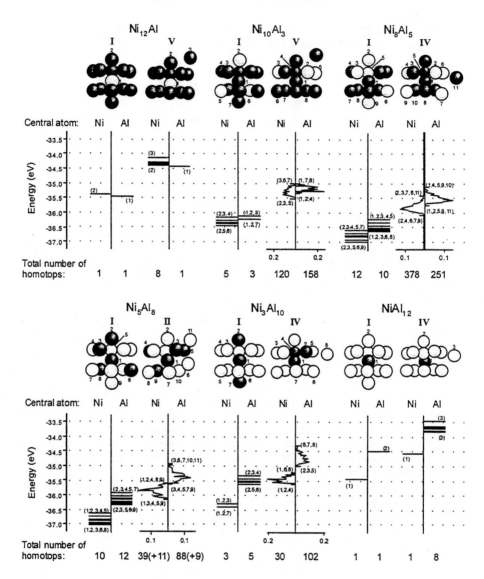

Figure 4. The same as Figs. 2 and 3, but for the first two (of the six considered) isomers of Ni$_{12}$Al, Ni$_{10}$Al$_3$, Ni$_8$Al$_5$, Ni$_5$Al$_8$, Ni$_3$Al$_{10}$, and NiAl$_{12}$. The numbers forming the labels of the lowest and the highest energy homotops in each class indicate the positions of the Al atoms in Ni$_{12}$Al, Ni$_{10}$Al$_3$, and Ni$_8$Al$_5$, and of the Ni atoms in Ni$_5$Al$_8$, Ni$_3$Al$_{10}$, and NiAl$_{12}$.

The peculiarities of structural and energy properties of two-component clusters are ultimately defined by the contributions to the configurational energy of the interactions between like atoms (in our case, Ni-Ni and Al-Al) vs. those between unlike atoms (in our case, Ni-Al). An analytical model that makes transparent the relative role of these interactions in defining the different properties (such, e.g., as the preferred type of the central atom, the propensity of two types of atoms to mix or to segregate, etc.) is presented in Refs. [7,16]. The model allows one to derive many of these properties directly from the parameters of the potential.

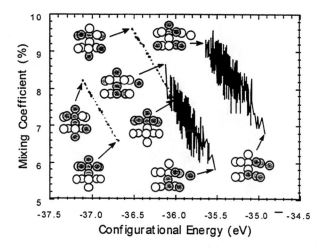

Figure 5. Mixing coefficient as a function of the equilibrium configurational energy of the homotops for four structural classes (see text) corresponding to the first two (of the six considered) isomers of Ni_6Al_7. The pictures show the lowest and the highest energy homotops in each class. The Al and Ni atoms are maximally mixed in the lowest energy homotops and segregated in the highest energy homotops.

4. DYNAMICS

The dynamical properties for the six compositions of the mixed Ni-Al 13-mer considered in Fig. 4 are displayed in Fig. 6, which shows graphs of the caloric curve, rms bond length fluctuation δ, and specific heat C, all considered as a function of the cluster energy (per atom). The features of the graphs - changes in the curvature of the caloric curves, abrupt change(s) in the δ graphs, and peak(s) in the C graphs - are typical indications of a solid-to-liquid-like transition induced by an increase of the cluster energy. Examination of the peculiarities of the graphs for different compositions shows that the details of the transition depend, in general, on the stoichiometry of the cluster.

As discussed in many earlier studies (see, e.g., [15,17] and references therein), the meltinglike transition in clusters takes place over a finite energy (or temperature) range, and in that it is different from the first-order melting phase transition in crystalline bulk materials. The transition range is associated with stages, such as isomerization

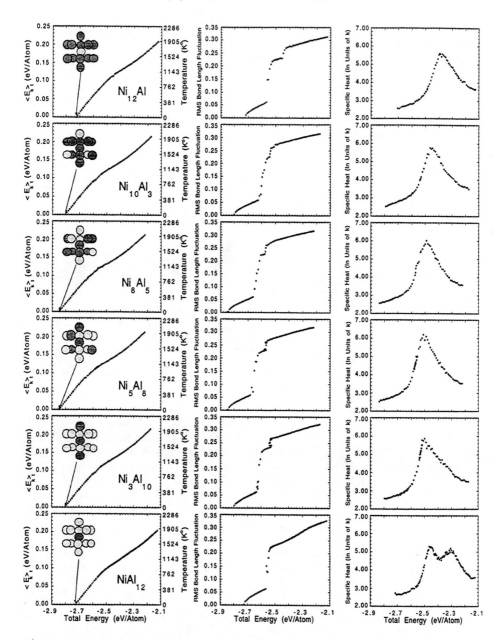

Figure 6. Caloric curves, rms bond length fluctuations, and specific heats (in units of the Boltzmann constant k) as functions of the cluster energy (per atom) for $Ni_{12}Al$, $Ni_{10}Al_3$, Ni_8Al_5, Ni_5Al_8, Ni_3Al_{10}, and $NiAl_{12}$. The pictures show the lowest energy structure for each stoichiometry.

transformations (or, alternatively, coexistence of isomers), partial melting, surface melting, etc. The nature of these stages, as well as their characteristics (e.g., the rate of isomerizations) depend on the cluster size, material(s), and energy (or temperature). In the case of two-component clusters, the structural transformations involve both isomeric and homotopic transitions.

The δ graph for $Ni_{12}Al$ shows two well-separated abrupt increases. The first one corresponds to the onset of the so-called surface isomerizations, which involve only the surface atoms of the cluster; in the course of these isomerizations the central Al atom remains in the central position. The reason for the existence of this stage is that the energy barrier for a replacement of the central Al atom by a surface Ni atom in the most stable icosahedral geometry (considered as the zero-temperature structure) is very high. It is easier first to isomerize the icosahedron into structures with a "defect", or, alternatively, an incomplete shell, keeping the Al atom in the center. The energies of such structures are close (cf. Fig. 2) and the barriers for their interconversions are lower than those for replacing the central Al atom by a surface Ni atom. These structures are sampled at energies corresponding to surface isomerizations.

When the cluster energy becomes sufficiently high to allow for an exchange in the positions of the central Al atom and a surface Ni atom (the barriers to this exchange in isomers with an incomplete shell are lower than in the icosahedron), global isomerizations (the ones that involve also the central atom) get "switched on". The onset of global isomerizations is the reason for the second abrupt rise in the δ graph. We confirmed the above interpretations by performing thermal quenching from many configurations sampled at different energies. Further increase of the energy allows the cluster to sample more of its isomeric and homotopic forms, and the rate of transitions between these forms increases. The cluster eventually attains its liquidlike state. The peak in the graph of the specific heat is a signature of the transition to this state.

Inspection of Fig. 6 shows that the overall picture of the stagewise solid-to-liquid-like transition is also valid for other compositions of the Ni-Al 13-mer. Since in other compositions Ni is the preferred central atom, the isomers and homotops involved in the stage of surface isomerizations have Ni in the center, and the number of the participating homotopic forms increases as the composition of the 13-mer approaches 50/50%. As the number of Al atoms in the cluster becomes dominant, the details of the graphs change. For $NiAl_{12}$ (and Ni_2Al_{11}, not shown) the clear distinction between the two abrupt changes in the δ graph vanishes. Instead, a change in the curvature emerges as a remnant of the second rise. On the other hand, the graph of the specific heat for $NiAl_{12}$ (and N_2Al_{11}, not shown) exhibits two peaks.

The dynamical features of a system depend, in general, not only on the minima, but also on the barriers (saddles) of its potential energy surface. The changes in the dynamics of the mixed Ni-Al 13-mer caused by the changes in its composition can be correlated with and understood in terms of the corresponding changes in the energy spectra of its isomers and homotops (cf. Fig. 4). As discussed in section 3, increase of the number of Al atoms leads, within each isomer, to an increase of the energy gap between the homotops with Ni in the center, on the one hand, and those with Al in the center, on the other. In isomers of $NiAl_{12}$ the gap between the single homotop with Ni in the center and those with Al in the center is especially large. The stage of surface isomerizations involves the single homotops of the different isomers with Ni in the center. Because of the mentioned large energy gap the cluster can absorb a substantial amount of extra energy and still sample only structures with the Ni atom in the center. The effect of the extra energy is to increase the rate of transitions between these structures. Eventually this rate becomes so high that the state of the cluster is best described as a liquidlike shell of twelve Al atoms that encage the single Ni atom. This is the stage of surface melting. Its signature is the first (lower energy) peak in the corresponding graph of the specific heat. Further increase of the cluster energy leads to the onset of global isomerizations, which allow for changes in the identity of the central (or most coordinated) atom, and, eventually, to a complete melting of the cluster. The second (higher energy) peak in the graph of the specific heat represents the transition to the fully developed liquidlike state.

Similar arguments explain the analogous (although exhibited in a somewhat less distinct form) meltinglike behavior of the Ni_2Al_{11} cluster. The fact that the peculiarities of the dynamics of the mixed Ni-Al 13-mer, as defined by the cluster composition, can be correlated with and explained in terms of the composition-specific energy spectra of its isomeric and homotopic forms suggests that the barriers (saddles) of the potential energy surface of the cluster change with its composition in a manner similar to that of the energy spectra of its structural forms. Further details on the dynamical properties of the Ni-Al 13-mer, including isomer-specific features and temperature-labeling of the different stages of the meltinglike transition, are given in Refs. [6-8,16].

5. SUMMARY

A brief review of structural and dynamical properties of 13-atom Ni-Al alloy clusters, as obtained from a semiempirical many-body Gupta-like potential, is presented. It is shown that the large manifold of the possible structural (isomeric and homotopic) forms of two-component clusters of fixed size and materials can be systematized by subdividing it into classes of homotops defined by the relative concentration of the components (stoichiometry), geometric structure (isomer), and the type of the central (or most coordinated) atom. Systematic changes in the patterns of the homotop energy spectra for the different classes of Ni_nAl_m, n+m=13, are described and analyzed as a function of the cluster composition (n, m). New general definitions of mixing energy and mixing coefficient, as quantitative measures of the degree of mixing of two components, are introduced. These are applicable to mixed systems described by both many-body (e.g., Gupta-like, embedded-atom, etc.) and pairwise-additive (e.g., Morse, Lennard-Jones, etc.) potentials. It is shown that the mixing coefficient is the parameter that defines, exactly (when the energy spacing between neighboring homotops is larger than the relaxation energy) or globally (when this spacing is smaller than the relaxation energy), the energy ordering of the homotops within the individual classes. It is mentioned that an analytical model has been developed that allows one to derive fundamental structural and energy properties of two-component clusters directly from the parameters of the potential. An additional merit of the model is that it makes transparent the relative roles played by the interactions between like atoms vs. those between unlike atoms in defining the different properties.

The dynamical features of the mixed Ni-Al 13-mer are analyzed as functions of the cluster energy and composition. It is shown that for all compositions the cluster undergoes a solid-to-liquid-like transition as its internal energy is increased. The mechanism(s) of the transition, however, depends on the composition. The stages involved in the meltinglike transformation of Ni_nAl_m, n=3-12, m=13-n, include surface isomerizations, global isomerizations, and, eventually, complete melting of the cluster. In the case of $NiAl_{12}$ and Ni_2Al_{11} the surface isomerizations become so rapid that they give rise to a qualitatively different stage of surface melting. The composition-specific peculiarities of the dynamics correlate with the composition-dependent patterns of the isomer and homotop energy spectra. Work is in progress on extending the studies to Ni-Al alloy clusters of other sizes. The results will be reported in forthcoming publications. The analyses introduced and used in this study are general and should be applicable to a broad variety of two-component systems.

ACKNOWLEDGMENTS

This work was performed under the auspices of the Office of Basic Energy Sciences, Division of Chemical Sciences, US-DOE under contract number W-31-109-ENG-38. EBK was also supported by the NIS-IPP Program.

References

[1] See, e.g., *Physics and Chemistry of Small Clusters*, P. Jena, B. K. Rao, and S. N. Khanna (Eds.), NATO ASI, Series B: Physics (Kluwer Academic Publishers, Dordrecht, 1987), Vol. 158; *Elemental and Molecular Clusters*, G. Benedek, T. P. Martin, and G. Pacchioni (Eds.) (Springer, Berlin, 1988); *Clusters of Atoms and Molecules*, H. Haberland (Ed.), Springer Series in Chemical Physics (Springer, Berlin, 1994), Vols. 52 and 56.

[2] See, e.g., *Nanophase and Nanocomposite Materials*, S. Komarneni, J. C. Parker, and G. J. Thomas (Eds.), Materials Research Society Symposium Proceedings (Materials Research Society, Pittsburgh, PA, 1993), Vol. 286; *Molecularly Designed Ultrafine/Nanostructured Materials*, K. E. Gonsalves (Ed.), Materials Research Society Symposium Proceedings (Materials Research Society, Pittsburgh, PA, 1994), Vol. 351; *Nanophase Materials, Synthesis, Properties, Applications*, G. C.Hadjipanayis and R. W. Siegel (Eds.), NATO ASI, Series E: Applied Sciences (Kluwer Academic Publishers, Dordrecht, 1994), Vol. 260; and references therein.

[3] See. e.g., *Clusters and Cluster-Assembled Materials*, R. S. Averback, J. Bernholc, and D. L. Nelson (Eds.), Materials Research Society Symposium Proceedings (Materials Research Society, Pittsburgh, PA, 1991), Vol. 206, and references therein; B. C. Guo, K. P. Kerns, and A. W. Castleman Jr., Science **255**, 1411 (1992); S. N. Khanna and P. Jena, Phys. Rev. Lett. **69**, 1664 (1992); P. Jena and S. N. Khanna, Material Science & Engineering A **217**, 218 (1996).

[4] See, e.g., A. P. Sutton, *Electronic Structure of Materials* (Clarendon Press, Oxford, 1993).

[5] J. Jellinek and E. B. Krissinel, Chem. Phys. Lett. **258**, 283 (1996).

[6] E. B. Krissinel and J. Jellinek, Int. J. Quantum Chem. **62**, 185 (1997).

[7] E. B. Krissinel and J. Jellinek, Chem. Phys. Lett. **272**, 301 (1997).

[8] J. Jellinek and E. B. Krissinel, in *Nanostructured Materials: Clusters, Composites and Thin Films*, V. M. Shalaev and M. Moskovits (Eds.), ACS Symposium Series (American Chemical Society, Washington, DC, 1997), Vol. 679, p. 239.

[9] R. P. Gupta, Phys. Rev. B **23**, 6225 (1981).

[10] F. Cleri and V. Rosato, Phys. Rev. B **48**, 22 (1993).

[11] G. E. Lopez and D. L. Freeman, J. Chem. Phys. **98**, 1428 (1993).

[12] W. C. Swope, H. C. Andersen, P. H. Berens, and K. R. Wilson, J. Chem. Phys. **76**, 637 (1982).

[13] E. M. Pearson, T. Halicioglu, W. A. Tiller, Phys. Rev. A **32**, 3030 (1985); S. Sawada and S. Sugano, Z. Phys. D **14**, 247 (1989).

[14] J. Jellinek and I. L. Garzon, Z. Phys. D **20**, 239 (1991); M. S. Stave and A. DePristo, J. Chem. Phys. **97**, 3386 (1992).

[15] J. Jellinek , in *Metal-Ligand Interactions*, N. Russo and D. R. Salahub (Eds.), NATO ASI, Series C: Mathematical and Physical Sciences (Kluwer Academic Publishers, Dordrecht, 1996), Vol. 474, p. 325.

[16] J. Jellinek and E. B. Krissinel, to be published.

[17] J. Jellinek, T. L. Beck, and R. S. Berry, J. Chem. Phys. **84**, 2783 (1986); R. S. Berry, T. L. Beck, H. L. Davis, and J. Jellinek, in *Evolution of Size Effects in Chemical Dynamics*, I. Prigogine and S. A. Rice (Eds.), Advances in Chemical Physics (John Wiley, New York, 1998), Vol. 70, Part 2, p. 75.

FIRST PRINCIPLES MOLECULAR DYNAMICS STUDIES OF ADSORPTION ON CLUSTERS

Vijay Kumar

Material Science Division
Indira Gandhi Centre for Atomic Research
Kalpakkam 603102, India

ABSTRACT

We consider aluminum clusters as a model system for understanding adsorption on metal clusters and present results of our studies of Na, Mg, Al, Si, P, S and Cl atoms on Al_7 and Al_{13} using the *ab initio* molecular dynamics method within the local density functional theory. Our results show a marked variation in the bonding behaviour of some atoms as a function of the size of the cluster which leads to a large variation in the binding energy. Also in some cases the variation in the binding energy of an adsorbate from one site to another is found to be much larger (≈ 1 eV) than it is known on flat surfaces. This could have important implications for reactions on clusters. We discuss the relaxation and reconstruction of clusters due to adsorption. The results are compared with the binding energies of Al-X dimers as well as X on semi-infinite Al surfaces where available.

1. INTRODUCTION

Studies of adsorption on clusters are important for understanding heterogeneous catalysis and in developing novel cluster assembled materials [1]. In the past 20 years or so, much efforts have been made to understand interaction of atoms and molecules on semi-infinite surfaces as well as on small clusters to develop a microscopic basis for catalytic processes. Our understanding of semi-infinite surfaces has improved very significantly over the years and often experimental results could be compared successfully with theoretical predictions. However, such detailed information is not easily available for clusters. Most of the theoretical studies have been done for selected geometries of clusters. However, small clusters, in general, have structures which could be very different from the corresponding bulk and this could have important consequences for their electronic and other related properties. Further, adsorption could lead to reconstruction of a cluster and affect its electronic structure. Such effects would need to be taken into account for a proper understanding of reactions. Aluminum surfaces have played a key role in the understanding of adsorption phenomenon and have helped to clarify the role of $s-p$ and d - electrons. Here, in the same spirit, we consider aluminum clusters as a model system to understand adsorption behaviour on metal clusters.

Qualitatively, adsorption of an atomic specie on metal surfaces has been shown [2] to give rise to a broadening of the adatom electronic level into a resonance due to interaction with the $s-p$ electrons of a metal. In the case of transition metals, this resonance interacts with the d-states of the metal and can give rise to bonding and antibonding states. The filling of this resonance or the antibonding state is important in determining the adsorption behaviour. In the case of clusters, however, the size plays a crucial role as the reaction rate in some cases has been found to vary by several orders of magnitude depending on the number of atoms in a cluster [3]. Developing an understanding of the variation in the adsorption properties with size is, therefore, very important for a proper understanding of reactions on clusters. Here we present results of our studies of the adsorption of Na, Mg, Al, Si, P, S and Cl atoms on Al_7 and Al_{13} clusters using the ab initio molecular dynamics method [4] within the local density functional theory. We have used a plane wave basis set with a cut-off of 11.5 Ry and an fcc unit cell of side 38.5 a.u.. The electron-ion interaction is represented by norm-conserving pseudo-potentials [5] with s non-locality in the Kleinman and Bylander [6] separable form. In the following section we present our results. A discussion and our conclusions are given in section 3.

2. RESULTS

We have considered Al_7 and Al_{13} clusters as these are abundant in the mass spectrum [7]. Also metal clusters with 20 and 40 valence electrons have closed electronic shell configurations similar to rare gas atoms. Al_7 (Al_{13}) cluster with 21 (39) valence electrons has one electron more (less) than the shell closing. One can, therefore, think Al_7 (Al_{13}) cluster analogous to an alkali (halogen) atom. This is likely to lead to interesting changes in the adsorption behaviour as a function of size.

The structures of Al_7 and Al_{13} clusters have been studied [8] and have been shown to be a capped octahedron and a slightly distorted icosahedron respectively. For studying adsorption, among the various possible sites on these clusters, we considered the 3-fold site of the Al_7 octahedron as shown in Fig. 1a. However, in the case of Na and Cl, the adatom drifted to the top site as shown in Fig. 1b. In some specific cases, other sites were also explored and these will be discussed. For Al_{13}, a 3-fold as well as a bridge site were considered. The clusters were fully relaxed to minimize energy. The calculated binding energies of different adsorbates are given in table 1 along with the binding energies for dimers. These were corrected [9] for spin polarization corresponding to the ground state electronic configuration of the adatom. In the case of clusters, however, we expect spin polarization effects to be small and these are not likely to effect significantly the conclusions presented here. In the following we present these results in some details.

Table 1: Binding energies E of Al-X dimer and adsorbate, X, on Al_7 and Al_{13} clusters. R is the dimer bond length.

X	Al	Al	E(Al$_7$) (eV)	E(Al$_7$) (eV)	E(Al$_{13}$) (eV)	E(Al$_{13}$) (eV)
	E (eV)	R (a.u.)	3-fold	Top	3-fold	Bridge
Na	1.13	5.66	1.78	1.92	2.86	2.84
Mg	0.63	6.45	1.25	1.10	2.00	1.98
Al	1.59	4.75	3.05	–	3.36	–
Si	2.38	4.28	4.45	–	4.61	4.64
P	2.99	3.94	5.01	4.86	5.47	5.08
S	5.51	3.89	5.85	5.14	5.76	5.84
Cl	6.30	4.12	4.46	5.89	4.33	4.61

For Na on aluminum clusters the binding energy increases from dimer to 13 atom cluster and for the latter it is much larger than the value [10] of 1.58 eV on Al(111) surface. Also the binding energy on Al_{13} is $\approx$ 1 eV larger than the value for Al_7. The HUMO-LUMO gap for $Al_{13}Na$ is 1.83 eV which is more than the value of 1.56 eV for Al_{13} cluster. This is in contradiction with the known behaviour of alkali adsorption on metals which leads to a lowering of the work function but agrees with the experimental result [11] of an increase in the ionization potential for $Al_{13}Na$ cluster. This different behaviour is due to the completion of the 2p electronic shell in this cluster. The Na-Al bond length for Al_7Na and $Al_{13}Na$ clusters is 5.66 and 5.55 a.u. respectively and is comparable to the value in Al-Na dimer.

The binding energy of Mg on aluminum clusters also increases with size. Al-Mg dimer is weakly bonded and the bond length is 6.45 a.u.. For Al_7Mg, we have studied adsorption on three different sites of a capped octahedron Al_7 cluster. When Mg atom is adjacent to the capping Al atom, then the energy is 0.028 eV lower than the case where Mg atom is on the opposite face (Fig. 1a). This configuration differs slightly from the capped pentagonal bipyramid reported by Röthlisberger and Andreoni [12] for Na_7Mg and can be treated as an analogue of an edge site on a surface. There is considerable elongation of the underlying Al-Al bond and this could be important to understand various processes occuring on clusters. Heating the cluster at around 500 K makes the cluster probe various positions including the Mg atom interacting with a pentagonal bipyramid. All these structures lie within $\approx$ 0.1 eV energy range and Mg atom is very mobile. Interaction of Mg atom with Al_7 remains weak but increases significantly for Al_{13}. This increase is not monotonic as a function of the size of the aluminum cluster. It has been shown [13] that shell closing for Al_2Mg and Al_6Mg clusters leads to a higher binding energy as also found for $Al_{13}Na$ cluster. For Al_{13} cluster Mg atom favours a 3-fold site. The Al-Mg bond lengths are 6.33 and 6.63 a.u. for 3-fold sites on Al_7 and Al_{13} clusters respectively and are close to the value in Al-Mg dimer.

Adsorption of an Al atom is favorable on a 3-fold site on both the clusters and the variation in the binding energy is relatively small. Incidently, the structures of Al_8 and Al_{14}, thus obtained, also correspond to the lowest energy geometries for these clusters.

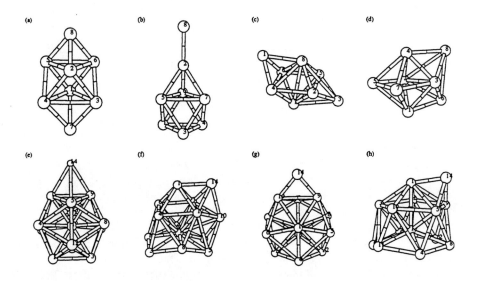

Figure 1. Structures of a) 3-fold site (X = Al, P, S) and b) top site (X = Na and Cl) and c) Si on Al_7 cluster. d) shows the lowest energy site for Mg atom on Al_7. e) corresponds to the 3-fold (Na, Mg, Al, Si) and f) the reconstructed structure of Si g) bridge site (Si, S and Cl) adsorption on Al_{13}. h) shows the distorted 3-fold site for P on Al_{13}.

Adsorption of a Si atom on Al_7, however, reconstructs the aluminum cluster. Si atom is bonded to five aluminum atoms. For Al_{13} cluster, the lowest energy isomer has a Si atom adsorbed on a bridge site. However, simulations at finite temperatures led to a totally different structure shown in Fig. 1f which seems to have origin in fcc or hcp type structure. In this case Si atom is 4-fold coordinated. For $Al_{13}Si$, Kumar and Sundararajan [14] have studied the case where Si is substituted at the center of the icosahedron and have shown that this substitution leads to a lower energy as compared to the case where a Si atom is adsorbed on Al_{13}. So the formation of aluminum-silicide is energetically favorable than adsorption of a Si atom. Viewing the $Al_{13}Si$ cluster as an Al atom interacting on $Al_{12}Si$ cluster which has a closed electronic shell structure, the interaction energy is found to be about 2.2 eV which is much smaller than the value of 3.4 eV for Al on Al_{13}. This interaction is, however, not weak as one may expect for inert gases.

Adsorption of P and S leads to significant relaxations in the two clusters and induces covalency in Al. Both favour a 3-fold site on Al_7 cluster. When P is adsorbed on a site adjacent to the capping Al atom, the cluster reconstructs and converges to the 3-fold site. For S there is ,however, a local mimimum and the S atom adsorbs on to the 3-fold site adjacent to the capping Al atom. Adsorption of P on Al_{13} leads to a reconstruction of the Al_{13} cluster and the reconstructed structure has a similarity with the one obtained for Si on Al_{13}. However, P is 3-fold coordinated. The binding energy of S on both the clusters is nearly the same but it is about 2.3 eV higher than the value of 3.5 eV reported by Feibelman [15] for S on Al(111) surface. The bridge site is favorable for S and there is a significant relaxation in the bond lengths on the opposite side of Al_{13} cluster.

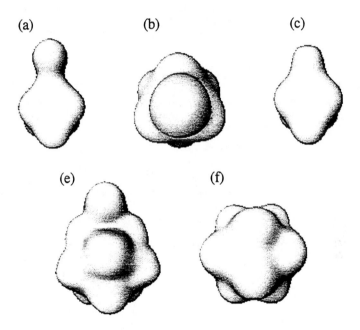

Figure 2. Electron charge density surface with 0.01 e/Ω for a) Al_7Cl cluster with Cl at the top site and b) 3-fold site. c) shows the same for Al_7Na at the top site, d) density surface at 0.012 e/Ω for $Al_{13}Cl$ and e) for Na on Al_{13} cluster. Na gives away its charge to the Al_{13} cluster.

For Cl on Al_7 cluster, we find the binding energy to decrease by 1.43 eV in going from the top site to the 3-fold site. Such large differences are normally not found for adsorption on semi-infinite surfaces and this should be very important for understanding the active sites as well as diffusion of adatoms on clusters. The binding energy decreases significantly on Al_{13} cluster and it is comparable to the value obtained for Al_7 on the 3-fold site. The bonding of Cl on the top site of Al_7 is predominently ionic as it can be noticed from the electronic charge density distribution shown in Fig. 2.

The Al-Cl bond length is 4.04 a.u. for the top site adsorption in contrast to 5.1 a.u. on the 3-fold site. This is, however, comparable to the value of 4.12 a.u. for Al-Cl dimer. The large difference is likely due to the poor screening on the top site as well as in Al-Cl. In both the cases of the top as well as 3-fold sites, there is about 20% increase in the Al-Al bond lengths on the opposite side of the adatom. On the other hand, bonding of Na on the top site of Al_7 is not ionic as it can be seen from Fig. 2. The binding energy difference on the two sites is also small

and is typical of what is observed on surfaces. The electronic charge density for $Al_{13}Cl$ cluster shows that Cl introduces covalency in Al-Al bonds. Since Cl poisons chemical reactions and is used for etching aluminum surfaces, our results on Cl interaction with Al clusters could be of help in understanding these processes.

3. DISCUSSION AND CONCLUSIONS

We have studied adsorption of Na, Mg, Al, Si, P, S and Cl atoms on Al_7 and Al_{13} clusters. We find a large variation in the binding energies with size for Na, Mg and Cl. Further, in some cases the variation of the binding enery from site to site is also of the order of 1 eV which is normally not found on semi-infinite surfaces. For Na, S and Cl the binding energies on Al clusters are about 1-2 eV more than the value on Al surfaces. This is expected to have very important consequences for reactions on clusters as these are some of the common species present in many of the catalytic reactions. In the case of Mg on Al_7 cluster, we find adsorption to be favorable on a high coordination site similar to a step site at a surface. The large changes in the bonding behaviour of the two clusters are due to different electronic structure of the two clusters. Al_7 is found to behave like an alkali atom whereas Al_{13} like a halogen. The screening at different sites in small clusters is likely to be significantly different and lead to large changes in the binding energies as we found here. These effects would need to be further studied in detail and we are currently pursueing it.

References

[1] V. Kumar in *Frontiers in Materials Modelling and Design*, Eds. V. Kumar, S. Sengupta and B. Raj, Springer Verlag, Heidelberg (1997).

[2] J.K. Norskov, Rep. Prog. Phys. **53**, 1253 (1990); B. Hammer and J.K. Norskov, preprint; R.A. Van Santen and M. Neurock, Catal. Rev. Sci. Eng. **37**, 557 (1995).

[3] R.L. Whetten, D.M. Cox, D.J. Trevor and A. Kaldor, Phys. Rev. Lett. **54**, 1494 (1985).

[4] R. Car and M. Parrinello, Phys. Rev. Lett. **55**, 2471 (1985); V. Kumar in *Lectures on Methods of Electronic Structure Calculations*, Eds. V. Kumar, O.K. Andersen and A. Mookerjee, World Scientific, Singapore (1994).

[5] G.B. Bachelet, D.R. Hamann and M. Schlüter, Phys. Rev. **B26**, 4199 (1982); R. Stump, X. Gonze and M. Scheffler, Fritz-Haber-Institut Research Report, February 1990.

[6] L. Kleinman and D.M. Bylander, Phys. Rev. Lett. **48**, 1425 (1982).

[7] M.F.Jarrold, J.J.E. Bower and J.S. Kraus, J. Chem. Phys. **86**, 3976 (1987).

[8] V. Kumar, Bull. Mater. Sci., in press.

[9] The spin polarization correction for Na, Al, Si, P, S, and Cl atoms was taken to be 0.249, 0.136, 0.612, 1.555, 0.721 and 0.188 eV respectively.

[10] J. Neugebauer and M. Scheffler, Phys. Rev. Lett. **71**, 577 (1993).

[11] A. Nakajima, K. Hoshino, T. Naganuma, Y. Sone and K. Kaya, J. Chem. Phys. **95**, 7061 (1991).

[12] U. Röthlisbeger and W. Andreoni, in *Clusters and Fullerenes*, Eds. V. Kumar, T.P. Martin and E. Tosatti, World Scientific, Singapore (1993).

[13] V. Kumar, to be published.

[14] V. Kumar and V. Sundararajan, Phys. Rev. B15 in press.

[15] P.J. Feibelman, Phys. Rev. **B49**, 14632 (1994).

NANOSTRUCTURED THIN FILMS OF CARBON, SILICON, AND MIXED CLUSTERS

M. Broyer,[1] C. Ray,[1] M. Pellarin,[1] J.Lerme,[1]
B. Prevel,[2] P. Melinon,[2] P. Keghelian,[2] and A. Perez[2]
[1] Laboratoire de Spectrométrie Ionique et Moléculaire, Univ. Claude Bernard Lyon I
[2] Départment de Physique des Matériaux, Univ. Claude Bernard Lyon I
69622 Villeurbanne, France

ABSTRACT

The formation of new cluster-assembled films of covalent materials using the Low Energy Cluster Beam Deposition technique is discussed. For carbon, a clear memory effect of the hybridization in free clusters is observed in the cluster assembled film. Moreover for carbon and silicon, the structure of the films obtained from small clusters (typically 30 atoms) is very close to the amorphon type structures and is very different from conventional amorphous structures. Very interesting properties are found for these films, as for example, a strong visible luminescence effect for silicon. Finally, preliminary results are found for Si-C clusters indicating very interesting structures for free clusters, such as mixed fullerenes with one or two Si atoms.

1. INTRODUCTION

Carbon and Silicon clusters have been extensively studied [1-5], over the ten past years. The discovery of C_{60} and other fullerenes has beautifully illustrated how atoms in very small systems may be arranged in a new way different from the known structure of the bulk. For silicon also the cluster geometries exhibit new atomic arrangements which may be interpreted in terms of surface reconstruction, with the occurrence of a large number of pentagons. On the solid state point of view, the small particles or nanoparticles appear as very promising systems both for the fundamental study of quantum size effects and for potential applications, as it is illustrated by the luminescence effect in nano-porous silicon or for transition metals by the interesting properties of magnetic nanostructures.

In this respect the Low Energy Cluster Beam Deposition (LECBD) offers unique possibilities [6] for the synthesis of new nanostructured materials : firstly it may be possible in some circumstances to keep the specific properties of the clusters in the gas phase (memory effect) and secondly the obtained nanostructured materials may be controlled, by varying and the grain size and the distance between the grains. These cluster assembled materials represent an original class of nanostructured solids, intermediate between amorphous and crystalline materials : the short range order is controlled by the grain size and no long range order exists due to the random stacking of incident clusters. Moreover, these cluster assembled films are highly porous with densities as low as about one half of the corresponding bulk density.

In this paper, we present recent results obtained on carbon and silicon cluster films and we compare them to illustrate how their similarities and their differences are related to different cluster structures in the gas phase. These results may be used as guide lines to investigate the new system of mixed SiC clusters and we present preliminary results obtained both in the gas phase and in thin films, and also for various relative concentrations of silicon and carbon. Considering the interest of SiC as a large gap semiconductor, nanostructured films constituted of Si_nC_m clusters having a controlled stoichiometry are very promising for the future.

2. EXPERIMENTAL TECHNIQUES

The experimental set up has been described in previous publications [6]. Clusters are produced in a laser vaporization source : a laser pulse is focused onto a rod or a disk, which is driven in a slow screw motion. In synchronism with the laser pulse, a short intense helium burst (several bars) delivered by a fast-pulsed valve cools the laser-produced plasma. Silicon clusters are formed by laser ablation of an undoped silicon wafer. High purity rods have been used for carbon and stoichiometric SiC. For non stoichiometric clusters, a small carbon film is deposited on a Silicon wafer, or small disks of defined Silicon and Carbon composition are prepared from compressed powders of the two elements. The cluster beam is analyzed by a high resolution time of flight mass spectrometer in the reflectron geometry, mounted perpendicular to the beam. For this analysis, we detect cluster ions merging directly from the source, or neutral clusters photoionized by an excimer ArF laser at 193 nm (6.4 eV). Cluster ions may also be size selected by a mass gate and then photodissociated by another laser (excimer XeCl at 308 nm).

After characterization of the free clusters, ions are deflected electrostatically and only neutral clusters are deposited on substrates at room temperature. For pure carbon clusters, they are not too reactive and they may be deposited in a chamber of moderately high vacuum (10^{-7} Torr). For Silicon and SiC, we use a Ultra High Vacuum Chamber (10^{-10} Torr). After cleaning ex-situ, the substrates are introduced in the UHV Chamber and placed on a rotating system allowing few configurations : preparation of the sample (heating), deposition or spectroscopic measurements (Auger and REED). The nature and the preparation of the substrate depends on the type of analysis to be performed on the film.

For Raman spectroscopy, the LiF platelets are freshly cleaved and introduced in the deposition chamber through the transfer chamber. The substrates are then heated up to 500°

C for five hours to drive off any adsorbates such as water. During the substrate heating, the desorbed species are monitored using a gas analyzer (Hiden quadrupole mass analyzer, having a partial pressure limit of 10^{-14} Torr in the size range 0 to 100 amu). After heating, the sample is placed into the cluster beam axis. After cluster deposition onto the substrate at room temperature, the sample can be transferred in the preparation chamber where a coating film can be evaporated from a Knudsen cell. The cluster deposition rate is controlled by a quartz microbalance. Since the beam is well collimated, the deposition on the substrate and the deposition rate measurement cannot be monitored together. A mechanical device allows the substitution of each element. The stability of the source is rather good and the deposition rate is controlled in each 15 minute interval. The vacuum is controlled during the deposition by the gas analyzer. The partial pressure of this reactive atmosphere is about 2×10^{-10} Torr. For example, we consider that during deposition each cluster undergoes an exposure of 10^{-1} to 1 Langmuir. To avoid any contamination during the air transfer, we evaporate a coated film at the surface of the clusters after the deposition. The thickness of the coated film measured by a quartz microbalance depends on the nature of the evaporated element. In the case of silver coating which presents a strong barrier to the oxygen diffusion, the coating thickness was fixed about 20-100 nm. After air transfer, all the samples are kept in vacuum atmosphere before analysis. For Raman spectroscopy, silicon clusters are deposited on LiF substrate and are coated by a silver film. The Raman study is performed through the LiF sample which is optically transparent. Moreover, the silver enhances the Raman signal allowing a good signal to noise ratio at low laser power.

For X-ray Photoelectron Spectroscopy (XPS) measurements, the procedure is more sophisticated. After cleaning of a monocrystalline silicon platelet *in situ*, we transfer the sample in the preparation chamber where a continuous thin film of silver (20nm) is evaporated. After that, clusters are deposited on the silver substrate. Finally, the film is transferred again in the preparation chamber and coated by a silver film (50nm thickness). The first silver film is used as reference and the second one as an oxygen barrier.

For microscopy (TEM), silicon clusters are deposited onto a copper grid (1000 mesh) coated by a thin film of carbon (5nm thickness). After film deposition (about 20 nm thickness), we evaporate a thin film of LiF (20 nm thickness). LiF is transparent to electron beam and forms a crystalline phase avoiding an additional signal corresponding to an amorphous state in the TEM pattern. Because of the LiF reactivity towards oxygen during air transfer, we evaporate a very thin film of carbon (thickness < 2 nm) onto the LiF. Earlier experiment where carbon has been evaporated directly onto silicon film has revealed by XPS, the formation of SiC bonds. Thus, the LiF acts as a barrier to the Si-C bond. However, most of the ionic elements (such as LiF) present damage irradiations under electron beam. Consequently, the observation is allowed during a small time interval (a few minutes) which limits the observation at large magnification for lattice imaging.

3. CARBON AND SILICON CLUSTER FILMS

Both carbon and silicon are tetrahedrally covalent materials but exhibit different behaviors related to their own electronic structures. In the case of carbon, since the 2p orbital is as compact as the 2s one, this element presents a wide variety of sp-mixing called hybridization.   In pure bulk carbon, the most stable configuration corresponds to four independent bonds (graphite). Three of them are orthogonal sp^2 states and the latter (p_z orbital) is perpendicular to the basal plane. Another structure is formed by four orthogonal hybrids leading to the sp^3 hybridization (diamond phase). Silicon presents two main differences with carbon : firstly, the main stable configuration corresponds to sp^3 hybridization. Secondly, Si atom can promote in 3d shell (carbon has no 2d shell). Thus, the d-state allows a mean coordination greater than four. This corresponds to the insulator-metallic transition observed in the high pressure phase of silicon.

Consequently, the structure of carbon clusters is dominated by a mean hybridization labelled sp^n with $1<n<3$, roughly related with the carbon basal plan curvature [7], while silicon clusters tend to have rather compact structures [5,8]. Four-fold bondings exist in both carbon and silicon, but carbon clusters promote preferentially two- and three-fold bondings corresponding to linear chains and cyclic or cage structures (fullerenes). It is generally admitted [2] that C_N clusters with $N<10$ atoms consist of linear or planar geometries with low coordination and above $N = 20$ atoms, a transition between two and three dimensional geometry may occur. Cage like ion-fullerene structures become the most stable isomers above $N = 32$ atoms [2], the dodecahedron C_{20} being the first fullerene (I_h symmetry) having the twelve pentagons required by Euler's rules.

For silicon, the sp^2 hybridization being highly unfavorable, the fullerene-like structures with an empty core are unstable for large Si_N ($N \geq 28$ atoms) clusters. For example, Si_{45} looks like a fullerene but with an inner core [5,9]. However, structures including a large amount of five-fold rings, and especially Si_N with $N = 20, 24, 28$, which are nearly purely sp^3 hybridized, are predicted. It is interesting to mention that these small fullerenes (Si_{20} and Si_{24}) exist in certain polymorph phases so called clathrates (i.e. Na_8Si_{46}) [10]. These structures consist in a cubic arrangement of Si_{20}-dodecahedra with Si_{24} in between. The electronic exchange with alkaline atoms located inside each polyhedron increases the stability of such structures.

3. 1. Carbon cluster films

On the basis of the above mentioned structures and properties of carbon neutral free clusters, LECBD experiments of size controlled distributions on various substrates at room temperature have been performed [6,11]. In the particular case of carbon, distributions centered around C_{20}, C_{60}, and C_{900} were chosen for their strong sp^3, intermediate sp^2-sp^3, and sp^2 characters, respectively. The corresponding films (≈ 100 nm thick) were systematically analyzed using various complementary techniques such as : Raman spectroscopy, Electron Energy Loss Spectroscopy (EELS), and X-ray Absorption Near Edge Structure (XANES). All these measurements converged to confirm the sp^3, sp^2 and intermediate characters of the C_{20}, C_{900}, and C_{60} films, respectively. This is a good evidence of a memory effect of the free cluster structures in the films. Moreover, the properties of the films (electrical conductivities, optical absorption spectra, hardnesses) [12] are in agreement with the more or less diamond or graphite characters predicted, emphasizing the capability of the LECBD technique to produce materials with adjustable structures and properties. AFM-observations of the deposits [12] and density measurements confirm the highly porous nanostructured morphologies resulting from the nearly random stacking process of incident clusters characteristic of the LECBD technique.

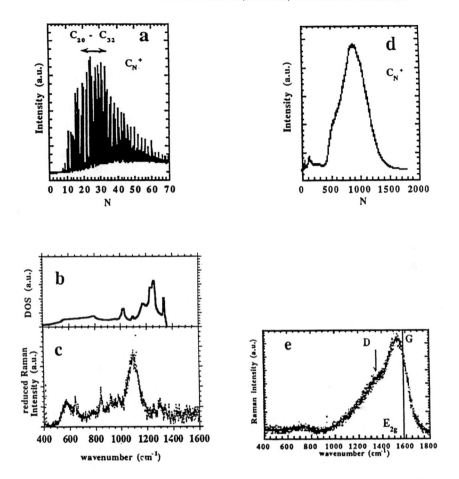

Figure 1. Mass abundance spectra of carbon clusters centered around C_{20}-C_{32} (a) and C_{900} (d). The reduced first order Raman scattering spectra measured with C_{20} and C_{900} films are given respectively in figures (c) and (e). Fig. (b) gives the phonon density of states of diamond calculated by Giannozzi *et al.* (see text).

To illustrate these results, Fig. 1 shows the Raman spectra of two different sizes of carbon clusters: a distribution centered around C_{20}-C_{32} (Fig. 1(a) and 1(c)) and another centered around C_{900} (Fig. l(d) and l(e)). In the case of C_{20} films, the Raman Spectrum exhibits a band around 1140 cm^{-1} which has been early interpreted as a signature of diamond nanocrystallites [13]. We do not significantly observe the diamond sharp peak at 1332 cm^{-1} and the graphite bands located around 1300 cm^{-1} and 1580 cm^{-1} (E_{2g} mode). The Raman spectrum of Fig. 1(c) is in rather good agreement with the diamond vibrational Density of States calculated by

Giannozzi *et al.* [14], except a small red shift in energy. These results can be understood if we consider the random stacking of very small fullerenes with quasi-pure sp³ hybridization. This film corresponds to a new kind of amorphous carbon formed of small cages with a large number of pentagons. Since Mott and Grigorovici [14], this kind of amorphous structure is called amorphon model, the small cages being the amorphons and our films may be considered, as an example of amorphon type structure. Finally, the red shift in energy may be explained by a lower energy binding in small fullerenes, responsible for lower vibrational frequencies. The lack of selection rules in such disordered structure of amorphon type explains the observation, for the first time, of the complete diamond vibrational density of states. Very recent calculations [15,16] performed on this kind of structure has confirmed both the red shift and the occurrence of the complete density of states by Raman spectroscopy.

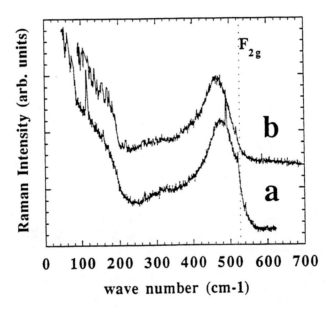

Figure 2. Raman spectrum in HV polarization (a) LECBD-film, (b) amorphous silicon produced by ion silicon implantation into a silicon wafer. The laser power was fixed at 10 mW (514.5 nm). F_{2g} indicates the Raman line in diamond silicon.

In conclusion, carbon cluster films can be really considered as a text book example of the memory effect: the hybridization in these cluster assembled materials varies from pure sp² to pure sp³ in direct relationship with the free cluster structure. Moreover, the structure of our film is of amorphon type with a large number of pentagonal rings.

3.2. Silicon cluster films

Comparable sets of deposition experiments have been performed with silicon clusters having a size distribution centered around Si_{50}, since interesting structures are predicted for this materials in this size range [5,9,17,18] : fullerenes with an inner core, (as Si_{33} or Si_{45}), and a large number of pentagonal rings at the surface. Fig. 2(a) shows the Raman spectrum of a silicon cluster thin film. Fig. 2(b) shows, for comparison, the Raman spectrum of conventional amorphous Silicon (a-Si). Both spectra are very similar, however, for the cluster film, a strong visible luminescence is observed in the red (Fig. 3). After heating under laser, the Raman spectrum of the cluster film gives other features close to those observed in diamond silicon near the melting point or in clathrates. After a total annealing, the film crystallizes under diamond phase giving the characteristic sharp peak located at 522 cm^{-1}.

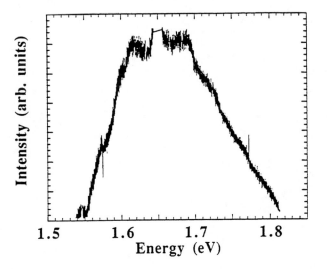

Figure 3. Photoluminescence spectrum of the silicones obtained under laser irradiation (514.5 nm).

We believe that our silicon cluster films have a structure close to the amorphon type silicon described by Grigorovici [19], which may be looked as an amorphous phase of the clathrate crystal. This structure contains a large number of odd-membered rings (mainly pentagonal rings). In such rings, a dissymmetry in atomic orbitals occurs due to the impossibility of alternant structures (electronic frustration). This dissymmetrization has been already observed in Si (111) 2 x 1 and 7 x 7 reconstruction where core level shifts of the silicon atoms (Si_{2p}) located at the surface are observed in X-ray Photo emission Spectroscopy [20]. Moreover, the modification of the electronic structure induced by the presence of pentagonal rings may also play a role in the luminescence effect (opening of the gap which may become almost direct).

In order to investigate the structure of our Silicon films we have systematically performed X-ray Photoelectron Spectroscopy (XPS) measurements. In fact, due to the highly reactive character of our films and despite UHV conditions used, the mean-stoichiometry is $SiO_{0.45}$. This problem somewhat complicates the analysis of Si-2p core level shift. However, our results may be interpreted if we consider a pure Si core embedded in SiO. No SiO_2 is observed. Moreover for pure Si, a core level shift $\Delta E = 0.6 \pm 0.3$ eV is observed. Such core level shift could be attributed to initial states as in reconstructed 7 x 7 silicon [20] (where the shift is 0.3 eV) or to final states (screening effect or increasing of the band gap).

We conclude that a new amorphous structure (amorphon type) is observed in the present silicon cluster films as well as very promising luminescence properties. New experiments are presently carried out with best vacuum and transfer conditions, in order to eliminate the oxygen in the film and to investigate the possible role of oxygen in the luminescence process.

4. Si_nC_m CLUSTERS

In the context of the above discussion concerning carbon and silicon films, the mixed Si-C clusters appear very interesting. As compared to pure silicon and carbon, the dissymmetrization (or frustration) in SiC pentagons is expected to be larger, and the question arises if this amorphon type structure may exist in Si-C cluster films. Moreover, SiC is a very interesting large gap semiconductor. In some aspects it is close to germanium or silicon but with a ionocovalent structure it is close to GaAs. Therefore, nanostructured SiC materials seem very promising for the future, especially if the grain size and stoichiometry can be controlled. It would be perhaps possible to vary and control the gap, or to stabilize the amorphon type carbon films by introducing a small amount of silicon in the carbon clusters.

The structure of free SiC clusters is not well known except for very small sizes [21]. Before performing experiments on deposited clusters, we have studied free Si_nC_m clusters of various compositions. The first run of experiments was carried out using a $Si_{0.5}C_{0.5}$ rod. The observed mass spectra (Fig. 4) is very complicated, and due to natural isotopic composition of carbon and silicon, one peak is obtained for each atomic mass unit. The periodicity of four a.m.u. observed in the mass spectrum (Fig. 4) is due to the most abundant isotopes (28 a.m.u for Si and 12 a.m.u) which are commensurable (four is the greatest common divisor of 12 and 28). The relative abundances in mass spectra well agree with simulations assuming cluster formation with fifty percent of Si and C, indicating that in average the free clusters have the same composition than the rod.

An approach to the determination of the structure of Si_nC_m consists in the photodissociation of size selected clusters. Fig. 5 shows the results of these photofragmentation experiments for different cluster sizes. Even for quite large clusters, as $Si_{11}C_{25}$, the main fragments correspond to the loss of Si_2C and at a less extend Si_3C. In bulk SiC, the main fragment corresponds to Si [22]. For very small clusters, the theoretical calculations predict that the Si atoms stay at the edges sites. Our photodissociation results seem to preclude a segregation between Si and C and are in favor of a network where Si and C atoms are strongly mixed.

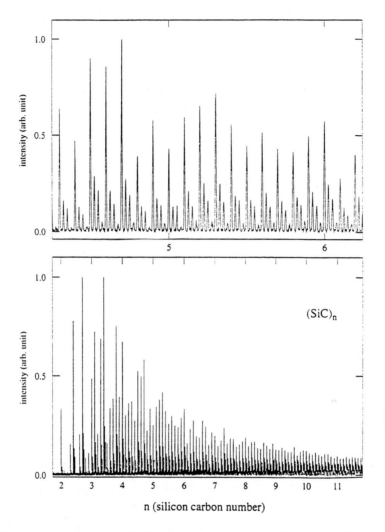

Figure 4. Mass spectra of Si_nC_m clusters obtained with a $Si_{0.5}$ $C_{0.5}$ rod. The mass unit on the horizontal axis corresponds to 40 a.m.u. (SiC mass for respectively isotopes 28 and 12 of Si and C). The lower part shows the whole mass spectrum. The upper part enlarges the details between $(SiC)_5$ and $(SiC)_6$.

Silicon doped carbon clusters have been produced using a disk consisting of compressed carbon and silicon powder with only five percents of silicon. Fig. 6 shows the mass spectrum obtained in these conditions. We observe from C_{36} only clusters with an even number of

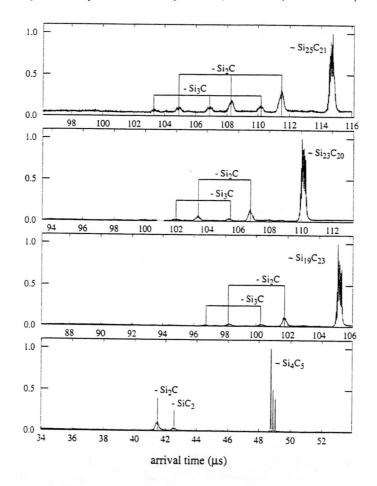

Figure 5. Photofragmentation mass spectra of size selected $Si_nC_m^+$clusters. Successive photo evaporation Of Si_2C is observed and at a less extend Of Si_3C.

carbon atoms which is an indication of the fullerene regime. Less intense peaks correspond to $SiC_{2n-1}Si$ and Si_2C_{2n-2}. If we compare the three series C_{2n}, SiC_{2n-1}, and Si_2C_{2n-2}, the relative peak intensities are very similar as a function of the total number of atoms. In the three series, X_{60} (C_{60}, SiC_{59}, Si_2C_{58}) is one of the most intense peaks. We believe that these clusters correspond to fullerenes where a silicon atom takes the place of a carbon atom, with probably a deformation of the cage [23]. This interpretation is confirmed by photodissociation experiments where the fragments correspond to SiC loss (or C_2 for the carbon fullerenes). These preliminary experiments on Silicon doped clusters are very encouraging for the future.

Increasing the silicon percentage will probably induce the presence of silicon atoms inside the cage or force carbon atoms to be in pure sp^3 hybridization.

Only preliminary deposition experiments have been performed on SiC cluster produced from a $Si_{0.5}C_{0.5}$ rod. The XPS analysis of these films reveals grains of SiC with the expected stoichiometry ($Si_{0.5}C_{0.5}$), but the detailed analysis is still under investigation.

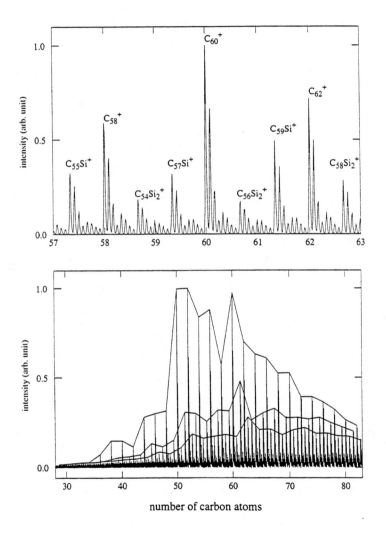

Figure 6. Mass spectra of C_{2n}, SiC_{2n-1}, Si_2C_{2n-2} fullerenes obtained with a small disk of carbon containing 5% of silicon.

5. CONCLUSION

We have shown that the films obtained by Low Energy Cluster Beam Deposition keep the memory of the free phase for covalent species. This is especially clear for carbon where hybridization varies from sp^2 to sp^3. For silicon, the hybridization remains close to sp^3, but different arrangements of the elementary tetrahedron lead to a different amorphous structure and we obtain silicon films close to amorphon type structure with very interesting luminescence properties. The novel sp^3 carbon films obtained by LECBD are also very close to the amorphon-type structure. For the future, mixed Si-C clusters appear as a very promising system for the synthesis of cluster assembled new materials. We have shown that we can control the cluster stoichiometry and determine the free cluster structure. Experiments on SiC films are in progress.

References

[1]   K.W. Kroto, J.R. Heath, S.C. O'Brien, R.F. Curl, R.E. Smalley, Nature **318**, 162 (1985).

[2]   J. Hunter, J.L. Fye, E.J. Roskamp and M.F. Jarrold, J. Phys. Chem. **98**, 1810 (1994).

[3]   V. Ray and M.F. Jarrold, J. Chem. Phys. **93**, 5709 (1990) and references therein.

[4]   W. Weltner and R.J. Van Zee, Chem. Rev. **89**, 1713 (1989) and references therein.

[5]   M.V. Ramakrishna and J. Pan, J. Chem. Phys. **101**, 8108 (1994).

[6]   P. Melinon, V. Paillard, V. Dupuis, A. Perez, P. Jensen, A. Hoareau, M. Broyer, J.L. Vialle, M. Pellarin, B. Baguenard, and J. Lermé, Int. J. Mod. Phys. B **9**, 339 (1995) and references therein.

[7]   R.C. Haddon, L.E. Brus and K. Raghavachari, Chem. Phys. Lett. **131**, 165 (1986).

[8]   J.L. Elkind, J.M. Alford, F.D. Weiss, R.T. Laaksonen, and R.E. Smalley, J. Chem. Phys. **87**, 2397 (1987).

[9]   V. Rothlisberger, W. Andreoni, and M. Parinello, Phys. Rev. Lett. **72**, 665 (1994).

[10] C. Cros, M. Pouchard, P. Hagenmuller, J of Sol State Comm. **2**, 570 (1970).

[11] V. Paillard, P. Melinon, J.P. Perez, V. Dupuis, A. Perez, and B. Champagnon, Phys. Rev. Lett. **71**, 4170 (1993).

[12] V. Paillard, P. Melinon, J.P. Perez, V. Dupuis, A. Perez, J.L. Loubet, H. Pascal, A. Tonck and M. Fallavier, Nanostruct. Mat. **4**, 759 (1994).

[13] R.J. Nemanich, J.P. Glars, G. Lucowsky, R.E. Shroder, J. Vac Sci Technal, A6, 1783 (1988).

[14] P. Giannozzi, S. de Gironcoli, P. Pavone, S. Baroni, Phys. Rev. B **43**, 7231 (1991).

[15] A. Canning, G. Galli and J. Kim (preprint).

[16] G. Benedek and L. Colombo, Materials Science Forum **232**, 247 (1996) [*Trans Tech. Publications* Ed. K. Sattler].

[17] E. Kaxiras, Phys. Rev. Lett. **64**, 551 (1989).

[18] M.F. Jarrold and W.A. Constant, Phys. Rev. Lett. **67**, 2994 (1991).

[19] R. Grigorovici in *Amorphous and Liquid Semiconductors*, Ed. J. Tauc (Plenum New York, 1974).

[20] A.L. Wachs, T. Miller, A.P. Shapiro, and T.C. Chiang, Phys. Rev. B **35**, 5514 (1987).

[21] H. Hunsicker and R.O. Jones, J. Chem. Phys. **105**, 5048 (1996) and references therein.

[22] J. Drowart, G de Maria, and M.G. Inghram, J. Chem. Phys. **29**, 1015 (1958).

[23] T. Kimura, T. Sugai, and H. Shinohara, Chem. Phys. Lett. **256**, 269 (1996).

SELF-ASSEMBLY AT METAL SURFACES

J. V. Barth, H. Brune, T. Zambelli,[#] and K. Kern
Institut de Physique Expérimentale
Ecole Polytechnique Fédérale de Lausanne, CH-1015 Lausanne, Switzerland
#Fritz-Haber-Institut der Max-Planck-Gesellschaft
Faradayweg 4-6, D-14195 Berlin, Germany

ABSTRACT

The recent development of temperature controlled scanning tunneling microscopy opened the door to a fascinating new world on the nanometer scale. The atomic scale information obtained provides unprecedented insight into the kinetics of surface phenomena such as epitaxial growth or chemisorption. In particular, a rich variety of surface structures could be observed, whose morphologies are determined by the self-assembly of the adsorbed particles. Temperature control allows for detailed investigations of the microscopic processes which are at the origin of the self-assembly. Some illustrative examples are discussed, particularly the initial stages in heteroepitaxial growth of metals (Ag, Cu) on fcc single crystal metal substrates with different symmetry (Pd(110), Pt(111), Ni(100)) with emphasis on the island shapes as well as a novel pattern formation in dissociative adsorption of molecular oxygen on Pt(111).

1. INTRODUCTION

In a world of continuing miniaturization, the science of nanostructures plays a vital role for the optimization of existing and the development of future materials and technologies. Extensive experimental and theoretical efforts are undertaken to explore the physical and chemical properties of materials at the nanometer scale [1]. Novel scientific approaches for the controlled formation and the characterization of nanostructured materials are in high demand. The advent of variable temperature and 4 K scanning tunneling microscopy (STM) certainly represent milestones in the development of techniques which can be used to investigate or influence ordering processes of particles on surfaces, be it directly by manipulation of individual atoms at the lowest temperatures [2], or indirectly by exploiting the laws governing their self-assembly [3]. The objective of the present contribution is to illustrate the capabilities of variable temperature STM in the domain of nanostructuring on surfaces via self-assembly by a discussion of a few selected examples.

Firstly, we concentrate on heteroepitaxial growth of metals on single crystal transition metal substrates with different symmetry. In an experiment, where metal atoms are deposited on a surface, aggregates will evolve, corresponding to a lowering of the system's free energy [4]. However, the formation of aggregates (islands) is subjected to kinetic limitations [5], which depend on the specific characteristics of the system and can be regulated by the substrate temperature and deposition flux. Agglomeration necessitates surface diffusion of the adsorbed metal atoms, initially randomly distributed on the substrate. The average adatom diffusion length Λ_a is strongly temperature (and flux) dependent (at constant flux: $\Lambda_a \sim \exp(-\chi E_a^*/2kT)$, where E^* is a corresponding activation energy and χ a scaling exponent). Once a minimum number of atoms find each other, their growth becomes more likely than their dissolution and a stable nucleus forms [5, 6]. The critical number of atoms which forms a stable nucleus upon incorporation of exactly one additional atom is referred to as the critical nucleus size i, which is the temperature dependent decisive parameter for the scaling exponent χ [7]. Attachment of additional diffusing atoms leads to the growth of the nucleus. The shape of the island formed, will depend on the one hand on the substrate symmetry. It may be used to induce anisotropies in the surface diffusion and the sticking of mobile atoms to islands, respectively. On the other hand, the island shape is determined by the mobility of attached atoms along the perimeter of the islands Λ_1, which is again temperature dependent ($\Lambda_1 \sim \exp(E_1^*/kT)$, and can be similarly influenced by the substrate geometry. At the lowest temperatures, where Λ_1 is very small, mechanisms with very limited edge mobility prevail, resulting in the formation of fractal islands, resembling those known from diffusion limited aggregation (DLA) studies [8] (note, however, that with all metal on metal systems investigated so far, edge mobilities in island formation never could be completely suppressed as would be the case for an ideal DLA growth). With increasing temperatures, the influence of Λ_1 becomes more pronounced, i.e., the adatom migration along island edges strongly influences the island morphology. A further parameter which influences the island shape, whose importance in heteroepitaxial growth was realized only recently, is the strain present in the islands due to the mismatch between the substrate and the adsorbate lattice. This can lead to symmetry breaking in growth and corresponding formation of ramified islands even in thermodynamic equilibrium, due to a more efficient strain relief associated with such structures.

Secondly, the application of variable temperature STM to growth processes which involve adsorption of gas molecules and surface chemical processes, will be demonstrated. With such systems, adsorbed particles do not necessarily reach the minimum of the chemisorption well directly, as opposed to the case of metal deposition. Rather, the intermediate population of metastable precursor states frequently occurs, leading to very different growth characteristics. By means of temperature controlled STM a comprehensive characterization of such processes becomes accessible, as will be demonstrated below. Additionally, it allows for an elucidation of the local chemical reactivity of the surface, which can play a decisive role in the self-assembly of the surface chemical reaction products.

2. EXPERIMENTAL

Experiments were performed in ultrahigh-vacuum chambers (base pressure ≈ 1 x 10⁻¹⁰ Torr) with home-built beetle-type STMs cooled by liquid He [9, 10]. Sample preparation and characterization followed standard procedures. The data presented were obtained in the constant-current mode. Metals were deposited by vapor-phase epitaxy with commercial Knudsen cells. Oxygen was exposed by back filling the chamber. Coverages are given in terms of monolayers (ML, where 1 ML corresponds to one adsorbed metal atom or O_2 molecule per substrate atom, respectively).

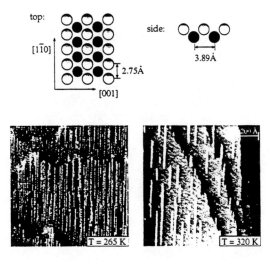

Figure 1. Model of the anisotropic Pd(110) surface and STM data of Cu islands grown on Pd(110). At T = 265 K monoatomic one-dimensional Cu chains forming the $[1\bar{1}0]$ channels on the substrate, whereas at T = 320 K two-dimensional anisotropic growth prevails (Θ_{Cu} ≈ 0.1 ML).

2.1. One-dimensional islands on anisotropic surfaces: Cu/Pd(110)

A conceptually simple way to synthesize islands with a desired shape is to take advantage of the anisotropy provided by a substrate with low symmetry. This can be used to obtain 1-dimensional aggregates, as demonstrated by the data reproduced in Fig. 1 for growth of Cu on a Pd(110) surface [3, 11]. For this system, the diffusion barriers along the close-packed $[1\bar{1}0]$ and open [001] direction were determined by a recent detailed analysis using kinetic Monte Carlo simulations to ≈ 0.30 eV and ≈ 0.45 eV, respectively [12]. Hence surface diffusion is predominantly one-dimensional at low temperatures and monoatomic chains of Cu atoms form in the $[1\bar{1}0]$ channels of the Pd(110) substrate. The average length of these

124   J.V. Barth, H. Brune, T. Zambelli and K. Kern

chains decreases with deposition temperature, allowing for controlled fabrication of monoatomic wires of a desired length. The simulations revealed that the monoatomic wires even evolve in the temperature regime where cross-channel diffusion is allowed.However, anisotropic irreversible sticking of diffusing atoms to the islands must exist, i.e., atoms can only be effectively bound to an island if they are attached at the energetically favorable next-neighbor site in the [1$\bar{1}$0] direction, whereas they are free to move if they are located in a [1$\bar{1}$0] channel next to an island [12]. With deposition temperatures exceeding a critical temperature of ≈ 270 K, this sticking irreversibility is lifted, which results in the formation of rectangular islands, whose average width can be controlled via the temperature.

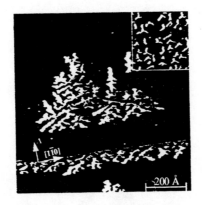

Figure 2. Dendritic islands formed in epitaxial growth of Ag on Pt(111) at 130 K and 80 K (inset); Θ_{Ag} = 0.12 ML, deposition flux 1.1 x 10^{-3} ML/s.

2.2. Dendritic islands by kinetic limitations in growth on a trigonal surface: Ag/Pt(111)

At low temperature branched islands evolve in epitaxial metal growth on a close-packed fcc(111) metal substrate.This is demonstrated by the STM image reproduced in Fig. 2 for the Ag/Pt(111) system [13, 14]. With this system the individual Ag atoms, initially present on the surface, are mobile even at low temperatures since the energy barrier for single atom migration is small on a close-packed substrate (160 meV [6]). In the example shown, the adatom aggregation leads to formation of dendritic islands with a trigonal symmetry reflecting the threefold symmetry of the Pt(111) surface. The inset in Fig. 1 demonstrates the initial branching for small islands obtained at T = 80K. These islands are Y-shaped with three branches rotated by 120°. More detailed investigations allow for the identification of the growth direction with the crystallographic [$\bar{1}\bar{1}$2] direction [6, 15]. Therefore the Y's exclusively appear in one orientation. The larger dendrites formed at 130 K retain the preferential growth direction, which leads to an overall triangular envelope of the islands.

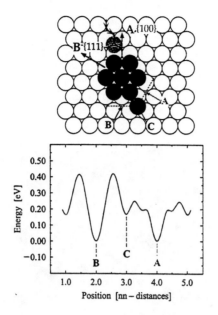

Figure 3. Model illustrating the mechanisms favoring formation of dendritic islands on a fcc(111) substrate and difference in total energy for the diffusion of a Ag atom from a corner site (C) of a Ag heptameron Pt(111) to the two possible step types (A and B).

The basic mechanisms leading to preferential growth directions, which are linked to the substrate symmetry, are illustrated by the model and results of effective medium theory (EMT [16]) calculations shown in Fig. 3 [15]. The starting point is a stable close-packed Ag-heptamer consisting of 7 atoms located at the preferred fcc substrate hollow sites. It is important to note that there are two different step types present on an fcc(111) surface, labeled A and B in the model. The A- or B-type step forms a {100}or {111} microfacet with the underlying substrate, respectively. An atom which reaches the preferential capture site of the island, a protruding corner (site C), faces two possibilities for edge diffusion along the heptamer, leading to its attachment to an A or B type step with increased coordination and hence higher binding energy. The EMT calculations, where the energetics of these two pathways were calculated, clearly demonstrate that the energy barrier from C to A is much smaller than that from C to B leading to the preferential population of A-type sites [15]. This energy difference can be rationalized by inspection of the respective paths. Diffusing to site B involves coming very close to the energetically unfavorable top position of a substrate atom.In contrast, site A can be easily reached via diffusing over a neighboring hcp site, without completely loosing the coordination to the heptamer. In a similar way, a randomly diffusing adatom occupying the same similar hcp site close to the heptamer, is attracted towards an A-type step. This is a second reason for the directional growth. B-type steps cannot be reached in the same way. Both processes are equally important for the preferred population of A-type steps and the resulting directional growth.Nevertheless, there is a small statistical chance for an attachment of diffusing atoms at B steps. This leads to some randomness apparent in the islands. It is interesting to note, that for very low deposition rates with the same system the growth behavior changes and fractal rather than dendritic islands form [13, 14].

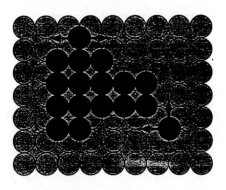

Figure 4. Model for edge diffusion mechanisms in epitaxial growth on a square substrate.

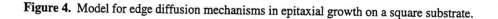

2.3. Strain-induced island ramification in equilibrium on a square substrate: Cu/Ni(100)

The discussion in the preceding section demonstrated that the dendritic growth of islands on a close-packed substrate is possible since kinetic limitations prevent the realization of the compact equilibrium shape. The symmetry of the fcc(111) surface plays a crucial role in this process, since it provides the possibility to reach sites with higher coordination by a single atom movement from a corner site to an edge. The anisotropy of this movement with respect to the two different types of atomic steps results in distinct growth directions. The situation is quite different for the quadratic fcc(100) substrate, as illustrated by the model in Fig. 4. Due to the substrate geometry there is no direct next-neighbor lateral binding at the island corner site and in addition, atoms arriving there face an equal choice for two diffusion processes towards similar island edges. Furthermore, once the atom is bound to the island edge, its only chance two find an energetically favorable site with higher coordination is diffusion to a kink site. Energetically favorable sites can usually not be reached by a single atom hop.From this it was inferred that in epitaxial growth on an fcc(100)square substrate a sharp transition between two growth regimes should prevail [17] : (i) at the lowest temperatures, where edge mobilities are possibly suppressed, fractal islands might form;(ii) if the temperature is high enough to allow for motions along the edge, compact islands should immediately evolve, since atoms at the edges remain mobile until the more stable kink sites are reached. So far these ideas seem to agree with experimental evidence and exclusively compact islands have been observed in metal epitaxial growth on square substrates (e.g., [18-20]).

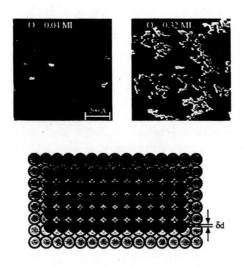

Figure 5. Island ramification in heteroepitaxial growth of Cu on a square Ni(100) substrate (deposition at 345 K, flux 1.5 x 10⁻³ ML/s) and ball model visualizing the local edge relaxation due to the lattice mismatch of the materials, which induces the island shape transition.

The STM data in Fig. 5 demonstrate, however, that island ramification in heteroepitaxial growth on square substrates can definitely exist even for growth at elevated temperature, where the edge mobilities are high enough to ensure equilibrium shapes. This holds for a wide range of substrate temperature (250 - 370 K) and deposition flux(6 x 10⁻⁵ - 3 x 10⁻² ML/s) [21]. Moreover, a statistical analysis of the observed island shapes reveals, that the ramification is size-dependent and sets only in if the number of atoms forming Cu islands exceeds a critical value of ≈ 480 atoms, quite independent of the growth conditions. Small islands remain compact, as expected, whereas the larger ones become irregular. These findings cannot be associated with kinetic limitations in the island growth. Rather, they are rationalized as a consequence of the positive lattice mismatch of the materials(+ 2.6 %). Since the islands grow pseudomorphic in the coverage range investigated here, this mismatch creates strain within the islands. Additionally, small displacements from the Cu atoms from the ideal fourfold hollow substrate positions exist, which are assumed to be largest at the island edges, as illustrated by the model in Fig. 5, corresponding to an effective strain relieve there. The strain relieved at the island edges results in an energy gain by the island ramification, which compensates for the loss of energy due to the increased perimeters of the irregular as compared to compact islands. Indeed, these findings confirm earlier theoretical considerations [22], where a similar strain driven, island size dependent shape transition was derived : small islands were predicted to be compact, whereas for larger islands a break of the symmetry allows for a more effective stress relaxation.

2.4. Pattern formation in precursor mediated dissociative chemisorption : O₂/Pt(111)

In contrast to metal deposition, for chemisorption of gases precursor states exist prior to the equilibration of the adsorbates. As a consequence the distribution of the adsorbed particles on a surface can be strongly affected by the precursor mobility on the surface, which will be in general very different from that of the final state. Dissociative adsorption of oxygen on Pt(111) proceeds via sequential population of precursor states [23-26], i.e., a physisorbed and a chemisorbed molecular oxygen species, which can be stabilized on the surface by reducing the crystal temperature to values below 30 K and 160 K, respectively. Our data show, however, in agreement with electron energy-loss spectroscopy [27] and recent STM observations [28], that for small coverages oxygen adsorbs dissociatively down to at least ≈ 95 K.

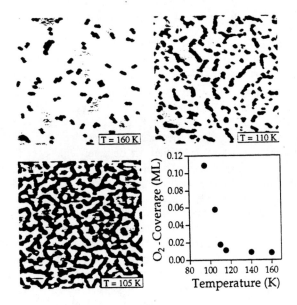

Figure 6. STM images of Pt(111), recorded after adsorption of 1 L of O₂ at the indicated temperatures and plot of the corresponding O₂-coverages as a function of temperature.

Thermally activated surface diffusion of chemisorbed O atoms is suppressed on the time scale of the experiments at temperatures below 160 K [10] and hence the STM images reproduced in Fig. 6, recorded after exposing the Pt(111) surface to an identical dose of 1 L O_2 (1 L = 10^{-6} Torr s) at sample temperatures between 105 K and 160 K, reflect the distributions of adatoms right after dissociation [29]. At the highest temperature, T =160 K, the O-atoms always appear as pairs (imaged as dark spots) which are randomly distributed across the surface. In agreement with previous findings, atoms in these pairs exhibit a preferential separation of twice the lattice constant, 2a = 5.6 Å, due to non-instantaneous release of the excess energy with dissociative adsorption ('hot atom' mechanism [30]) [10]. Upon lowering the substrate temperature to 115 K, cluster formation of the oxygen pairs is predominant. At T = 110 K, these clusters have grown quasi-one-dimensionally along the close-packed [1$\bar{1}$0] directions on the surface, with lengths between 10 Å and 50 Å. At 105 K the atoms are almost exclusively arranged in such chains, which are partly interconnected now forming an irregular network with about 40 Å periodicity. Oxygen islands were observed beginning with the lowest coverages and during *in situ* experiments at appropriate temperatures. These observations rule out the possibility that island formation is due to a locally enhanced sticking coefficient at the island edges. Rather, a *mobile* precursor state for the oxygen molecules occupied prior to their dissociation must be invoked to account for this process. The island formation is rationalized by the role of already chemisorbed oxygen as an active site for the dissociation of this precursor. This behavior is similar to the O_2/Ag(110) system, where mobile 'hot' O_2-precursors can be trapped by chemisorbed O_2-molecules [31]. The oxygen coverages included in Fig. 6 indicate in agreement with earlier observations a marked increase of the sticking probability at temperatures below 120 K [26]. It is obvious that oxygen island formation and the increase of the sticking probability are directly correlated. The temperature dependence of the sticking is understood as the result of the competition between mobility, allowing to reach the active sites for equilibration, and thermal desorption of oxygen molecules from their precursor state. This mechanism is substantiated with the help of the atomic resolution STM data in Fig. 7. Initially, at 160 K, O_2 molecules may statistically dissociate and form pairs. Already at 140 K some of the precursor molecules have a chance to find their way to pairs of atoms where they dissociate (see Fig. 7). At even lower temperatures this process is more pronounced leading to the larger agglomerates. With decreasing temperature the lifetime of the precursor at the surface increases. Since the activation energy for surface diffusion is generally merely a fraction of the adsorption energy, this causes an increase of the diffusion length of the precursor on the surface. (Note that this contrasts the behavior of nonevaporating particles like the metal-on-metal systems discussed above, where the diffusion length increases with the temperature.)

The mean free path of the precursor state is determined by its mean residence time τ on the surface and its hopping rate υ_{hop} on the surface, parameters determined by Arrhenius equations. The mean free path of the molecules is then given by the expression $\lambda = a\sqrt{\tau \upsilon} = a$ $[\upsilon_{hop}/\upsilon_{des} \exp(E_{des}(1-\alpha)/kT)]^{1/2}$, where E_{des} is the activation energy for desorption, α the ratio between the diffusion and the desorption barrier and υ_{hop} and υ_{des} the corresponding prefactors. Assuming $\upsilon_{hop} \approx \upsilon_{des}$ and $\alpha \approx 0.2$, a rule of thumb in surface diffusion, this expression reduces to $\lambda \approx a \exp(2 E_{des}/5 kT)$. Taking $E_{des} = 100$ meV, which is the binding energy of physisorbed molecules [24, 32], the temperature dependence of the mean free paths matches very well the requirements for the oxygen uptake curve and the pattern formation (e.g., $\lambda \approx 20$ a, 100 a at

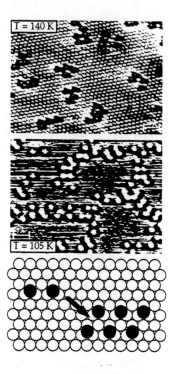

Figure 7. Atomic resolution images of oxygen islands illustrating the enhanced precursor dissociation probability near adsorbed atomic oxygen and the resulting quasi-one-dimensional growth for T = 140 K : small clusters form (3 L dose), T = 105 K : chain-shaped islands in <110> corresponding to strings of oxygen pairs (1 L dose) and model visualizing the directional growth of O_{ad} islands (black spheres) on Pt(111).

160 K, 100 K, respectively). The increased dissociation probability of the diffusing precursors near already chemisorbed oxygen is attributed to an increase of the adsorption energy and corresponding lowering of the activation barrier for dissociation near adsorbed oxygen atoms which leads to the formation of 4 atom-clusters and continues with the creation of chains. The quasi-one-dimensional growth of the clusters is very surprising in view of the threefold symmetry of the substrate lattice. It cannot be explained by the diffusion of the precursor alone and strongly contrasts the dendritic growth of metal islands on the same surface. It must be related to an interaction anisotropy between the atoms in the chains and the molecular precursor, which is associated with a higher chemical activity of O atoms at the chain ends due to their lower coordination to neighboring O (see model in Fig. 7).

3. CONCLUSIONS

The processes of self-assembly described above are believed to be of general importance for the synthesis of nanostructures on surfaces. The knowledge of the parameters governing the self-assembly of adsorbed particles may be transformed to obtain versatile tools for the tailoring of islands with a desired shape : for growth systems in the kinetic regime via diffusion-controlled aggregation and in the thermodynamic equilibrium regime via strain-induced island shapes, respectively. In addition we demonstrated the complexity for growth processes involving precursors and surface chemical reaction, resulting from the precursor mobility and the heterogeneity created by the reaction products. The pathways revealed have to be considered in catalytic reactions or chemical vapor deposition if diffusion lengths reach certain critical values determined by the distribution of adsorbed particles, and might be thus similarly useful for formation of nanostructures.

ACKNOWLEDGEMENTS

J.V.B. wishes to express his sincere thanks to the organizers, Profs. B. Rao, P. Jena, and S.N. Behera, for the invitation to a remarkable workshop on novel materials. The work of T.Z. was supported by the Deutscher Akademischer Austauschdienst (DAAD). Stimulating discussions with J. Wintterlin and G. Ertl are gratefully acknowledged.

References

[1]   *Physics and Chemistry of Finite Systems : From Clusters to Crystals*, Eds. P. Jena, S. N. Khanna, and B. K. Rao(Kluwer, Dordrecht, 1992).

[2]   D. M. Eigler and E. K. Schweizer, Nature **344**, 524 (1990).

[3]   H. Röder, E. Hahn, H. Brune, J. P. Bucher, and K. Kern, Nature **366**, 141 (1993).

[4]   R. Kern, G. L. Gay, and J. J. Metois, in *Current Topics in Material Science* Vol. 3, E. Kaldis, Eds.(North-Holland, 1979), pp. 134.

[5]   J. A. Venables, G. D. T. Spiller, and M. Hanbücken, Rep. Prog. Phys. **47**, 399 (1984).

[6]   H. Brune, H. Röder, C. Borragno, and K. Kern, Phys. Rev.Lett. **73**, 1955 (1994).

[7]   J. Villain, A. Pimpinelli, L. Tang, and D. Wolf, J. Phys.(Paris) I **2**, 2107 (1992).

[8]   T. A. Witten and L. M. Sander, Phys. Rev. B **27**, 5686(1983).

[9]   H. Brune, H. Röder, K. Bromann, and K. Kern, Thin Sol. Films **264**, 230 (1995).

132   J.V. Barth, H. Brune, T. Zambelli and K. Kern

[10] J. Wintterlin, R. Schuster, and G. Ertl, Phys. Rev. Lett. **77**, 123 (1996).

[11] J. P. Bucher, E. Hahn, P. Fernandez, C. Massobrio, and K. Kern, Europhys. Lett. **27**, 473 (1994).

[12] Y. Li, M. C. Bartelt, J. W. Evans, N. Waelchli, E. Kampshoff, K. Kern, and A. dePristo (1997); submitted.

[13] H. Brune, C. Romainczyk, H. Röder, and K. Kern, Nature **369**, 469 (1994).

[14] H. Brune, K. Bromann, K. Kern, J. Jacobsen, P.Stoltze, K. Jacobsen and J. Nørskov, Mat. Res. Soc. Symp. Proc. **407**, 379 (1996).

[15] H. Brune, K. Bromann, K. Kern, J. Jacobsen, P. Stoltze, K.Jacobsen, and J. Nørskov, Surf. Sci. Lett. **349**, L115(1996).

[16] K. W. Jacobsen, J. K. Nørskov, and M. J. Puska, Phys. Rev. B **35**, 7423 (1987).

[17] Z. Zhang, X. Chen, and M. G. Lagally, Phys. Rev. Lett. **73**,1829 (1994).

[18] E. Kopatzki, S. Günther, W. Nichtl-Pecher, and R. J. Behm, Surf. Sci. **284**, 154 (1993).

[19] J. A. Stroscio, D. T. Pierce, and R. A. Dragoset, Phys. Rev.Lett. **70**, 3615 (1995).

[20] E. Hahn, E. Kampshoff, N. Wälchli, and K. Kern, Phys. Rev.Lett. **74**, 1803 (1995).

[21] B. Müller, L. Nedelmann, B. Fischer, H. Brune, J.V. Barth, and K. Kern, Phys. Rev. Lett. (1997); submitted.

[22] J. Tersoff and R. M. Tromp, Phys. Rev. Lett. **70**, 2782 (1993).

[23] A. C. Luntz, M. D. Williams, and D. S. Bethune, J. Chem. Phys.**89**, 4381 (1988).

[24] A. C. Luntz, J. Grimblot, and D. E. Fowler, Phys. Rev. B **39**, 12903 (1989).

[25] W. Wurth, J. Stöhr, P. Feulner, X. Pan, K. R.Bauchspiess, Y. Baba, E. Hudel, G. Rocker, and D. Menzel, Phys. Rev.Lett. **65**, 2426 (1990).

[26] C. T. Rettner and C. B. Mullins, J. Chem. Phys. **94**, 1626 (1991).

[27] H. Steininger, S. Lehwald, and H. Ibach, Surf. Sci. **123**,1 (1982).

[28] B. C. Stipe, M. A. Rezaei, W. Ho, S. Gao, M. Persson, and B. I. Lundqvist, (1997); submitted.

[29] T. Zambelli, J. V. Barth, J. Wintterlin, and G. Ertl, (1997); submitted.

[30] H. Brune, J. Wintterlin, R. J. Behm, and G. Ertl, Phys. Rev.Lett. **68**, 624 (1992).

[31] J. V. Barth, T. Zambelli, J. Wintterlin, and G. Ertl, Chem.Phys. Lett. (1997); in press.

[32] A. N. Artsykhovich, V. A. Ukraintsev, and I. Harrison, Surf.Sci. **347**, 303 (1996).

LATTICE RELAXATIONS AROUND IMPURITIES IN METALS

P. H. Dederichs[1], N. Papanikolaou[1], N. Stefanou,[2] and R. Zeller[1]

[1] *Institut für Festkörperforschung, Forschungszentrum Jülich,*
D-52425 Jülich, Germany
[2] *Section of Solid State Physics, University of Athens, Panepistimioupolis,*
GR-15784 Athens, Greece

ABSTRACT

We review first-principle calculations of the size effect in dilute transition metal alloys. The calculations apply local density functional theory and a Green's function method based on the KKR-multiple scattering formalism. In each cell the full, anisotropic potential is included and the forces on the atoms are calculated by the Hellmann-Feynman theorem. The method is applied to predict the atomic positions around $d+sp$ impurities in Cu and Al. The results compare favorably with experimental data from extended X-ray-absorption fine-structure and lattice-parameter measurements.

1. INTRODUCTION

A point defect in a crystal, such as a vacancy or an impurity atom, presents not only a potential inhomogeneity, but also induces a displacement of the neighboring host atoms from their ideal lattice position. These displacements are in fact long ranged, varying with the inverse of the square of the distance from the impurity, and lead to a volume change of the crystal. Very complete information about this displacement field can be obtained by diffuse X-ray or neutron scattering, but unfortunately very few systems have been measured. Detailed information about many more systems has been obtained by extended X-ray-absorption fine-structure (EXAFS) measurements [1], yielding reliable data for the nearest neighbor shifts. In addition lattice parameter measurements [2, 3] are available for many systems, giving direct information about the volume changes induced by the impurities.

From the theoretical point of view the treatment of structural relaxation due to defects in crystals is a difficult task. In the past this problem has been mostly dealt with on a phenomenological basis, e.g. by applying models of lattice statics or continuum theory [4]. A reliable microscopic description of lattice relaxation effects based on first-principles electronic structure calculations requires very accurate total energies or forces and has mostly been attempted so far for simple metals and semiconductors on the basis of pseudopotential treatments. For instance, the pseudopotential supercell approach has been used to describe structural distortion in simple-metal [5] and semiconductor [6] systems. Lattice relaxation effects around defects in semiconductors have also been treated by the pseudopotential Green function method [7, 8].

In the paper we want to review the progress, which has been achieved recently, in calculating forces and lattice relaxations for transition metal systems [9, 10]. These calculations use the full-potential Korringa-Kohn-Rostoker (KKR) Green function method which offers an elegant and efficient framework for treating the impurity problem. In the following section we will shortly describe this method and details about the calculation of forces and lattice relaxations. In section 3 we will then present results obtained by this method for transition metal impurities and sp impurities in Cu. In section 4 similar results are given for the Al host. In both cases we discuss mostly $3d$ impurities and the connection between local moments and structural distortions. Moreover, we give also a detailed comparison with the experimental information obtained from EXAFS and lattice parameter measurements.

2. THEORETICAL METHOD

2.1. The Full Potential KKR Green Function Method

Within the framework of the KKR multiple scattering theory a crystalline solid is divided into non-overlapping space-filling cells around each atomic cite $\mathbf{R}^n$. The effective one-electron potential is written as a collection of individual potentials $V^n(\mathbf{r})$, which in each cell n are non-spherical and have a faceted cut-off structure. Using a site-centered expansion, the crystal Green function can be written in the form [11]

$$G(\mathbf{r} + \mathbf{R}^n, \mathbf{r}' + \mathbf{R}^{n'}; E) = \delta_{nn'} G_s^n(\mathbf{r} + \mathbf{R}^n, \mathbf{r}' + \mathbf{R}^{n'}; E)$$

$$+ \sum_{L,L'} R_L^n(\mathbf{r}; E) G_{LL'}^{nn'}(E) R_{L'}^{n'}(\mathbf{r}'; E) \qquad (1)$$

The vectors $\mathbf{r}$ and $\mathbf{r}'$ are restricted to the Wigner-Seitz cells around the atomic positions $\mathbf{R}^n$ and $\mathbf{R}^{n'}$, while $L = (\ell, m)$ denotes the angular momentum quantum numbers. G_s^n is the Green function for a single scattering potential in cell n in an otherwise free space. All multiple scattering contributions are contained in the second term through the so-called structural Green function $G_{LL'}^{nn'}(E)$. The wave function $R_L^n(\mathbf{r}, E)$ is the solution of the single-potential-scattering problem with a spherical wave $j_\ell(\sqrt{E}r)Y_L(\hat{\mathbf{r}})$ of angular momentum L incident on a general potential $V^n(\mathbf{r})$, where $j_\ell(x)$ is a spherical Bessel function and $Y_L(\hat{\mathbf{r}})$ is a real spherical harmonic.

The electronic structure of a crystal with a localized perturbation, induced by the presence of a point defect for instance, can be obtained in two steps. We first calculate the ideal host Green function G^0 following a band structure calculation and obtain the host structural Green function $G_{LL'}^{0nn'}$ from (1). The Green function of the defect system can be then calculated in a second step from expansion (1), with the structural Green function given by the solution of the algebraic Dyson equation [11]

$$G_{LL'}^{nn'}(E) = G_{LL'}^{0nn'}(E) + \sum_{n'',L'',L'''} G_{LL''}^{0nn''}(E)\Delta t_{L''L'''}^{n''}(E)G_{L'''L'}^{n''n'}(E) \qquad (2)$$

The summations in (2) extend only over those cells and angular momenta where the difference $\Delta t_{LL'}^{n}(E)$ between the t-matrices of the defect and the host system is significant. Equation (2) can be abbreviated in matrix form: $\mathcal{G} = \mathcal{G}^0 + \mathcal{G}^0\Delta t\mathcal{G}$.

For a general potential $V^n(\mathbf{r})$ the above t-matrices are given by

$$t_{LL'}^{n}(E) = \int d\mathbf{r} \; j_\ell(\sqrt{E}r)Y_L(\hat{\mathbf{r}}) \; V^n(\mathbf{r})R_{L'}^{n}(\mathbf{r},E) \qquad (3)$$

Since the potentials include non-spherical contributions, we expand both the wave function as well as the potential in real spherical harmonics [12]

$$R_L^n(\mathbf{r}; E) = \sum_{L'} R_{L'L}^n(r; E)Y_{L'}(\hat{\mathbf{r}}) \qquad (4)$$

$$V^n(\mathbf{r}) = \sum_{L} V_L^n(r)Y_L(\hat{\mathbf{r}}) \qquad (5)$$

The non-spherical components of the potential couple the angular momentum channels and one is faced with the problem of a system of coupled radial equations. This is solved by an iterative procedure, starting from the solution of the spherical potential [12].

The division of the space into space-filling, non-overlapping cells is described by the shape functions $\Theta^n(\mathbf{r})$ which equal 1 inside cell n and vanish outside. The shape functions are expanded in real spherical harmonics

$$\Theta^n(\mathbf{r}) = \sum_{L} \Theta_L^n(r)Y_L(\hat{\mathbf{r}}) \qquad (6)$$

The expansion coefficients $\Theta_L^n(r)$ are calculated with the algorithm of Stefanou et al. [13] following a semianalytical approach which can be used for any arbitrary Voronoi polyhedron.

The charge density is calculated from the Green function by integrating over all occupied energies. This integration can be performed efficiently by a contour integral in the complex energy plane [11, 12].

2.2. Dyson Equation for Displaced Atoms

Let us now suppose that we introduce a distortion in the crystal, displacing a number of atoms by $\mathbf{s}^n$ from their ideal lattice sites $\mathbf{R}^n$. The structural Green function matrix $\mathcal{G}^0 = \{G^{0nn'}_{LL'}\}$ of the ideal host can be expanded around the shifted positions, with the help of the matrix transformation [14, 15]

$$\tilde{\mathcal{G}}^0 = \mathcal{U}\mathcal{G}^0\mathcal{U}^{-1} \tag{7}$$

The transformation matrix $\mathcal{U} = \{U_{LL'}(\mathbf{s}^n)\}$ is local in the site index and is given by

$$U_{LL'}(\mathbf{s}^n; E) = 4\pi \sum_{L''} i^{l+l''-l'} C_{LL'L''} j_{l''}(s^n\sqrt{E})Y_{L''}(\hat{\mathbf{s}}^n) \tag{8}$$

where $C_{LL'L''} = \int d\hat{\mathbf{r}} Y_L(\hat{\mathbf{r}})Y_{L'}(\hat{\mathbf{r}})Y_{L''}(\hat{\mathbf{r}})$ are the Gaunt coefficients.

The defect structural Green functions $\mathcal{G}$, describing the real system with perturbed atoms located at the shifted positions $\mathbf{R}^n + \mathbf{s}^n$, is naturally expanded in the shifted coordinate system and related to the host structural Green function by the following Dyson equation

$$\mathcal{G} = \tilde{\mathcal{G}}^0 + \tilde{\mathcal{G}}^0(t - \tilde{t}^0)\mathcal{G} \tag{9}$$

where $\tilde{t}^0 = \mathcal{U}t^0\mathcal{U}^{-1}$ is the ideal host t-matrix in the expansion around the shifted position, and t is the defect t-matrix. We have used a tilde to denote quantities that are obtained using the coordinate system transformation.

A problem occurring in the transformation of the Green function $\mathcal{G}^0$ or the t-matrix t^0 is the angular momentum convergence. However for small shifts, to first order in the displacement, the only non-vanishing off-diagonal elements of the transformation matrix (8) are those with $|l - l'| = 1$ [15]. Therefore, for moderate lattice distortions, accurate calculations up to l_{max} can be done if angular momentum components up to $l_{max} + 1$ are included. The full-potential KKR Green function method has been found to give accurate, well-converged total energies and forces using a cut-off at $l_{max} = 3$. Therefore in the present calculation we use $l_{max} = 4$ in order to obtain reliable results.

2.3. Calculation of Forces and Lattice Relaxations

Calculation of interatomic forces is an important and difficult problem in electronic-structure calculations. According to the Hellmann-Feynman theorem the force on an atom is given by the electric field at the nuclear position due to all electrons and all other nuclei. The application of this theorem is however severely limited, since it requires a full-potential treatment of the core electrons. This problem can be overcome by making a spherical ansatz $\rho^n_c(r)$ for the core density entering in the total energy expression. The force is then calculated as the derivative of the total energy with respect to the nuclear position assuming that the Kohn-Sham equations are solved exactly for the valence electrons only. The resulting expression for the force $\mathbf{F}^n$ on the atom $\mathbf{R}^n$ is given by [16, 17]

$$\mathbf{F}^n = Z^n \partial_{\mathbf{r}} V^n_C(\mathbf{r})|_{\mathbf{r}=0} - \int d^3r \rho^n_c(r)\partial_{\mathbf{r}} V^n(\mathbf{r}) \tag{10}$$

where V^n_C is the Coulomb part of the effective one-electron potential in cell n due to all electrons and all other nuclei. Clearly the first term is the force on the nucleus, evaluated with spherical core charge densities and the second term is the force on the core electrons. Since V^n is the effective Kohn-Sham potential, the latter term includes both electrostatic and exchange-correlation contributions arising from the exchange and correlation between core and valence electrons.

Using the angular momentum expansion (5) for the potential, one finally obtains for the $i = x, y, z$ component of the force on atom n

$$F_i^n = Z^n \sqrt{\frac{3}{4\pi}} \left. \frac{V_{C;1i}^n(r)}{r} \right|_{r=0} - \sqrt{\frac{4\pi}{3}} \int dr\, \rho_c^n(r) \frac{\partial}{\partial r} \left(r^2 V_{1i}^n(r) \right) \tag{11}$$

As we see from this result, within the KKR Green function method the force calculation requires no additional effort, since the force is readily calculated from the $l = 1$ component of the potential which is anyhow evaluated self-consistently. Furthermore one can show that Pulay corrections to the Hellmann-Feynman force formula (10), arising from the finiteness of the basis used in the calculations, vanish within the full potential KKR formalism [17].

Knowing the forces, the determination of the displacements s^n describing the equilibrium structure is another non-trivial task. In order to accelerate convergence we use here a lattice statics procedure known as Kanzaki method [4, 9]. This requires the calculation of the Kanzaki forces

$$\mathbf{F}_K^n = \mathbf{F}^n - \sum_{n'} \Delta\underline{\phi}^{nn'} \mathbf{s}^{n'} \tag{12}$$

where $\mathbf{F}^n$ are the Hellmann-Feynman forces (10) and $\Delta\underline{\phi}$ are the changes of the Born-von-Karman coupling parameters in the vicinity of the defect, being estimated from calculations with given $\mathbf{s}^{n'}$. Within lattice statics the displacements are then determined by

$$\mathbf{s}^n = \sum_{n'} \underline{G}^{nn'} \mathbf{F}_K^{n'}(\mathbf{s}) \tag{13}$$

where $\underline{G}^{nn'}$ is the lattice Green function of the ideal harmonic crystal (without the defect). In this way a new displacement pattern $\mathbf{s}^n$ for all atoms around the impurity is generated which serves as input for another electronic-structure and force calculation, etc. In practical cases only one or two self-consistency cycles are needed.

3. LATTICE RELAXATIONS IN Cu

3.1. Consistency of Force and Total Energy

In order to check the reliability of the calculated forces, we allow only a relaxation of the first neighbors of the impurity, by fixing the more distant ones at their regular positions. The equilibrium positions can then either be calculated from the minimum of the total energy, or from the condition of vanishing forces. The good agreement between results represents a stringent test of the accuracy of the force calculation.

3.2. Importance of Semicore States

In the force formula (10) the role of the core and valence electrons is very different, since the core charge density is assumed to be spherically symmetric, while the valence charge density is calculated exactly. For the semicore states the question therefore arises, if they are sufficiently localized to be treated as core states. For the 3d-impurities in Cu, we find that this is not the case for the 3p semicore states. For instance, for a Ti impurity the force on the neighboring Cu atoms is about 50 % too large, if the semicore charge density of the impurity is considered as spherical. Thus for the early transition metal impurities in Cu the 3p semicore states have to be treated as valence states, since they give an important binding contribution to the forces.

3.3. Displacements of Nearest Neighbors

For the case of 3d impurities in Cu, Fig. 1 shows the displacements of the NN Cu atoms calculated by the Kanzaki procedure. Also included are experimental values form EXAFS measurements by Scheuer and Lengeler [1] which agree with the calculated results within the experimental uncertainty. In most cases the Cu lattice is dilated by the impurities, except for the cases of Fe, Co or Ni impurities.

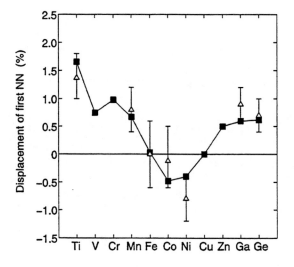

Figure 1. Relative displacements of the nearest neighbor Cu atoms (with respect to the n.n. distance) for 3d and 4sp impurities in Cu. The triangles with error bars give the results of EXAFS measurements [1].

Since the 3d impurities in Cu are magnetic, it is interesting to see how the local moments are affected by the lattice displacement. We find that this influence is vanishingly small. Our calculation gives for the local moments of V, Cr, Mn, Fe, and

Co impurities: 0.98 (0.85), 2.96 (2.91), 3.42 (3.39), 2.53 (2.53), and 0.96 (1.01) Bohr magnetons, respectively, in the relaxed (unrelaxed) geometry. These results show that outward relaxations slightly increase the moments (V, Cr, and Mn), while inward relaxations lead to a small decrease of the magnetic moment (Co). For Fe there is no relaxation of the first NN so that, in this case, there is no moment change. This vanishingly small influence of the local lattice distortion on the impurity moments can be understood, if we compare the spin-polarization energies of the impurities [18] with lattice relaxation energies. While the relaxation energies in the whole series from V to Ge are smaller than 0.03 eV, the spin-polarization energies of the magnetic impurities are huge, e.g. 0.7 eV for Mn, 0.36 eV for Fe and 0.39 eV for Cr. Therefore in Cu the local moments are very stable and practically not affected by the relaxations.

3.4. Induced Volume Change

For a statistical distribution of many point defects the lattice expansion or compression due to the individual defects results in a change of the crystal volume. In the case of cubic metals the volume change due to a single defect is given by the first moment of the Kanzaki forces [4]

$$\Delta V = V - V_0 = \frac{1}{3K} \sum_n \mathbf{F}_K^n \cdot \mathbf{R}^n \tag{14}$$

where V and V_0 are the atomic volumes of the defect system and the ideal host, respectively, and K is the bulk modulus of pure Cu: $K = 1.55$ Mbar. As an example we present in Fig. 2 the calculated relative volume changes $\Delta V/V_0$ for $4d$ impurities in Cu, together with the experimental data as obtained from lattice-parameter measurements [2, 3]. For cubic crystals the volume change is related to the change of the lattice parameter by $\Delta V/V_0 = 3\Delta a/a_0$. The agreement with the experiment is good.

The relative sizes of the nearest neighbor displacements and the volume changes can be estimating from a simple elasticity model. Assuming that the asymptotic displacement field, varying as $1/r^2$ in elasticity theory (r = distance from impurity), is already valid at the nearest neighbor sites, the volume change of a sphere with radius equal to the nearest neighbor distance a_{NN} is given by $\Delta V \cong 4\pi a_{NN}^2 s_{NN}$, where s_{NN} is the nearest neighbor displacement. Inserting the values $a_{NN} = a/\sqrt{2}$, $V_0 = a^3/4$ for the fcc lattice and correcting for the image expansion [4] we obtain in this model: $\Delta V/V_0 \approx 27 s_{NN}/a_{NN}$. The simple model explains the similarity between the ΔV and s_{NN} curves found in the calculation. Also the prefactor 27 is reasonable.

The dashed line in Fig. 2 refers to the volume expansion as estimated by Vegard's law. This indeed shows the qualitative trends as observed for the sp and early 3d impurities. However the bump in the curve for Fe, Mn and Cr impurities cannot be explained in this way. It results from the large local moments of the impurities, leading to a sizeable magneto-volume expansion. The same effect can also be seen for the displacements in Fig. 1.

4. LATTICE RELAXATIONS IN Al

Al is a three-valent simple metal with a rather high electron density. Nevertheless its lattice constant is about 10 % larger than the one of Cu. Both facts are important for understanding the defect relaxations in Al as compared to Cu.

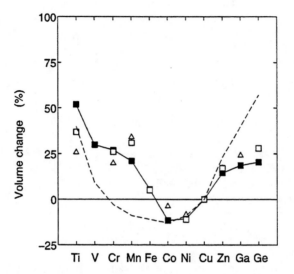

Figure 2. Volume changes (in units of the elementary volume) due to 3d and 4sp impurities in Cu. The squares and triangles denote experimental values from lattice parameter measurements [2, 3]. The dashed line is obtained using Vegard's law.

Fig. 3 shows the predicted displacements of the first nearest neighbors for $3d$ and $4sp$ impurities in Al. The results are in good agreement with the available experimental data from EXAFS measurements [1]. Upon alloying with $3d$ elements the Al lattice contracts. Across the $3d$ series the resulting displacements show a parabolic behavior with a minimum at Fe, while in the case of $4sp$ impurities we find only very small distortions. As compared to Cu the displacements in Al are much larger, which can only partially be explained by the differences between the host lattice constants. For instance, in Cu the difference between the NN relaxations of Co and Ge is about 1 %, while in Al this difference amounts to 4.5 %. Clearly the different bonding in both hosts is very important. In particular the formation of the strong and compact bonds between the impurity d- and Al p-electrons leads to the large contraction in the $3d$ series. As a consequence the relaxation energies, varying quadratically with the distortions, are very large (Fig. 4). For instance, for Co and Fe in Al, these amount to 0.42 and 0.44 eV which is more than an order of magnitude larger than the corresponding values in Cu.

The strong d-p hybridization between the impurity d-states and the Al p-electrons dramatically affects the local moments. In the unrelaxed configuration only Cr, Mn and

Fe impurities are magnetic, with moments of 2.2, 2.6 and 1.7 μ_B. The corresponding spin-polarization energies are shown in Fig. 4. They are considerably smaller than the spin polarization energies in Cu, which amount to e.g. 0.7 eV for Mn and 0.36 eV for Fe. More important is that these energies are smaller than the energies being gained in Al by lattice relaxations. The corresponding relaxation energies are also shown in Fig. 4. Since the inward relaxation increases the p-d hybridization, this has the tendency to strongly reduce the local moments. From energetic point of view the large gain in relaxation energy can overcome the loss of spin-polarization energy. For the equilibrium configuration we find a moment of 0.7 μ_B for Cr, 1.3 μ_B for Mn and a vanishing moment for Fe. Experimentally Cr and Mn in Al are spin-fluctuation systems with Kondo-temperatures of 500 K and 1500 K [19], being in qualitative agreement with our results. The result for Fe is in accordance with the experimental finding that Fe is non-magnetic in Al. In fact in the calculation the Fe-moment vanishes already at a relaxation of 2.8 %, being appreciably smaller than the equilibrium value of 4.2 %.

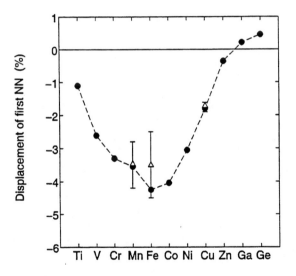

Figure 3. Relative displacements of the n.n. Al atoms due to 3d and 4sp impurities in Al. The triangles denote results form EXAFS measurements [1].

Thus when we compare the magnetic and structural properties of $3d$ impurities in Cu and Al, we find a very different behavior. In Cu the spin-polarization energies are large and the relaxation energies very small, typically 0.03 eV. Therefore the local moments are very stable and practically not affected by the relaxations. On the other hand the relaxations themselves are small, but depend strongly on magnetism. In Al we are in the opposite limit of large relaxation energies and small spin polarization energies (Fig. 4). Therefore the relaxations are large, with only a small magnetic anomaly appearing for the Mn impurity, while the moments are relatively small and strongly depend on relaxations.

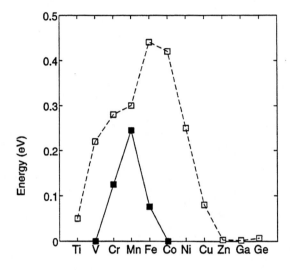

Figure 4. Lattice relaxation energies (dashed line) and spin polarization energies (full line, for the unrelaxed configuration) for 3d and 4sp impurities in Al.

5. SUMMARY

We have presented an *ab initio* method to calculate lattice distortions around point defects in metals. The method was applied to $3d$ and $4sp$ impurities in Cu and Al. We have shown that the full potential KKR Green function method allows an accurate and easy calculation of forces, since Pulay-type corrections to the Hellmann-Feynman force vanish. The results for the NN shifts and the volume changes induced by the defect are in very good agreement with data from EXAFS and lattice parameter measurements. We believe that the present method can be successfully applied to many other structural or dynamical properties of solids, e.g. relaxations of ideal surfaces or of surfaces with adsorbate atoms, phonons of transition metals or of transition metal surfaces etc.

References

[1] U. Scheuer and B. Lengeler, Phys. Rev. B **44**, 9883 (1991)

[2] H. W. King, J. Mat. Sci. **1**, 79 (1966)

[3] W. B. Pearson, *A Handbook of Lattice Spacings and Structures of Metals and Alloys*, Vol. 1 (Pergamon Press, London, 1958), p. 570; Vol. 2 (Pergamon Press, Oxford, 1967) p. 868

[4] G. Leibfried and N. Breuer, *Point Defects in Metals I* (Springer-Verlag, Berlin, 1978)

[5] R. Pawellek, M. Fähnle, C. Elsässer, K. M. Ho and C. T. Chan, J. Phys.: Condens. Matter **3**, 2451 (1991); R. Benedek, L. H. Yang, C. Woodward, B. I. Min, Phys. Rev. B **45**, 2607 (1992)

[6] Y. Bar-Yam and J. D. Joannopoulos, Phys. Rev. Lett. **52**, 1129 (1984); Phys. Rev. B **30**, 1844 (1984)

[7] M. Scheffler, J. P. Vigneron, and G. B. Bachelet, Phys. Rev. Lett. **49**, 1765 (1982); Phys. Rev. B **31**, 6541 (1985); M. Scheffler, Physica **146B**, 176 (1987)

[8] G. A. Baraff and M. Schlüter, Phys. Rev. B **30**, 1853 (1984); Phys. Rev. Lett. **55**, 1327 (1985)

[9] N. Papanikolaou, R. Zeller, P. H. Dederichs and N. Stefanou, Phys. Rev. B **55**, 4157 (1997)

[10] N. Papanikolaou, R. Zeller, P. H. Dederichs and N. Stefanou, Comp. Mat. Science (accepted)

[11] P. J. Braspenning, R. Zeller, A. Lodder, and P. H. Dederichs, Phys. Rev. B **29**, 703 (1984)

[12] P. H. Dederichs, B. Drittler, and R. Zeller, Mat. Res. Soc. Symp. Proc. **253**, 185 (1992)

[13] N. Stefanou, H. Akai, and R. Zeller, Comput. Phys. Commun. **60**, 231 (1990); N. Stefanou and R. Zeller, J. Phys.: Condens. Matter **3**, 7599 (1991)

[14] A. Lodder, J. Phys. F **6**, 1885 (1976)

[15] N. Stefanou, P. J. Braspenning, R. Zeller, and P. H. Dederichs, Phys. Rev. B **36**, 6372 (1987)

[16] J. Harris, R. O. Jones, and J. E. Müller, J. Chem. Phys. **75**(8), 3904 (1981)

[17] K. Abraham, Diploma thesis, RWTH Aachen (1991) (unpublished)

[18] P. H. Dederichs, T. Hoshino, B. Drittler, K. Abraham, and R. Zeller, Physica B **172**, 203 (1991)

[19] F.J. Kedves, M. Hordos, and L. Gergely, Solid State Commun. **11**, 1067 (1972); E. Babic, P.J. Ford, C. Rizzuto, and E. Salamoni, J. Low Temp. **8**, 219 (1972)

CONTROLLED GROWTH OF SEMICONDUCTORS WITH SURFACTANTS

Efthimios Kaxiras[a],† and Daniel Kandel[b]
[a] Institute for Theoretical Physics
University of California, Santa Barbara, CA 93106, USA
[b] Department of Physics of Complex Systems
Weizmann Institute, Rehovot, Israel

ABSTRACT

. Controlled growth of semiconductors at low temperature is essential in many applications of nanoscale structures in electronic and optical devices. In the last few years, the use of surfactants for enhancing layered growth of various elemental and compound semiconductors has been successfully demonstrated. We discuss a theoretical model that provides a comprehensive picture of growth of semiconductors through the use of various types of surfactants. The model is based on first principles calculations for the activation energies of key processes and kinetic Monte Carlo simulations of solid-on-solid growth that afford direct comparison with experimental observations.

1. INTRODUCTION

Growth of semiconductor thin films, both in homoepitaxy (material A on substrate A) and heteroepitaxy (material A on substrate B), has become an integral step in the fabrication of many devices with submicron features. There are usually two important requirements in this process. First, high quality crystalline material must be obtained and second, low growth temperatures must be maintained. The need for the first requirement is self evident, since a highly defected crystal will perform poorly in electronic applications. The second requirement arises from the need to preserve the characteristics of the substrate during growth, such as doping profiles and sharp interfaces between layers, which can be degraded due to atomic diffusion at high temperature. These two requirements seem to be incompatible: In order to improve crystal quality, atoms need to have sufficient surface mobility so that they can find the proper crystalline sites to be incorporated into a defect-free crystal. On the other hand, low enough temperature must be maintained so that the substrate characteristics are not altered due to excessive atomic mobility in the bulk. These problems are exacerbated in the case of semiconductor growth, where the strong directional covalent bonds inhibit diffusion at low temperatures and lead to the trapping of atoms at positions that do not correspond to crystalline sites.

Several years ago Copel *et al.* [1] demonstrated that the use of surfactants improves the growth of semiconductor crystals at temperatures significantly lower than would be required in their absence. Since that original demonstration, a number of experiments have been reported exhibiting surfactant mediated growth in homoepitaxy and heteroepitaxy of semiconductors [2, 3, 4, 5, 6, 7, 8, 9, 10]. Despite the progress in this field, several puzzles remain in understanding how surfactants work at the microscopic level. The understanding of surfactant mechanisms is crucial in making it possible to select successful surfactant–substrate combinations for a given application. One of the most intriguing observations is that surfactant mediated growth often proceeds through the nucleation of enhanced density of small islands [2, 3]. This has been interpreted as an indication that the surfactant decreases the surface diffusion length. A reduction of the diffusion length seems counter-intuitive, in light of the passivation of the surface by the surfactant layer.

We note here that this behavior is not universal, since in many cases of surfactant mediated growth, the growth occurs trough step flow [7]. This case is easier to rationalize, because enhanced diffusion on terraces and enhanced reactivity at step edges naturally lead to step flow growth. This is precisely the type of behavior expected of a surfactant layer, which typically provides a chemically passive environment on which the newly deposited atoms can easily diffuse until they find a step edge where they become incorporated into the substrate. Nevertheless, growth which exhibits enhanced density of islands is just as as common as step flow growth, and needs to be explained from an atomistic perspective, if a comprehensive picture of surfactant mediated growth is to be attained.

In this paper we review our recent work toward explaining these questions. We discuss a model which accounts for the two aforementioned types of surfactant mediated growth of semiconductors. The model is based on first-principles calculations and stochastic simulations of growth, which can successfully account for essentially all the important experimental observations reported on this topic.

2. SURFACTANT MECHANISMS

Before we embark on the construction of the theoretical model we briefly review the available experimental information. It appears that a full monolayer of surfactant coverage is required for growth of high quality semiconductor crystals. This is different from the case of surfactant effects in the growth of metals, where a small amount of surfactant (typically few percent of a monolayer coverage) is sufficient. Direct evidence on this issue was provided by the experiments of Wilk *et al.* [10], who studied homoepitaxial growth of Si on Si(111) using Au as a surfactant. These authors report that the density of defects in the film correlates well with the surfactant coverage, with the minimum defect density corresponding to full monolayer coverage by the surfactant. This is a physically appealing result, and can be interpreted as evidence that the better the passivation of the surface by the surfactant, the more effective the surfactant is in promoting high quality growth. In the following we will assume that full monolayer coverage of the substrate is the standard condition for successful surfactant mediated growth of semiconductors.

In order to understand the microscopic aspects of surfactant mediated growth, one has to consider all the possible atomic processes involved in the phenomenon. A schematic representation of such processes is shown in Fig. 1. The simplest process is of course diffusion of the newly deposited atoms on top of the surfactant layer [Fig. 1(a)]. A second necessary process is the exchange of newly deposited atoms with the surfactant atoms, so that the former can be buried under the surfactant layer and become part of the bulk. This process can take place either on a terrace or at a step [Fig. 1(b)]. From thermodynamic considerations, we must also consider the process by which atoms de-exchange and become again equivalent to newly deposited atoms on top of the surfactant layer [Fig. 1(c)]. Again, this process can take place on terraces or at surface steps. Up to this point, we have assumed that the surfactant passivates equally well the terraces and the steps, so that all atomic processes are equivalent on terraces and steps, as suggested schematically in Fig. 1(a)-(c). Finally, we have to consider separately surfactants that *cannot* passivate step edges, in which case both the exchange [Fig. 1(d)] and de-exchange processes [Fig. 1(e)] will be different than at passivated steps, since at step edges these processes no longer involve actual exchange events between newly deposited atoms and surfactant atoms.

This last issue is crucially important to the morphology of growth. Accordingly, we discuss in some more detail the ability of a surfactant to passivate terraces or steps on the semiconductor surface. We distinguish these two possibilities according to the chemical nature of the surfactant. For instance, group V atoms (especially As and Sb) should be effective in passivating steps on the (111) and (100) surfaces of tetravalent semiconductors such as Si and Ge. This is because group V atoms prefer to have three-fold coordination, in which they form three strong covalent bonds with their neighbors using three of their valence electrons, while the other two valence electrons remain in a low-energy lone-pair state. This is precisely what is needed for passivation of both terrace and step geometries in the (111) and (100) surfaces of the diamond lattice, which are characterized by three-fold coordinated atoms [11]. On the other hand, it is expected that elements with the same valence as the substrate, or noble metals, will not be effective in passivating step edges. In the case of the tetravalent semiconductors Si and Ge, for example, the elements Sn and Pb have the same valence, and while they can form full passivating layers on top of the substrate, they clearly cannot passivate the step geometries since they have exactly the same valence as the substrate atoms and hence can only form similar structures. Analogously, certain noble metals can form a passivating monolayer on the semiconductor surface, but their lack of strong covalent bonding cannot affect the step structure. We note that not all noble metals behave in a similar manner: some of them form complex structures in which they intermix with the surface atoms of the substrate (such as Ag on the Si(111) surface), in which case it is doubtful that they will exhibit good surfactant behavior.

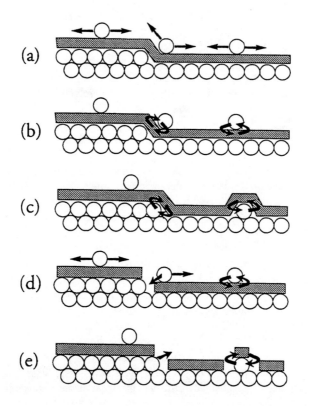

Figure 1. Schematic illustration of important mechanisms in surfactant mediated growth on a substrate (represented by white circles) with a full monolayer surfactant coverage (represented by continuous shaded area): (a) diffusion on terraces and steps for surfactant that passivates step edges; (b) exchange at terraces and passivated steps; (c) de-exchange at terraces and passivated steps; (d) diffusion on terrace and exchange at non-passivated steps; (e) de-exchange at terrace and at non-passivated steps.

For those surfactants that passivate the substrate terraces but not the substrate steps, we expect that the mode of growth will be step flow. The reasoning is that the newly deposited atoms will diffuse fast on the passivated terraces and be able to reach the nearest (unpassivated) step, where they quickly react and become incorporated into the substrate (as indicated in Fig. 1(d)). This picture seems to be supported by experimental evidence which suggests that Sn, Pb and Au, when used as surfactants lead to step flow growth on Si(111) surfaces [10, 7] (as argued above, these elements should passivate the terraces but not the steps on this substrates).

The opposite case, where the surfactant passivates equally well the terraces and the steps, is a more interesting one. As indicated in Fig. 1(a)-(c) a number of processes must be considered in this case, the balance of which will give the observed behavior in a physical system. In order to evaluate their relative contributions, the corresponding activation energies must be calculated. This is a difficult task because very little is known about the atomic configurations involved in these processes. In the next section we consider two idealized processes, discuss how the corresponding activation energies should be representative of growth mechanisms, and obtain their values from first principles calculations.

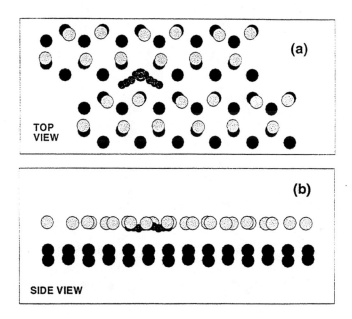

Figure 2. Representative surface diffusion pathway, top and side views. The dark circles represent the substrate atoms, the light circles the surfactant atoms. The smaller grey circle represents an extra atom deposited on top of the surfactant layer, at different positions.

3. REPRESENTATIVE ATOMIC PROCESSES

The first process we consider is diffusion on a surface covered by a surfactant monolayer. The representative system we chose to study consists of a Si(111) substrate, covered by a bilayer of Ge, with Sb as the surfactant. In this case, it is known that the structure of the Sb layer is a chain geometry with a periodicity of (2×1), as shown in Fig. 2 [11]. An additional Ge atom is then placed on top of the Sb layer and the total energy of the system is optimized for a fixed position of the Ge atom along the direction parallel to the Sb chains. All other atomic coordinates, including those of the Ge atom perpendicular to the Sb chain and vertical with respect to the surface, are allowed to relax in order to obtain the minimum energy configuration. The energy and forces are computed in the framework of Density Functional Theory and the Local Density Approximation (DFT/LDA), a methodology that is known to provide accurate energetic comparisons for this type of system. By considering several positions of the extra Ge atom along the chain direction and calculating the corresponding total energy of the system we obtain a measure of the activation energy for diffusion in this direction. We find that the activation energy for diffusion along this path is 0.5 eV.

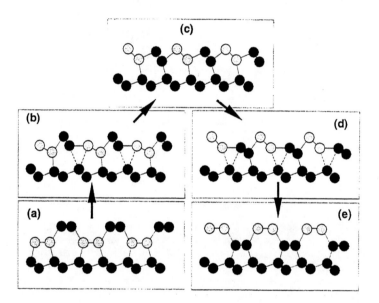

Figure 3. Representative exchange pathway. The color scheme is the same as in Fig. 2. (a) Structure with one layer of newly deposited atoms on top of the surfactant layer. The geometries depicted in (b), (c), (d) are the intermediate structures during a concerted exchange that brinks the surfactant layer on top of the newly deposited layer, shown as the final configuration in (e). Structure (c) is metastable, while structures (b) and (d) are saddle-point configurations.

We next consider a possible exchange mechanism in the same system, through which the newly deposited Ge atoms can interchange positions with the surfactant atoms and become buried under them. To this end, we model the system by a full monolayer of Ge deposited on top of the surfactant layer [Fig. 3(a)]. We studied a concerted exchange type of motion for the Ge-Sb interchange. In the final configuration [Fig. 3(e)] the Ge layer is below the Sb layer, and the system is now ready for the deposition of the next Ge layer on top of the surfactant. The middle configuration, Fig. 3(c), corresponds to a metastable structure, in which half of the newly deposited Ge layer has interchanged position with the Sb surfactant layer. The configurations between the initial and middle geometries and the middle and final geometries, Fig. 3(b) and Fig. 3(d) respectively, correspond to the saddle point geometries which determine the activation energy for the exchange.

From our DFT/LDA calculations we find that the energy difference between structures 3(a) and 3(b) is 0.8 eV, and the energy difference between structures 3(c) and 3(d) is the same to within the accuracy of the results. Similarly, the energy difference between structures 3(c) and 3(b) and 3(e) and 3(d) is 1.6 eV. The activation energy for exchange (going from 3(a) to 3(c) through 3(b), or going from 3(c) to 3(e) through 3(d)) is then 0.8 eV, while the activation energy for de-exchange (going from 3(c) to 3(a) through 3(b), or going from 3(e) to 3(c) through 3(d)) is 1.6 eV for the hypothetical process described in Fig. 3.

We discuss next why these calculations give reasonable estimates for the activation energies involved in surfactant mediated growth. As far as the diffusion process is concerned, it is typical for semiconductor surfaces to exhibit anisotropic diffusion constants depending on the surface reconstruction, with the fast diffusion direction along channels of atoms that are bonded strongly among themselves. This is precisely the pathway we examined in Fig. 2. As far as the exchange process is concerned, it is believed that the only way in which atoms can exchange positions in the bulk is through a concerted exchange type of motion, as first proposed by Pandey for self diffusion in bulk Si [12]. This motion involves the breaking of the smallest possible number of covalent bonds during the exchange, which keeps the activation energy relatively low. In the case of bulk Si, the activation energy for concerted exchange is 4.5 eV. In the present case the activation energy is only 0.8 eV, because, unlike in bulk Si, the initial configuration [Fig. 3(a)] is not optimal, having the pentavalent Sb atoms as four-fold coordinated (they would prefer three-fold coordination) and the newly deposited Ge atoms as three-fold coordinated (they would prefer four-fold coordination). In the final configuration [Fig. 3(e)], which has lower energy than the initial one, all atoms are coordinated properly (three-fold for Sb, four-fold for Ge).

While we have argued that the atomic processes described above are physically plausible, we have not established neither their uniqueness, nor their supremacy over other possible atomic motions. This, however, is not necessary, for the following reasons: The activation energy for diffusion corresponding to the mechanism illustrated in Fig. 2 should be taken as an upper bound, since possible diffusion paths with lower activation energy cannot be excluded. On the other hand, the activation energy of the exchange mechanism illustrated in Fig. 3, should be taken as a lower bound, since the atoms were arranged in an optimal manner as far as the breaking of covalent bonds is concerned, that is, the exchange is facilitated by the geometry chosen. Interchange of atoms in other geometries would not benefit from the concerted exchange aspect built in the model of Fig. 3. Thus, through the above calculations we have determined a plausible upper bound for the activation energy for diffusion on top of the surfactant, and a plausible lower bound for the activation energy of exchange at terraces. We will also take the de-exchange energy barrier of the hypothetical mechanism in Fig. 3 as representative of this process, in analogy with the activation energy for exchange.

This exercise in determining plausible bounds for the activation energies involved in the possible processes during surfactant mediated growth (see Fig. 1), is useful because it already provides some clues on what the likely microscopic evolution of the system is. Specifically, based the fact that the upper bound to the activation energy for diffusion (0.5 eV) is almost half of the lower bound to the activation energy for exchange (0.8 eV), we expect that diffusion will be a much faster process than exchange in this system. This is already interesting, in that it seems to be incompatible with the simple explanation of the observed enhanced density of islands during growth in this system, which invoked the notion that the surfactant suppresses the surface diffusion [2, 3]. On the other hand, fast diffusion on the surfactant, is promising, since it will be beneficial to high-quality crystal growth.

The proper way to evaluate the effect of the rates of the various mechanisms on the morphology of the system is by performing kinetic Monte Carlo simulations, with all the possible processes occurring randomly with rates determined by the corresponding activation energies, as described in the next section.

4. KINETIC MONTE CARLO SIMULATIONS

We consider a system in which the processes examined above are operative, and the corresponding activation energies are the ones obtained from the DFT/LDA calculations for the hypothetical cases illustrated in Fig. 2 and Fig. 3. Before reporting on the simulations, certain subtleties need to be clarified.

Since the hypothetical system we are considering consists of a Si(111) substrate with Sb as a surfactant, we will assume that the island edges are passivated by the surfactant, as argued earlier for the case of group V surfactants on a tetravalent substrate. We will therefore assume that activation barriers for diffusion and exchange are the same at terraces and at step edges. There is however an important difference between processes that take place at step edges of small islands and edges of larger islands. Specifically, consider an island consisting of one atom buried under the surfactant. This atom will undergo de-exchange at a certain rate. An atom at the edge of a large island will undergo de-exchange at a much slower rate, since that process involves the breaking of additional bonds that the atom forms *laterally* to its neighbors under the surfactant layer (see Fig. 1). This difference is crucial, since it will affect atoms at the edge of islands of different sizes in a different manner. Consequently, we will assume that the activation energy for de-exchange at the edge of islands that are bigger in size than one atom will be higher than the the de-exchange activation energy of an isolated buried atom. The difference in de-exchange energies is proportional to the number of broken covalent bonds involved in the de-exchange process. This is an empirical correction factor that must be introduced in the model for physical reasons.

With these considerations, we performed kinetic Monte Carlo (KMC) simulations using the activation energies discussed above, for the case of homoepitaxial growth, i.e.

when there is no strain involved in the system due to a lattice mismatch between the substrate and the newly deposited material [13]. For simplicity, we considered a square lattice. The results of these simulations are displayed in Fig. 4, for temperatures of (a) 600°C, (b) 700°C, and (c) 850°C. The flux in the simulations was taken to be the same as in experiments, and the amount of deposition was 0.15 monolayers for direct comparison with experimental results [2]. It is clearly seen from these simulations that at low temperature there is enhanced density of small islands, which decreases with increasing temperature, precisely as observed experimentally. Moreover, the shape of the islands is fractal-like with rough edges, again very closely matching the experimental observations.

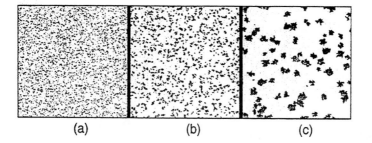

(a)   (b)   (c)

Figure 4. Kinetic Monte Carlo simulation of homoepitaxial surfactant mediated growth. A total of 0.15 monolayer of new material has been deposited at (a) 600°C, (b) 700°C and (c) 850°C. The high density of small islands at low temperature is evident, as well as the decrease of the island density with increasing temperature.

Thus, we find that our simulations reproduce qualitatively the experimental results, without having to introduce any artificial assumptions about the restriction of the diffusion length of newly deposited atoms on top of the surfactant layer. In fact, in our model the diffusion on top of the surfactant is much faster than exchange. What leads to the formation of many small islands is the fact that island step edges are passivated, and therefore do not act as sinks of the newly deposited atoms.

What is equally important is that when we continue the simulation of growth in this model, we find that growth remains flat in the sense that each layer is essentially completed before the next higher layer. This is crucial in order to maintain the high quality of the crystal and to avoid a very rough surface. This type of growth is difficult to account for in a model which involves reduced diffusion length on the surfactant layer, while it comes as a natural consequence of our model, which does not involve such an assumption. What preserves the layered growth in our model is the fact that atoms are allowed to de-exchange, which enhances their mobility since once they get on top of the surfactant layer they can again diffuse very fast. In that sense, the edges of larger islands are more attractive, since atoms there de-exchange less frequently because they have to break more lateral bonds. This attraction, however, is an *a posteriori* effect due to de-exchange, rather than an *a priori* effect, as one might expect at reactive step edges.

Finally, we will consider the effects of strain in surfactant mediated growth. Strain is difficult to include in an atomistic calculation in a self consistent manner. Here we will rely on the theory developed by Tersoff *et al.* [14] for the elastic energy of strained islands on a substrate. According to this theory the elastic energy per unit volume is given by:

$$E_{el} = \epsilon \left(\frac{\ln s}{s} + \frac{\ln t}{t} \right), \qquad (1)$$

where s and t are the lateral dimensions of the island (measured in the lattice constant), and ϵ is a quantity that includes the stress involved in the lattice mismatch, as well as additional parameters related to the shape of the island [14]. For the present study, we will take ϵ to be a free parameter in our model. The effect of strain then is to alter the strength of the bonds in elastically strained islands according to the expression of Eq. (1), which depends on the island size through the values of s and t. The most important consequence of this effect is a change in the de-exchange activation energy for atoms that belong to an island: this energy will now depend on the island size. The other barriers, having to do with processes that take place on top of the surfactant (diffusion and exchange at terraces and step edges), will be unaffected to lowest order by the presence of strain. Therefore, the only important change in the kinetics comes from an island-size dependent de-exchange rate, given by

$$E'_{de-ex} = E_{de-ex} + \epsilon \left(\frac{\ln s}{s} + \frac{\ln t}{t} \right), \qquad (2)$$

with E_{de-ex} the de-exchange barrier in the absence of strain effects. With this modification, and using a value of $\epsilon = 3.0$ eV, which is a reasonable number for the typical strength of bonds and the amount of strain involved in the systems of interest (4% for the case of Ge on Si), we have repeated our KMC simulations. The results are shown in Fig. 5. This sequence of figures corresponds to growth of one monolayer of new material on top of a substrate of a different lattice constant at temperatures of (a) 350^0C, (b) 400^0C, (c) 450^0C, and (e) 600^0C. In the first two cases, growth is essentially indistinguishable from the case of homoepitaxy discussed earlier, with a high density of small islands. However, in case (c), despite the small change in temperature of only 50^0C relative to case (b), a dramatically different growth mode is evident, with a large number of tall 3D islands and a substantial amount of the substrate left uncovered. This trend is even more evident at higher temperature, as shown in case (d). This is precisely the type of abrupt transition from layer-by-layer growth at low temperature, to 3D island growth at higher temperature observed experimentally for the strained heteroepitaxy systems, such as Ge/Si.

5. DISCUSSION AND CONCLUSIONS

We have presented a theoretical model of surfactant mediated homoepitaxial and heteroepitaxial growth on semiconductor surfaces. The ingredients of the model include: (i) Activation energies for the important processes in the system, such as diffusion, exchange and de-exchange mechanisms, obtained from first-principles calculations on representative cases that involved a Si(111) substrate, Ge overlayers, and a group V surfactant (Sb). (ii) Kinetic Monte Carlo simulations of solid-on-solid growth, that include all these processes, and assume that steps and terraces are equally well passivated in the system of interest. (iii) Strain effects, manifested by the dependence of lateral bonds on the size of strained islands, as given by continuum elasticity theory [14].

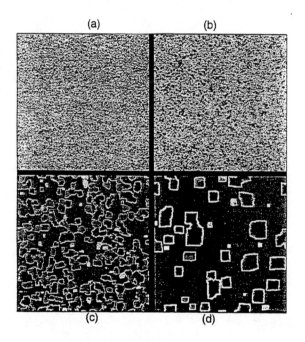

Figure 5. Kinetic Monte Carlo simulation of heteroepitaxial surfactant mediated growth. A total of 1 monolayer of new material has been deposited at (a) 350^0C, (b) 400^0C, (c) 450^0C and (d) 600^0C. The different shades indicate island height. The transition between layer-by-layer growth and 3D island growth takes place somewhere between (b) and (c). In (d) the predominance of tall 3D islands is striking.

This model is able to account for all the important experimental observations, including the presence of a high density of small islands, the persistence of layer-by-layer growth, and the transition to 3D island growth in heteroepitaxy at higher temperatures. We emphasize that no artificial assumptions, such as a reduction of the diffusion length on top of the surfactant layer, were required in order to account for the high density of small islands. Rather, this is a natural consequence of the balance of rates of different mechanisms as determined by the corresponding activation energies obtained from the first-principles calculations. Similarly, the persistence of layer-by-layer growth is a consequence of the ability of atoms to de-exchange, and thus to find optimal sites for incorporation since they can diffuse very fast once on top of the surfactant layer.

Nevertheless, this theory must be viewed as the minimal microscopic model that can account convincingly and without artificial assumptions for a wide range of experimental observations in surfactant mediated homoepitaxial and heteroepitaxial growth of semiconductors. There is no doubt that the actual physical system is considerably more complicated than the model described above. The importance of the model lies in the fact that simple mechanisms with theoretically determined rates can reproduce the behavior of a very complex physical system, making possible the exploration and prediction of growth phenomena which is necessary for the successful design and processing of nanostructures.

ACKNOWLEDGMENTS

This work was supported in part by the Office of Naval Research under Grant No. N00014-95-1-0350, and in part by the National Science Foundation under Grant No. PHY94-07194.

† On leave of absence from the Department of Physics and the Division of Engineering and Applied Sciences, Harvard University.

References

[1] M. Copel, M.C. Reuter, E. Kaxiras, and R.M. Tromp, Phys. Rev. Lett. **63**, 632 (1989).

[2] G. Meyer, B. Voigtlander, and N.M. Amer, Surf. Sci. Lett. **274**, L541 (1992); B. Voigtlander and A. Zinner, Surf. Sci. Lett. **292**, L775 (1993); B. Voigtlander and A. Zinner, J. Vac. Sci. Technol. A **12**, 1932 (1994).

[3] R.M. Tromp and M.C. Reuter, Phys. Rev. Lett. **68**, 954 (1992).

[4] M. Copel, M.C. Reuter, M. Horn von Hoegen, and R.M. Tromp, Phys. Rev. B **42**, 11682 (1990); M. Horn-von Hoegen, F.K. LeGoues, M. Copel, M.C. Reuter and R.M. Tromp, Phys. Rev. Lett. **67**, 1130 (1991); J. Falta, M. Copel, F.K. LeGoues, and R.M. Tromp, Appl. Phys. Lett. **62**, 2962 (1993); M. Copel and R.M. Tromp, Phys. Rev. Lett. **72**, 1236 (1994).

[5] M. Horn-von Hoegen, J. Falta, M. Copel, and R.M. Tromp, Appl. Phys. Lett. **66**, 487 (1995).

[6] G.S. Petrich, A.M. Dabiran, J.E. Macdonald, and P.I. Cohen, J. Vac. Sci. Techn. B **9**, 2150 (1991); G.S. Petrich, A.M. Dabiran, and P.I. Cohen, Appl. Phys. Lett. **61**, 162 (1992).

[7] S. Iwanari and K. Takayanagi, J. Cryst. Growth **119**, 229 (1990).

[8] H. Minoda, Y. Tanishiro, N. Yamamoto, and K. Yagi, Surf. Sci. **287**, 915 (1993).

[9] H. Nakahara and M. Ichikawa, Appl. Phys. Lett. **61**, 1531 (1992).

[10] G.D. Wilk, R.E. Martinez, J.F. Chervinsky, F. Spaepen, and J.A. Golovchenko Appl. Phys. Lett. **65**, 866 (1994).

[11] E. Kaxiras, Europhys. Lett. **21**, 685 (1993).

[12] K.C. Pandey, Phys. Rev. Lett. **57**, 2287 (1986).

[13] D. Kandel and E. Kaxiras, Phys. Rev. Lett. **75**, 2742 (1995).

[14] J. Tersoff and R.M. Tromp, Phys. Rev. Lett. **70**, 2782 (1993).

EXCITON CONFINEMENT IN III-V QUANTUM-WELL STRUCTURES: ROLE OF MAGNETIC FIELDS

J.R. Anderson,[a] M. Górska,[b] J.Y. Jen,[a] J. Kossut,[b] Y. Oka,[c] and M. Sakai[d]
[a] Dept. of Physics and Joint Program on Advanced Electronic Materials
University of Maryland, College Park, MD 20742, U. S. A.
[b] Institute of Physics, Polish Academy of Sciences
Al. Lotników 32/46, Warsaw, Poland
[c] Research Institute for Scientific Measurements
Tohoku University, Katahira 2-1-1, Sendai 980, Japan
[d] Dept. of Functional Materials Science
Saitama University, 255 Shimo-okubo, Urawa 338, Japan

ABSTRACT

Photoreflectance and photoluminescence have been studied in GaAs/AlGaAs single quantum wells doped with Si. Photoreflectance features were observed from the AlGaAs barriers, GaAs buffer layers, and also from the GaAs wells. Electro-optic energies, obtained from fits to the expression for Franz-Keldysh oscillations, have been used to estimate the built-in electric fields in these materials. From photoluminescence at high magnetic fields a cyclotron mass that is more than 1.5 times the bulk GaAs electron effective mass has been obtained. The contributions from non-parabolicity and from exchange and correlation corrections due to the high carrier density in the well do not contribute significantly to an increased mass. A more rigorous theoretical approach for a two-dimensional system including both the Coulomb potential and a magnetic field yields an increased mass but only about 20 % larger. Enhanced luminescence in high magnetic fields, which was observed from the screened quantum-well excitons, suggests that we have produced the "zero-dimensional exciton state" by magnetic-field confinement.

1. INTRODUCTION

An exciton created in a narrow semiconductor quantum well shows a markedly increased binding energy due to confinement. Moreover, a large magnetic field, applied parallel to the quantum-well growth direction, induces confinement in the two directions orthogonal to the direction of growth. Such confinement enhances the luminescence as an exciton recombines.

If donor (or acceptor) impurities are introduced into the well, mobile donor electrons tend to screen the exciton's Coulomb potential and cause a decrease in the exciton binding energy [1]. There is an additional complication: The charge distribution of the ionized donors modifies the potential of the quantum well.

We have studied photoreflectance (PR) and photoluminescence (PL) of heavily doped GaAs/AlGaAs single quantum wells in order to examine exciton confinement and the electron distribution in quantum wells with different widths. From the PR studies we have determined the electric fields in the buffer, well, and barrier for each quantum-well (QW) structure and have estimated the positions of the subbands and the Fermi level. High magnetic fields have been used with photoluminescence in order to investigate the Landau levels in these structures. The PL results show that the magnetic fields above about 10 T separate the broad luminescence peak into contributions from individual Landau levels. In addition, the "two-dimensional" electron-hole pair state, although highly screened, is confined by the magnetic field producing a "zero-dimensional" quantum-well exciton state with a finite binding energy.

Partial results of the PL measurements have been presented previously [2]. From the measurements of the magnetic-field dependence of the lowest four Landau levels, a cyclotron mass has been obtained, which is considerably larger than the mass in bulk GaAs. Although this difference is not understood at present, some contributions to the enhancement of the mass will be discussed in Section IV.

2. EXPERIMENT

Single quantum wells of $GaAs/Al_xGa_{1-x}As$ were grown by molecular beam epitaxy on Cr-doped insulating substrates, using a VG system. The wells were uniformly doped n-type with Si to a maximum concentration of approximately 3×10^{18} cm^{-3}. In Table I we show the characteristics, well width (L_w), barrier width (L_B), and doping concentration (n), of the five structures used in these experiments.

The photoreflectance experiments were carried out with a standard pump-probe technique. For most of the measurements the pump was a green He-Ne laser with a wavelength of 5435 Å and the probe was a tungsten-halogen lamp with the beam passed through a 0.5m spectrometer. The pump beam was chopped at 200 Hz. The detector was a Si diode. A computer was used to control the spectrometer and collect the data.

The pump for the PL studies was an Ar-ion laser and the detector was either a photomultiplier or a Si-diode. The high magnetic fields up to 23 T were produced by a hybrid magnet at the High-Field Laboratory for Superconducting Materials of the Institute for Materials Research of Tohoku University. For high magnetic fields the detector was a Si-diode array which spanned a range of 200 Å. Since the diode array spanned such a limited range of wavelengths, it was necessary at each magnetic field to divide the total wavelength range into segments and adjust the grating for each segment. Usually the range of measurement was from 7700 Å to 8300 Å. The grating settings were made so that there would be an overlap of 50 Å in order to fit the segments to make a complete, smooth curve. The data were taken with about 5 different segments at each value of magnetic field.

Table 1. GaAs/Al$_x$Ga$_{1-x}$As Sample Parameters.

Sample	x	L_W (Å)	L_B (Å)	Doping (10^{18}cm^{-3})
MB941	0.3	50	500	undoped
MB933	0.22	100	300	undoped
MB895	0.22	50	300	3
MB896	0.22	100	300	3
MB898	0.22	200	300	3

3. RESULTS AND DISCUSSION

The PR spectra ΔR *versus* probe energy for several temperatures are shown in Figs. 1, 2, and 3 for the 50Å, 100Å, and 200Å wells. The spectra can be divided into three energy regions, α, β, and γ, corresponding to the GaAs buffer, the GaAs quantum well, and the Al$_x$Ga$_{1-x}$As barriers, respectively. It is clear that the spectral features shift toward higher energies as the temperature decreases. This is primarily a reflection of the temperature dependence of the band gap. At the temperatures shown, the α and γ features represent mainly Franz-Keldysh (FK) oscillations due to large electric fields in the GaAs buffer and the AlGaAs barriers, while at lower temperatures excitonic spectra dominate. From the periods of these oscillations we have determined the built-in electric fields in the barriers and buffer by fitting the peak maxima and minima to the equation in asymptotic form [3,4]

$$\hbar\omega = E_g + (\hbar\Omega)F_j, \tag{1}$$

where $F_j = [1.5(j\pi-\phi)]^{2/3}$ and the electro-optic energy is given by

$$\hbar\Omega = [e^2\hbar^2E^2/(8\mu)]^{1/3}. \tag{2}$$

Here E is the electric field, $\hbar\omega$ is the energy of the probe light, e is the charge of the carrier, $\hbar$ is Planck's constant over 2π, and μ is the reduced mass. The parameter ϕ, which should be used as a fitting parameter, appears to depend on the situation [4], but typically has been taken as 0.5. For fits with a fixed value of ϕ, the intercept should be E_g and the slope $\hbar\Omega$, from which we obtain the electric fields at the surface and interfaces.

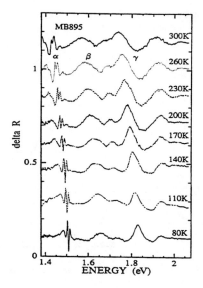

Figure 1. Photoreflectance versus probe energy at different temperatures for MB895.

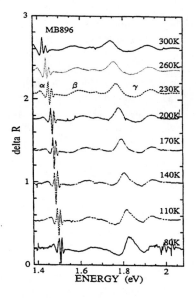

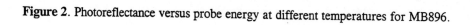

Figure 2. Photoreflectance versus probe energy at different temperatures for MB896.

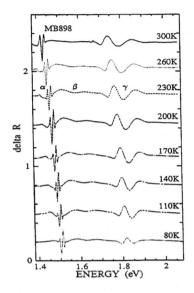

Figure 3. Photoreflectance versus probe energy at different temperatures for MB898.

There are three problems in using Eq. 1 with fixed ϕ for fitting the data: First, since the equation represents an asymptotic limit, it is not valid for small values of E_g-$\hbar\omega$, second, there are only a few peaks representing Franz-Keldysh oscillations and the positions of the maxima and minima are not well-defined, and third, there are actually two quantities, which are difficult to establish, E_g and ϕ. The fit using Eq. 1 should give good values for E_g if ϕ is selected correctly. However, we have found that, because of the ambiguity in choosing the energies corresponding to the peak maximum and minimum and because the choice of the first peak ($j=1$) is not obvious, a range of values for E_g was found at each temperature.

There is another approach: Fit $(\hbar\omega - E_g)^{3/2} = 1.5(\hbar\Omega)^{3/2}(j\pi-\phi)$. In this case, it is necessary to know E_g and ϕ can be determined from the intercept of this plot. As before, the electro-optic energy can be determined from the slope. The values of E_g were determined using the results of Bosio including the temperature dependence [5,6].

We have used both approaches and the results for the electric fields obtained from the electro-optic energy are plotted *versus* temperature in Figs. 4 and 5 for barriers and buffer. This approximate approach seems to give fairly consistent values for the electric fields at each temperature, and the electric field increases with temperature as expected [7], There is, however, considerable scatter in the plots due, at least in part, to our method of extracting the electro-optic energy from the PR. (In the future we plan to fit these spectra directly to an expression for FK oscillations in the asymptotic regime [4,7] or to a more complete formula using Airy functions.) We see that the electric fields in the barrier, assumed to be the one nearest the surface, are approximately the same for the 50-Å and 100-Å well samples. We do not know why the electric fields for the 200-Å well sample are lower. Again it may only reflect the inaccuracies of our fitting procedure. We expected the electric fields in the barriers of all three samples to be approximately the same if all the states at the surface are occupied, since the barrier width is 300 Å in each case. We have not attempted to

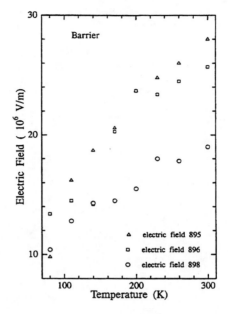

Figure 4. Electric field values in the surface barrier region versus temperature.

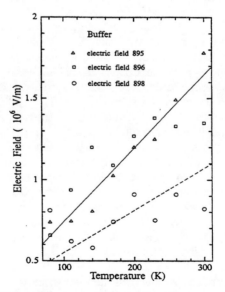

Figure 5. Electric field values in the buffer versus temperature. The solid line is a fit to the data for MB895 and MB896 taken together. The dashed line is a similar fit for MB898.

interpret the values of ϕ, but the values for electric fields have been used to calculate the positions of the Fermi level in the wells and to interpret photoluminescence spectra.

The characterization of the features at energies corresponding to the quantum wells (β) is more difficult. These features are broad and it is difficult to specify precisely peak maxima and minima. If the wells are doped, there should be an electric field in each well, but probably not a constant value. In fact, the potential in the well may be parabolic if the distribution of the charge is uniform. It is usually stated that FK oscillations will not occur unless the carriers can be accelerated by the electric field, but the features we observe seem to correspond more closely to FK oscillations than excitonic effects. Typical phonon energies in GaAs are about 30 meV, which is about half of the peak separations of the β structure. Therefore, phonons seem to be an unlikely mechanism to produce this structure. (In the undoped samples, MB941 and MB933, we did not observe PR features in the β region.)

Since our data warrant only a qualitative interpretation, we have made Franz-Keldysh fits to the well spectra. We note that Hughes *et al.* have fitted their PR spectra in undoped GaAs wells with a combination of FK oscillations and excitonic features [8], but this seems to require too many parameters for an unambiguous fit to our data. For the FK fits in the β region we have used the Aspnes and Studna expression [4], omitting the term giving a singularity at the critical point,

$$\Delta R/R \sim C \exp\{-\Gamma\Xi^{1/2}/(\hbar\Omega)\}\cos\{\theta+(2/3)\Xi^{3/2}\}, \qquad (3)$$

where $\Xi = (\hbar\omega-E_c)/(\hbar\Omega)$, E_c is a critical point energy, C and θ are fitting parameters, and Γ is a broadening parameter. For E_c we have used the calculated subband energies for our quantum wells, but we have ignored the fact that there is at least partial filling of these subbands. For all three well widths, 50 Å, 100 Å and 200 Å, we have included only one term corresponding to Eq. 3 in our fits, although it is clear from the data that more than one critical point is present. For MB 895 we find that the electric field increased from about 1.2×10^7 V/m to 1.34×10^7 V/m as the temperature increased from 80 K to 300 K. For both MB896 and MB898 the field did not appear to change monotonically with temperature but had an average value of $1.05 \pm .08 \times10^7$ V/m. The electric field in the well can be pictured as a result of a parallel plate capacitor with the dielectric constant of GaAs and a charge density (per unit area) equal to that of the ionized Si ions in the well. For an electric field of 10^7 V/m this implies a charge of about 7×10^{11} cm^{-2}, which is a reasonable value for the charge in surface states. In the future we plan to use these values for electric fields to determine the modification of the well potential due to uncompensated charge in the well.

Photoluminescence spectra for samples MB895 and MB896 are shown for magnetic fields of 0 T and 18 T in Figs. 6 and 7. The sharp lines at low energies are also present in the substrates and result from excitons bound to impurities. We will not discuss these features here. For the 50 Å well the zero-field PL spectrum appears as a broad band peaking at about 1.58 eV and extending to about 1.62 eV. For the 100 Å well the zero-field broad band peaks at 1.585 eV and extends to about 1.610 eV on the high-energy side. At high magnetic fields the broad peaks split into at least four components as shown in Figs. 6 and 7, but the high-energy edge of the broad structure, shown by arrows in Fig. 7, does not shift. The low-energy component at 1.535 eV in Fig. 7 is related to the bound exciton in the quantum well and will not be discussed here.

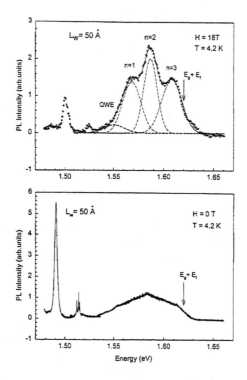

Figure 6. Photoluminescence spectra for MB895 at 4.2K for magnetic fields of 0 and 18 T. The dashed lines show Gaussian fits to the separate components and the solid line is the resulting fit to the data.

Fig. 8 shows the magnetic field dependence of the luminescence peak energies (solid rectangles and circles) for MB895 and MB896. At fields below about 10 T we were unable to resolve the structure in the broad peak, but above 10 T it was possible to determine that the peak positions shifted linearly to higher energies with magnetic field.

First we consider all the lines for n from 1 to 4 in Fig. 8. For MB 895 the peak positions extrapolate to a zero-field value of about 1.55 eV and for MB896 to a value of 1.560 eV. We denote these energies as E_s^{50} and E_s^{100}, respectively. Because of the linear dependence on field we assign these peaks to recombination by Landau level transitions with Landau quantum numbers n = 1-4. Such features were not seen in the undoped samples. The Landau level luminescence of MB895 and MB896 can be assigned to recombination by electrons in the quantum-well subbands with photoexcited holes. The converging energies, E_s^{50} E_s^{100}, correspond to the lowest intersubband transition.

From the slopes of peak energies versus magnetic field for both MB895 and MB896 we have determined a cyclotron mass ratio, $m^*/m_0 = 0.11 \pm 0.01$, a value which differs significantly from the electron mass ratio for bulk GaAs ($m^{GA}/m_0 = 0.067$) [9]. Perhaps this discrepancy is not too surprising since the problem of a hydrogen atom in an intermediate magnetic field is difficult to solve.

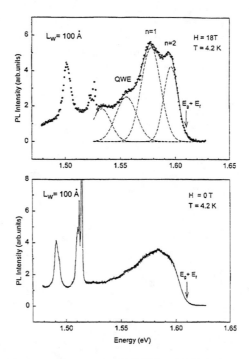

Figure 7. Photoluminescence spectra for MB896 at 4.2 K for magnetic fields of 0 and 18 T. The dashed lines show Gaussian fits to the separate components and the solid line is the resulting fit to the data.

There are several possible reasons for the difference. Akimoto and Hasegawa have addressed the problem of two-dimensional magnetoexcitons and sought a solution within the WKB approximation [10]. The eigenvalues of the system as functions of magnetic field consist of levels n = 0,1,2,... At a high magnetic field **H** these levels approach the normal Landau levels and n corresponds to a Landau level quantum number. As **H** → 0, the energies represent the levels of the two-dimensional hydrogen atom,

$$E_n = E_g - Ry^*/[n + 1/2]^2,$$

where Ry^* is the effective three-dimensional Rydberg constant $\{0.5m^*e^4/(4\pi\epsilon_0\epsilon\hbar)^2\}$, and E_g is the energy gap. Here m^* is the effective mass and ϵ is the dielectric constant. It is clear that the two-dimensional binding energy is four times the binding energy for the exciton in three dimensions, as expected.

Fig. 9 shows the energy in effective Ry units *versus* **H** (in effective Ry), calculated from the expressions of Akimoto and Hasegawa. (For our system 1 Ry is approximately 6.55 T.) We note that, over the field range shown, the n=0 level is affected very little by magnetic field while the other levels begin to approach the regular Landau levels shown by the dashed lines. From these results we have calculated a "cyclotron" mass defined as m^{**}

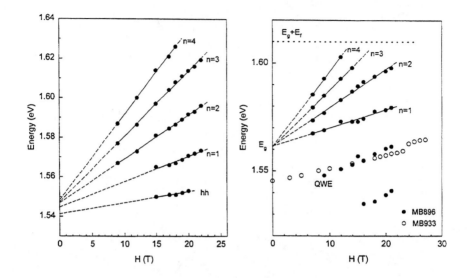

Figure 8. Photon energies at PL peaks versus magnetic field. n is the Landau level number and hh denotes the heavy-hole exciton. (a) MB895: The solid lines are linear least-square fits for the individual Landau levels and the dashed lines show extrapolations to zero magnetic fields. (b) MB896: The solid lines represent linear least-square fits for each Landau level. The open circles represent the undoped QW, MB933. The dotted line shows the position of the Fermi level, the arrow in Fig. 7.

$= e\hbar(n+1/2)/(dE_n/dH)$. It is clear that m^{**} is not a constant and increases to infinity as $H \to 0$. If, however, we take values for the lowest Landau levels extrapolated to high fields, we find $m^{**}/m_0 \approx 0.079$. Thus we find an increase, but only by about 20%, which is not enough.

Szmyd et al. [11] have used photoluminescence to determine the electron effective mass in bulk GaAs doped with Se, grown by metalorganic chemical vapor deposition. From fits to their photoluminescence peaks they find an empirical relation for the mass ratio at the conduction band minimum as a function of electron concentration n:

$$m^*/m_0 = 0.0635 + 2.06\times10^{-22}n + 1.16\times10^{-40}n^2.$$

This variation is attributed to band-gap shrinkage due to electron-electron exchange and correlation. For our doping density of 3×10^{18} cm^{-3}, we would expect an increase of less than 3% in the mass ratio, which is much too small.

Szmyd et al. also consider nonparabolicity in the form $m_F/m_0 = 1 + 7.5E_f/E_g$, where E_f is measured from the bottom of the conduction band. In our samples E_f is typically 50 meV and therefore nonparabolicity would be only about a 2.5% correction to the mass ratio, again much too small. Therefore, since we are unable to account completely for the cyclotron mass difference, we suggest that the Si$^+$ ions in the well are responsible in part.

We locate the Fermi energy, E_f, at the high-energy edge of the PL broad peak. For MB896 this edge is at about 1.610 eV for all magnetic fields up to our maximum of 21 T. The difference between E_f and E_s is 50 meV, which represents the filling of the quantum well

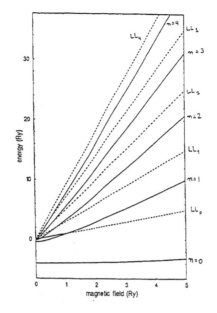

Figure 9. Magnetoexciton energy in Ry *versus* the magnetic field in Ry according to the theory of Akimoto and Hasegawa[10]. 1 Ry ≈ 6.55 T.

above the first subband level. Only one subband in the conduction band is occupied for the 100 Å well with a barrier potential of 0.21 eV. Since the two-dimensional (2D) density of states is $\rho_s = m^*/\pi\hbar^2$, we can calculate the number of 2D-electrons in the well (n_w) by using $\Delta E_f = 50$ meV. For $m^*/m_0 = .067$, we find that $n_w = 1.4 \times 10^{12}$ cm^{-2}. The doping electron density, determined from the growth conditions was 3×10^{12} cm^{-2}. This implies that only about 47% of the doping electrons remain in the well with the remainder migrating to surface states and/or the buffer. If we assume that electrons migrate to both the buffer and surface states and assume the electric fields in the barriers and buffer are those obtained from the fits to the asymptotic form of the Franz-Keldysh oscillations, we find that the occupation in the well is about 2×10^{12} cm^{-2}, in reasonable agreement with our measurements.

This simple explanation does not work for the 50 Å well, MB895. The luminescence peak (Fig. 6) is as broad as that for MB896, beginning about 1.56 eV, reaching a peak slightly above 1.58 eV, and cutting off about 1.62 eV. Even the zero-magnetic-field peak appears to be made up of several components. Our calculation for square wells with finite barriers would imply that the energy difference between the heavy-hole subband and the first electron subband would be about 1.60 eV with the Fermi energy only about 20 meV above the first electron. Thus, we are unable to predict either the position or the width of the luminescence spectra in this case. It may be, however that the presence of Si$^+$ ions in the well causes a lowering of the luminescence energy by roughly 40 meV, but the width of about 60 meV is more difficult to understand. Self-consistent calculations for this problem would be helpful.

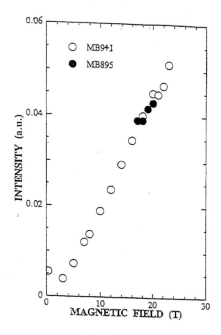

Figure 10. Photoluminescence intensity in arbitrary units *versus* magnetic field. The open circles represent the undoped and the solid circles the doped 50 Å wells.

We attribute the peak-energy components in the vicinity of 1.55 eV for both wells, shown in Figs. 6 and 7, to excitons in the quantum wells. The peak energies depend quadratically on magnetic field. Since this dependence is the same as that for the quantum-well exciton (QWE) observed for the undoped samples, MB941 and MB933, we conclude that the recombination of the QWE in the doped quantum well can be discriminated from the Landau sublevel transitions for fields greater than 10 T. At lower fields the QWE transition merges with the band-to-band recombination in the well. Screening by electrons in the QW smears out the bound state of the QWE, while the confinement of the exciton wave function by the magnetic field results in a finite binding energy of the screened QWE state for fields greater than 10 T.

Figs. 10 and 11 show the PL intensity of the QW excitons in MB895 and MB896 (closed circles) and the undoped samples (open circles) as a function of magnetic field. The QWE intensity increases by an order of magnitude in the undoped sample as the field increases from 0 T to 21 T. This result indicates that the oscillator strength of the QW exciton increases due to magnetic-field-induced confinement. The exciton luminescence intensity in the doped sample (MB896) between 10 T and 21 T has a dependence similar to that of the undoped sample, demonstrating the transformation of the screened electron-hole pair state to the QW exciton state by magnetic-field confinement.

For a two-dimensional system the Bohr radius $a_B^{[2]}$ is 1/2 of the three-dimensional radius $a_B^{[3]}$ and the two-dimensional Rydberg $Ry^{[2]}$ is four times the bulk value, $Ry^{[3]}$. We also note that the screening length due to an electron density n_W in the two-dimensional exciton state is given by

$$r_s = (1/2)\{(a_B^{[2]})^2/n_W\}^{1/4}, \tag{4}$$

where we use $a_B^{[2]}$ as the exciton Bohr radius in the QW. For $n_W = 1.4 \times 10^{12} cm^{-2}$, the screening length is 36 Å. In the undoped QW at 0 T, $a_B^* = 63$ Å for the QW exciton with a binding energy of 15.7 meV. The magnetic length $l_M = (\hbar/eB)^{1/2}$ is 81 Å at 10 T and 57 Å at 20 T. Thus, at fields of the order of 10 T, the magnetic length becomes smaller than the width of the quantum well and for somewhat higher fields the Bohr radius of the screened QW well exciton can be decreased to a value comparable with r_s. Magnetic field induced confinement perpendicular to the field causes the energy splitting of the screened QWE state from the Landau sublevel. Such a fully quantized screened QWE state is the "zero-dimensional exciton state", which has been reported for undoped quantum wells [12].

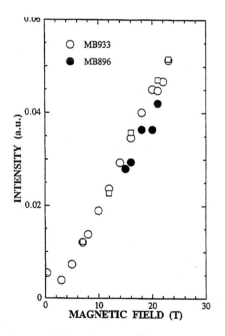

Figure 11. Photoluminescence intensity in arbitrary units *versus* magnetic field. The open circles and squares represent the undoped and the solid circles the doped 100 Å wells. The open squares and circles are the result of different fits to the data.

4. CONCLUSIONS

We have shown that photoreflectance can be used to determine the built-in electric fields in our undoped and doped quantum-well systems. Confinement in high magnetic fields is effective even in heavily doped wells, but more experimental and theoretical work is needed to explains the shapes of our PL and PR spectra from the quantum wells.

ACKNOWLEDGMENTS

We thank Dr. C.E.C. Wood who prepared many of the samples used in this research. We also thank Dr. D.R. Stone, Dr. W.B. Johnson, and Dr. W.T. Beard for their encouragement and helpful discussions.

References

[1]   De-Sheng Jiang, Y. Makita, K. Ploog, and H.J. Queisser, J. Appl. Phys. **53**,999 (1982).

[2]   J.R. Anderson, M. Górska, J.Y. Jen, Y. Oka, I. Mogi, and C.E.C. Wood, Materials Science & Engineering A **217-218**, 184-188 (1996).

[3]   R. Enderlein, Phys. Stat. Sol. **20**, 295 (1967).

[4]   D.E. Aspnes and A.A. Studna, Phys. Rev. **B 7**, 4605 (1973).

[5]   C. Bosio, J.L. Staehli, M. Guzzi, G. Burri, and R.A. Logan, Phys. Rev. **B 38**, 3263 (1988).

[6]   S. Adachi, J. Appl. Phys. **58**, R1 (1985).

[7]   C.R. Lu, J.R. Anderson, D.R. Stone, W.T. Beard, R.A. Wilson, T.F. Kuech, and S.L. Wright, Phys. Rev. **B 43**, 11791 (1991).

[8]   P.J. Hughes, T.J.C. Hosea, and B.L.Weiss, Semicond. Sci. Technol. **10**,1339 (1995).

[9]   L. Hrivnak, Appl. Phys. Lett. **56**, 2425 (1990).

[10] Okikazu Akimoto and Hiroshi Hasegawa, J. Phys. Soc. Japan **22**, 181 (1967).

[11] D.M. Szmyd, P. Porro, A. Majerfeld, and S. Lagomarsino, J. appl. Phys. **68**, 2371 (1990).

[12] H. Sakaki, Y. Arakawa, M. Nishioka, J. Yoshino, H. Okamoto, and N. Miura, Appl. Phys. Lett. **46**, 83 (1985).

STRONGLY CORRELATED FERMIONS ON SUPERLATTICES: MAGNETISM AND SUPERCONDUCTIVITY

Thereza Paiva and Raimundo R. dos Santos
Instituto de Física
Universidade Federal Fluminense
Av. Litorânea s/n, 24210-340 Niterói RJ, Brazil

ABSTRACT

We consider fermions in one-dimensional superlattices (SL's), modelled by site-dependent Hubbard-U on-site couplings, and use Lanczos diagonalization to investigate their properties. The "magnetic" SL is made up of a repeated pattern of repulsive (i.e., $U > 0$) and free ($U = 0$) sites. Depending on the band-filling, Spin–Density-Wave correlations are wiped out for some patterns, while they are enhanced for others. Also, below half-filling, the local-moment maxima are displaced from the repulsive to the free sites; the peak in the magnetic structure factor is also shifted from $q = \pi$, for the alternating configuration when $\rho = 1/3$ and $5/3$. The "superconducting" SL, on the other hand, consists of a repeated pattern of attractive (i.e., $U < 0$) and either free or repulsive sites. We find that some patterns give rise to ground states with dominant superconductivity, others with spin-density-wave (SDW), and others with coexistence of both; some ground states have no dominant correlations. A systematic study of the behavior with band filling and layer thickness reveals a sensitive dependence of these properties with the superlattice configuration. While coexistence is more frequent above half-filling, superconductivity is more frequent below. Further, since in the case of coexistence pairing occurs predominantly on the free sites, the substitution of repulsive for free sites wipes out superconductivity in these cases.

1. INTRODUCTION

The development of diverse and very accurate deposition techniques over the last two decades has generated a whole new class of materials, generically called heterostructures. In particular, systems made up of very thin layers – in some cases a few atoms thick – of materials with different properties have been grown; if the layered structure is periodic, it is referred to as a superlattice (SL). Exciting new phenomena have emerged from these heterostructures, demanding novel theoretical mechanisms to explain the observed data and giving rise to improved devices for technological applications.

As a first example, consider multilayers built up from magnetic transition metals such as Fe, Ni or Co, separated by non-magnetic metallic spacers such as Cu, Ag, Au, Mg, or Mn; see e.g. [1] for a survey of experimental data. It has been observed that the exchange coupling between magnetic layers oscillates as one varies both the spacer and the magnetic layer thicknesses [2, 3]. Also, a drop of unprecedented magnitude in the electrical resistance with an applied magnetic field (giant magnetoresistance) has been found in these magnetic multilayers [4], with outstanding implications for magnetic recording [5]. The first theories devised to explain the oscillatory exchange coupling, namely the quantum-well theory [6] and the so-called RKKY theory [7], were later realized [8] to correspond to extreme and opposite limiting cases. Indeed, while the former corresponds essentially to an *infinite* on-site (*i.e.*, Hubbard-U) repulsion on the magnetic layers, the latter is equivalent to U *small* on the same layers [8]. In order to interpolate between these two limits, one usually starts off with a band theory obtained within a spin-density functional theory [8], which incorporates the effects of electronic correlations in a simplified fashion. Being a one-particle approach, it yields results resembling those obtained through a Hartree-Fock approximation (HFA); in effect, the appearance of a ferromagnetic ground state for two-dimensional layers at moderate U has long been known to be an artifact of the HFA [9, 10]. Though many of the features related to exchange oscillation in magnetic multilayers can be accounted for by those theories, the search for a deeper understanding of the true many-body effects brought about by electronic correlations is clearly in order.

A different class of heterostructures is obtained by combining superconducting (active) layers and non-superconducting (spacer) layers, which have also unveiled interesting new phenomena. For a semiconducting spacer, such as the case of Nb/Ge superlattices, it was found [11] that the critical temperature (T_c) initially decreases as the spacer layer thickness (L_N) increases, for a fixed superconducting layer thickness (L_S), and stabilizes for large L_N; a qualitatively similar behavior has been observed with the roles of L_S and L_N interchanged. Interestingly, an analogous trend holds when the superlattice is made up of a high-T_c cuprate and an insulator, such as in $YBa_2Cu_3O_{7-x}/PrBa_2Cu_3O_{7-x}$ [12, 13]. For metallic spacers, as in Nb/Cu superlattices, a similar decrease in T_c with layer thickness followed by its saturation has also been detected [14]. If, on the other hand, the metal spacer is ferromagnetic (as in the case of Nb/Gd multilayers [15]) or a spin glass (as in the case of Nb/CuMn multilayers [16]), the critical temperature *oscillates* with spacer layer thickness, for a range of L_S [17]. From the theoretical point of view, crucial issues need to be clarified. Firstly, while proximity-effect theories [18] seem to explain the observed superconductivity and saturation of T_c for thick non-magnetic spacers, they cannot consistently account for the strong dependence of T_c on L_N when $L_N \sim L_S$ [13, 14]. Instead, this behavior has been attributed to a coupling between the superconducting layers [13], the nature of which is still obscure. Nonetheless, since this trend holds for both conventional (CS) and high-temperature superconductors (HTS), the coupling between the active layers is probably independent of the details of microscopic mechanism (i.e., electron-phonon, magnetic, and so forth) responsible for superconductivity. Secondly, in the case of magnetic spacers the oscillatory behavior of T_c is consistent with the predictions [19] from an extended quasiclassical theory of superconductivity [20], though for thin spacers several discrepancies were found, casting doubts on the general validity of the theory [15].

The above examples are indicative that considerable insight into both problems of magnetic and superconducting superlattices should be gained by considering microscopic models from the outset. With this in mind, here we review two recent studies [21, 22] of one-dimensional superlattice models in which electronic correlations are incorporated and treated non-perturbatively by means of Lanczos diagonalization on finite systems [23, 24, 25]. The magnetic SL consists of a periodic arrangement of L_U sites ("layers") in which the on-site coupling is repulsive, followed by L_0 free (i.e., $U = 0$) sites; see Fig. 1 for an example with $L_U = 1$ and $L_0 = 2$. The superconducting SL is similarly made up of a periodic arrangement of pairing and non-pairing sites; the correlation energy is negative (i.e., an attractive Hubbard-U coupling) on the pairing sites, thus favoring the formation of local pairs. The discussion about the underlying mechanism for pairing is therefore circumvented, in a way completely analogous to using the attractive Hubbard model to probe normal and superconducting properties of homogeneous systems [26, 27, 28, 29]; the model should then be applicable to both HTS and CS.

$$\bigcirc \text{\textbullet} \text{\textbullet} \bigcirc \text{\textbullet} \text{\textbullet} \bigcirc \text{\textbullet} \text{\textbullet} \bigcirc \text{\textbullet} \text{\textbullet}$$
$$\text{U} \quad 0 \quad 0 \quad \text{U} \quad 0 \quad 0 \quad \text{U} \quad 0 \quad 0 \quad \text{U} \quad 0 \quad 0$$

Figure 1. An example of a lattice with $N_s = 12$ atoms: open circles represent repulsive sites (i.e., those for which $U_i > 0$) and full circles represent free sites (i.e., those for which $U_i = 0$); the respective "layer" thicknesses are $L_U = 1$ and $L_0 = 2$.

In dealing with one-dimensional superlattices, two remarks are in order. First, in doing so one probes the influence of electronic correlations along the direction of superlattice growth, thus capturing the role played by relative layer thicknesses on the magnetic and/or superconducting properties. Secondly, recall that the ground state of the half-filled homogeneous Hubbard chain corresponds to an insulating Spin-Density–Wave (SDW) state, which resembles the Néel state, but with power-law decay of correlations [30]; its strong-coupling limit describes localized spins coupled antiferromagnetically through a Heisenberg exchange interaction. Away from half-filling the ground state is metallic [30, 31], with very short-range spin-spin correlations [32]. By the same token, what one means by a superconducting (SUC) state in one dimension actually corresponds to one with *quasi*-long range order, characterized by a power-law decay of correlations [26].

The layout of the paper is as follows: In Sec. 2 we introduce the one-dimensional magnetic superlattice and present the results; the superconducting SL is discussed in Sec. 3, and Sec. 4 summarizes our findings.

Thereza Paiva and Raimundo R. dos Santos

2. 1-D MAGNETIC SUPERLATTICES

Following the above discussion, we define the Hamiltonian as

$$\mathcal{H} = -t \sum_{i,\,\sigma} \left(c_{i\sigma}^{\dagger} c_{i+1\sigma} + H.c. \right) + \sum_{i} U_i \, n_{i\uparrow} n_{i\downarrow} \tag{1}$$

where, in standard notation, i runs over the sites of a one-dimensional lattice with periodic boundary conditions, $c_{i\sigma}^{\dagger}$ ($c_{i\sigma}$) creates (annihilates) a fermion at site i in the spin state $\sigma = \uparrow$ or $\downarrow$, and $n_i = n_{i\uparrow} + n_{i\downarrow}$, with $n_{i\sigma} = c_{i\sigma}^{\dagger} c_{i\sigma}$; the on-site Coulomb repulsion is taken to be site-dependent: $U_i = U > 0$, for sites within the repulsive layers, and $U_i = 0$ otherwise.

Since one should not expect a uniform distribution of the local moments, we will address the following questions: (1) Is the SDW state generally robust in the half-filled case? (2) Where are the magnetic moments preferentially located? (3) Is the period of short-range correlations sensitive to L_U and L_0?

In order to answer these questions, the ground state $|\psi_0\rangle$ of the Hamiltonian (1) for finite-sized systems is obtained with the aid of the Lanczos algorithm [23, 24, 25]. We then calculate spin-spin correlation functions

$$\langle S_i S_j \rangle = \frac{1}{4} \langle m_i m_j \rangle; \tag{2}$$

where the ensemble averages at zero temperature become ground state averages, $\langle \ldots \rangle \equiv \langle \psi_0 | \ldots | \psi_0 \rangle$, and $m_i \equiv n_{i\uparrow} - n_{i\downarrow}$ is the net spin on site i. As for the homogeneous case, the ground state corresponds to a singlet, with vanishing magnetization, $\langle m_i \rangle = 0$, at every site. We therefore assess the local moments through $\langle S_i^2 \rangle$, which is a measure of the degree of itinerancy of the system. For a completely localized homogeneous system at half-filling, one has $\langle S_i^2 \rangle = 3/4$ for any i, whereas the opposite limit of complete itinerancy ($U_i = 0$ for all i) yields $\langle S_i^2 \rangle = 3/8$.

We also probe the periodicity of the dominant magnetic correlations through the magnetic structure factor, defined as the Fourier transform of Eq. (2),

$$S(q) = \frac{1}{N_c} \sum_{i,j} e^{iq(r_i - r_j)} \langle (n_{i\uparrow} - n_{i\downarrow})(n_{j\uparrow} - n_{j\downarrow}) \rangle, \tag{3}$$

where one should have in mind a lattice made up of N periodic *cells* each of which with a *basis* of $N_b = L_U + L_0$ sites. Accordingly, the total number of *sites* is $N_s = N \cdot N_b$, and the total number of states in the full Hilbert space is 4^{N_s}, while the first Brillouin zone and the allowed values of q are determined by N. Not all configurations $\{U_i\}$ fit into all sizes and occupations considered. Nevertheless, we were able to perform a systematic check of our results by first comparing with those obtained from brute-force diagonalization for $N_s = 4$ or 6; then, a consistent trend was established as N_s is increased. In what follows, we present the results for the largest lattice size used, $N_s = 12$; we also consider the intermediate-coupling regime by taking $U = 4t$ in all cases.

Let us first discuss the case of a half-filled band, for which the number of electrons $N_e = N_s$. Fig. 2 shows the distribution of the local moment over the sites of a lattice with $N_s = 12$, for several thicknesses, L_U, of the repulsive layers; the "free" layers (i.e., those for which $U_i = 0$) in this case are chosen to be one-atom thick, or $L_0 = 1$, and the positions of these atoms correspond to the dips in the figure. As L_U increases, the local moment on each free layer sticks to its fully itinerant value, while the localization on the repulsive layers is enhanced, which should be expected as a "surface-to-bulk" effect. A similar behavior has been observed for free layers with $L_0 = 2$; details will be published elsewhere [33]. We have also examined the case of fixed L_U as L_0 is varied, and found that the local moment is still larger on the repulsive sites, but its magnitude now decreases as L_0 increases, again due to a "surface-to-bulk" effect [33].

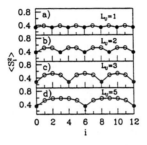

Figure 2. Local moment at different lattice sites, i, for a half-filled chain with 12 sites and a single-atom free layer ($L_0 = 1$). The moment at repulsive and free sites is represented by empty and full circles, respectively. Repulsive layers have thicknesses $L_U = 1$ (a), 2 (b), 3 (c), and 5 (d).

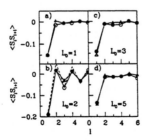

Figure 3. Correlation functions *vs.* intersite distance for a half-filled chain with 12 sites and a single-atom repulsive layer ($L_U = 1$). Circles (triangles) refer to the origin at a repulsive (free) site, and full (open) symbols correspond to free (repulsive) sites. The free layers have thicknesses $L_0 = 1$ (a), 2 (b), 3 (c), and 5 (d). The dashed lines are the results for the corresponding homogeneous system, scaled by a factor of two to stress the rate of decay.

The correlation functions $\langle S_i S_{i+\ell} \rangle$ shown in Fig. 3 are calculated for $L_U = 1$, taking i either as a repulsive site (circles) or as its neighboring free site (triangles). They indicate that a spin on any site (i.e., irrespective of being repulsive or free) is more likely to be surrounded by opposite spins in nearest-neighbor sites. According to Fig. 3(a), a spin arrangement starting from a repulsive site does not match the one starting from a free site; though quite weak, this frustration is sufficient to wipe out the characteristic power-law decay of spin correlations in the SDW state of the otherwise homogeneous system. Figs. 3(c) and (d) also correspond to frustrated cases. The situation is completely different for $L_U = 1$ and $L_0 = 2$, where there is no frustration due to a perfect matching of short range correlations. In Fig. 3(b) we compare the correlation functions for this $\{U_i\}$ configuration with the corresponding one (i.e., $N_s = 12$) for the homogeneous system. The rate of spatial decay being the same in both cases indicates the presence of SDW correlations in this particular superchain. For larger L_U there are also non-frustrated configurations, but now giving rise to a slow decay of *ferro*magnetic correlations on every other site [33]. Thus, if the repulsive layers are separated by an odd number of free layers, frustration does not allow SDW (or other slowly-decaying) correlations to set in, and inhomogeneities tend to favor itinerancy over localized moments through frustration. The magnetic structure factor, Eq. (3), shows a maximum at $q = \pi$ for all combinations of L_U and L_0 examined. Notice that this by no means implies long range antiferromagnetic order, but simply that short-range correlations are such that the magnetic arrangement of the basis is more likely to be repeated at every two cells (or at every $2(L_U + L_0)$ sites).

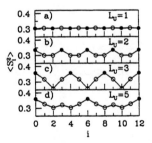

Figure 4. Same as Fig. 2, but for a quarter-filled band ($\rho = 1/2$).

We now discuss occupations smaller than half-filling. Fig. 4 shows the local moment for $\rho \equiv N_e/N_s = \frac{1}{2}$ on a 12-site lattice with $L_0 = 1$. For $L_U = 1$ the electrons are more evenly distributed throughout the lattice than for the half-filled case. As L_U increases, the local moment reaches its maximum values precisely on the free sites, contrary to the half-filled case; since there are fewer electrons than sites, the system lowers its energy by letting them spend more time near the free sites, thus decreasing the likelihood of double occupancy on the repulsive sites. The behavior for $\rho = 1/3$ is qualitatively similar to that for $\rho = 1/2$; quantitative differences, such as in the range of values of $\langle S_i^2 \rangle$, are attributed to the smaller number of electrons. One should stress that this behavior is also present with comparable amplitudes for $N_s = 6$, indicating that this result is very unlikely to be due to finite-size effects. Analogous features are observed as we increase the size of either the repulsive or the free layers, with the maxima of Fig. 4 sometimes developing into *plateaux* [33]. Thus, the usual idea that magnetic and non-magnetic sites should be respectively associated with repulsive and free sites has been proven not to be generally true: as the electron concentration decreases below half-filling, the largest magnetic moment is displaced from the repulsive sites to the free ones.

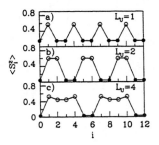

Figure 5. Local moment at different lattice sites, i, for a chain with 12 atoms with 18 electrons ($\rho = 3/2$) and a two-sites free layer ($L_0 = 2$). The moment at repulsive and free sites is represented by empty and full circles, respectively. Repulsive layers have thicknesses $L_U = 1$ (a), 2 (b), and 4 (c).

Correlation functions have also been calculated for both $\rho = 1/2$ and $1/3$, and, unlike the half-filled case, no evidence of dominant SDW correlations has been found for any configuration $\{U_i\}$. Thus, inhomogeneities do not change the correlations below half-filling; e.g., the ground state of the quarter-filled Hubbard model with homogeneous on-site repulsion is metallic [30, 31] without SDW correlations [32]. Nevertheless, the peak position of the magnetic structure factor is not so robust: While for $\rho = 1/2$ it lies at $q = \pi$ for all superlattice configurations examined, for $\rho = 1/3$ the peak shifts to $q_{max} = 2\pi/3$ when $L_0 = L_U = 1$; again, this shift from $q = \pi$ is not likely to be a finite-size effect since it is also present for $N_s = 6$.

Finally, we now examine the results obtained for occupations above half-filling. For both $\rho = 3/2$ and $\rho = 5/3$, the maxima of $\langle S_i^2 \rangle$ now lie on the repulsive sites, similarly to the half-filled case; Fig. 5, corresponding to $\rho = 3/2$ with $L_0 = 2$ and several values of L_U, is representative of the general behavior. The system now minimizes the energy by making it more likely to find two electrons of opposite spins located at each free site, since the local moment decreases to zero very fast with L_U; for the same reason, the maxima are much larger than their $\rho' = 1 - \rho$ counterparts. The local moments therefore behave qualitatively similar to the half-filled case. In some cases, however, the presence of inhomogeneities affect the correlation functions in a fundamental way.

While away from half-filling in the homogeneous system there are no SDW correlations, on a 'superchain' these are favored for spins on a subset of sites. As shown in Fig. 6, there are strong SDW correlations for spins on the repulsive sites; for comparison, the correlations for the corresponding homogeneous cases are depicted as dashed lines in the same figure. The analysis of the magnetic structure factor also shows a shift in the peak position for $L_0 = L_U = 1$: $q_{max} = \pi/3$ for $\rho = 5/3$, whereas $q_{max} = \pi$ for all other cases considered.

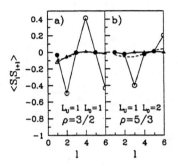

Figure 6. Same as Fig. 2, but for (a) $\rho = 3/2$ with $L_U = L_0 = 1$, and (b) $\rho = 5/3$ with $L_U = 1$; $L_0 = 2$. The dashed lines are the results for the corresponding homogeneous systems.

3. 1-D SUPERCONDUCTING SUPERLATTICES

The model for a superconducting SL follows along the same lines as those for the magnetic case, discussed in Sec. 2. That is, it consists of a repeated pattern of L_A sites with negative on-site correlation energy – the "attractive layer" – followed by L_0 sites with zero on-site energy – the "free layer". However, it is also interesting to consider the case of a "repulsive layer" of L_R sites with positive on-site energy, instead of the free one. We will therefore address the following questions: (1) Is the SUC state generally robust? (2) Is there coexistence between SUC and SDW states? (3) How do the superlattice properties depend on the nature of the non-attractive layer (i.e., whether $U > 0$ or $U = 0$ on its sites)?

The Hamiltonian is the same as Eq. (1),

$$\mathcal{H} = -t \sum_{i,\,\sigma} \left(c_{i\sigma}^{\dagger} c_{i+1\sigma} + H.c. \right) + \sum_{i} U_i\, n_{i\uparrow} n_{i\downarrow} \qquad (4)$$

with the difference that now $U_i = -|U|$ for sites within the attractive layers; otherwise U_i is either 0 or $|U|$. The ensuing heterostructures will then be referred to as attractive/free (A/F) or attractive/repulsive (A/R) superlattices.

As before, the ground state of Eq. (4) is obtained with the aid of the Lanczos algorithm [23, 24, 25]; we consider lattices with $N \equiv N_s/(L_A+L_{0/R})$ periods. Again, a SDW ground state is characterized by a slow spatial decay of the spin-density correlation function, Eq. (2). Singlet superconducting (SS) correlations are similarly probed by

$$C(i;\ell) = \frac{1}{2}\langle c_{i+\ell\downarrow}c_{i+\ell\uparrow}c_{i\uparrow}^{\dagger}c_{i\downarrow}^{\dagger} + \text{H.c.}\rangle; \qquad (5)$$

other pairing correlations, such as triplet, have been examined, but it turned out that SS were always the dominant ones. In what follows, we present the results for the largest lattice size used, $N_s = 12$; in all cases, we consider the intermediate-coupling regime by taking $|U| = 4t$.

We have calculated the above correlation functions for different band fillings and superlattice configurations, and examined their behavior against distance. Before discussing the overall behavior with band filling and superlattice configurations, let us take a closer view at the different outcomes of our calculations. As is well known [26], for a homogeneous lattice the SS correlation function displays a power law decay with distance, and SDW correlations have very short range; these facts are well reproduced in our analysis using finite-sized systems, as shown by the triangles in Fig. 7, corresponding to a half-filled band, $\rho = 1$. Turning to the inhomogeneous A/F case, still at half-filling, a typical plot is also shown in Fig. 7, for $L_A = 2$ and $L_0 = 1$. The SS correlations between fermions on the attractive sites decay slowly, comparably to the homogeneous case, therefore indicating the presence of a superconducting state; spin-density correlations, on the other hand, have very short range, as in the homogeneous system.

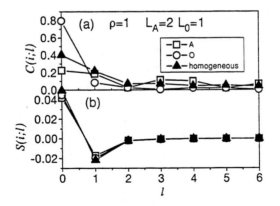

Figure 7. Singlet superconducting (a) and spin density (b) correlation functions vs. intersite distance for a half-filled attractive/free superlattice with 12 sites and $L_A = 2$ and $L_0 = 1$. Circles and squares refer to the origin being taken at a free and at an attractive site, respectively; for comparison, the triangles denote the corresponding data for the homogeneous attractive case.

Other configurations display exactly the opposite behavior, such as the one shown in Fig. 8 for $L_A = 1$ and $L_0 = 2$: a substantial increase in the range of spin-density correlations for fermions on free sites with strongly damped superconducting correlations. A third group of configurations, a representative example for $\rho = 1$ being $L_A = L_0 = 1$, displays no quasi-long range correlations at all. These different behaviors can be understood qualitatively as follows. In the homogeneous case, the energy is lowered by forming pairs irrespective of their position, so they can coherently hop around the lattice at virtually no extra cost in energy; when $L_A = L_0 = 1$ all fermions can still be paired but their mobility is hindered since a pair moving on to a free site would increase the energy by $|U|$. When $L_A = 2$ and $L_0 = 1$, the ground state corresponds to each attractive site accommodating one pair and the remaining pairs resonating between empty attractive sites belonging to different cells, thus increasing the range of SS correlations; the lack of unpaired fermions is clearly responsible for the absence of SDW correlations. When $L_A = 1$ and $L_0 = 2$, one finds paired and unpaired fermions: the former do not superconduct due to their reduced mobility, whereas the latter tend to distribute themselves through the free sites, forming a SDW state to take advantage of the decrease in hopping energy.

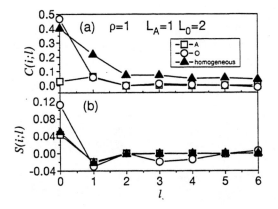

Figure 8. Same as Fig. 7, but for $L_A = 1$ and $L_0 = 2$.

Another possible outcome is the one depicted in Fig. 9, corresponding to a superlattice with $L_A = 2$, $L_0 = 1$ (i.e., $-U - U\ 0\ -U - U\ 0 \ldots$) and filling factor $\rho = 5/3$: *both* SS and SDW correlations *on the free sites* have long range, implying a *coexistence* of spin and superconducting quasi-order. This comes about as a result of the dominant contributions to the ground state in this case being given by

$$| \uparrow\downarrow\ \uparrow\downarrow\ 0\ \uparrow\downarrow\ \uparrow\downarrow\ \uparrow\downarrow\ \uparrow\downarrow\ \uparrow\downarrow\ 0\ \uparrow\downarrow\ \uparrow\downarrow\ \uparrow\downarrow \rangle + \text{C.p.}, \qquad (6)$$

and, with the same weight, by

$$| \uparrow\downarrow \ \uparrow\downarrow \ \uparrow \ \uparrow\downarrow \ \uparrow\downarrow \ \downarrow \ \uparrow\downarrow \ \uparrow\downarrow \ \uparrow \ \uparrow\downarrow \ \uparrow\downarrow \ \downarrow \rangle + \text{C.p.}, \qquad (7)$$

where C.p. stands for circular permutation; the former favors superconductivity while the latter favors a SDW arrangement. This analysis also explains why coexistence is accompanied by a shift of the peak, from the attractive to free sites, in both SS and SDW correlation functions. A similar effect was also found for the local magnetic moment in the repulsive/free case [21], which illustrates the unexpected features resulting from the interplay between strong electronic correlations and a superlattice structure.

We are now in a position to discuss the "phase diagrams" ensuing from a systematic analysis of correlation functions. In Fig. 10 we display the nature of the ground state for different occupations and lengths of the attractive layer; the length of the free layer is kept fixed. For a given occupation, say below half-filling, as the length of the attractive layer increases, one goes through a sequence of SUC and non-SUC ground states. The latter arise whenever the number of pairs is commensurate with the number of attractive layers, giving rise to an even distribution of pairs throughout the lattice. One can also infer from Fig. 10 that superconductivity is least robust (against the imposition of a superlattice structure) in the half filled band case; it is also the only band filling for which a SDW state (without coexisting with a SUC state) is induced by the superlattice structure. Though for the cases shown in Fig. 10 the ground states with SUC-SDW coexistence only occur above half-filling (and they actually consist in the majority of cases) they can also be found below half-filling, such as when $\rho = 2/3$ and $L_A = 1$, $L_0 = 3$.

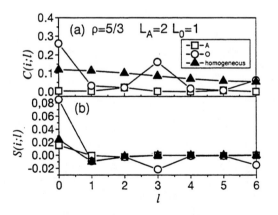

Figure 9. Same as Fig. 7, but for $\rho = 5/3$ and for $L_A = 2$ and $L_0 = 1$.

Another interesting feature in the diagrams of Fig. 10 is that they are *almost* particle-hole symmetric. As $\rho \to 2 - \rho$, a SUC state is generally mapped onto a state with coexistence of SUC and SDW, and vice-versa, while an orderless state is preserved; for the cases shown, the exceptions occur for both $L_A = 3$ and $L_A = 5$, in the sense that there are SUC states not being mapped onto coexistence states. It is straightforward to verify that a mapping $\rho \to 2 - \rho$ for the inhomogeneous Hamiltonian (4) cannot be generated by the usual particle-hole transformation $c_{i\sigma}^\dagger \to (-1)^i c_{i\sigma}$, and so forth. Further, such mapping, if it existed, should also take SUC correlation functions onto SUC+SDW ones, which would impose severe constraints on their forms. As a further check on this "quasi"-symmetry, we have compared the spectra of (4) for densities ρ and $2 - \rho$, in search of a trend indicative of a hidden symmetry, such as the successive energy gaps being constant; again, no signature of an exact symmetry was found [34].

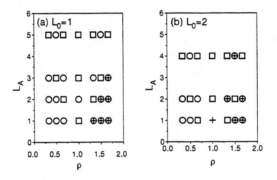

Figure 10. Phase diagram in the parameter space attractive layer length (L_A) versus occupation (ρ), for fixed free layer length, $L_0 = 1$ (a) and $L_0 = 2$ (b). Crosses, circles, and squares respectively denote ground states with spin–density-wave quasi-order, superconducting quasi-order, and none of these. Coexistence is indicated by a circle superimposed to a cross.

Let us now discuss the case of a superlattice made up of attractive and repulsive sites. Below half-filling, and for the cases without SUC-SDW coexistence, the results are qualitatively very similar to those for the A/F case. This is simply because the paired electrons can always fit in the attractive sites, irrespective of whether the non-attractive sites are repulsive or free. Since in the A/F case superconducting correlations at half-filling are peaked on the attractive sites, no qualitative changes occur in the A/R case; similarly, since in the A/F case the SDW correlations are peaked on the free sites, here they are also peaked (and larger) on the repulsive sites. As expected, the cases of coexistence are the ones most drastically affected by replacing repulsive for free sites: SS correlations are wiped out, while SDW correlations are preserved and, in most cases, significantly enhanced.

4. CONCLUSIONS

In conclusion, we have singled out the effect of strong electronic correlations in superlattices through a simple model based on an inhomogeneous, but periodic, site-dependent on-site coupling U_i. In the "magnetic" case (i.e., a regular array of repulsive and free sites) novel features were found due to the nonuniform distribution of local moments. Firstly, at half-filling frustration sets in for some configurations $\{U_i\}$, manifested by a mismatch of the spin arrangements in different sublattices. Consequently, SDW quasi-order, which is a trademark of the half-filled homogeneous Hubbard chain, is wiped out by frustration; in the absence of frustration, slowly decaying correlations persist. Secondly, the dominant local magnetic moment is displaced from the repulsive sites to the free ones, as the occupation decreases below half-filling, thus invalidating the idea that local moments would always be concentrated on the repulsive sites. Thirdly, for some configurations above half-filling we have found a slow decay of correlations (i.e., SDW-like) within the repulsive sublattice, a situation reminiscent of superexchange in insulators. And finally, for the "dimerized" configuration (i.e., alternating free and repulsive sites) the peak in the magnetic structure factor was found to shift from $q_{max} = \pi$ to $2\pi/3$ and $\pi/3$, if $\rho = 1/3$ and $5/3$, respectively. This model corresponds to the simplest situation incorporating electronic correlations in superlattices. In addition to generalizing to higher dimensions [35] – i.e., sites becoming lines or planes, so that SDW quasi-order may become true antiferromagnetic long-range order – we may extend the concept to generalized objects such as CH groups in conducting polymers (e.g., *trans*-polyacetylene), where correlations increase dimerization [36]. Further, the occupation-induced displacement of the local moment found below half-filling could provide a mechanism for magnetic switching, with the density of carriers being varied by e.g., optical injection, temperature, or even through direct contact with a different crystal acting as a particle reservoir.

In the case of a "superconducting" superlattice, and for a mixture of attractive and free sites, we have found that superconductivity is more robust (relative to the homogeneous case) against the imposition of a superlattice structure away from half-filling. We have also established that some superlattice configurations give rise to ground states with coexisting quasi–long-ranged superconducting and spin-density–wave correlations, predominantly above half-filling. However, the peaks in the correlation functions are shifted to the free sites whenever coexistence occurs. As a consequence, the strong superconducting correlations in this case are wiped out when repulsive sites replace the free ones, whereas they are hardly affected in the absence of coexistence. Finally, we should comment on the extent to which contact can be made with the experimental systems referred to earlier on. We recall that the layers are macroscopic in those systems, in the sense that they can support either true superconductivity and/or magnetic long range order. This is clearly not the case of the model treated here, for which the "layers" are quite thin – and one-dimensional – to sustain local order. Nonetheless, our results indicate that strong correlations provide a mechanism for coexistence of magnetism and superconductivity. Also, it is interesting to notice that, for fixed band filling, pair coherence is very sensitive to the number of free layers in the case of attractive/free superlattices, a feature that can be regarded as reminiscent of the oscillatory behavior of T_c with the spacer thickness; further work at higher dimensions and at finite temperatures is clearly needed in order to make quantitative contact with actual experimental data.

We hope the intriguing results reported here will stimulate further experimental studies in both magnetic and superconducting superlattices. A particularly interesting question is the influence of electron concentration on the measurable properties.

ACKNOWLEDGMENTS

The authors are grateful to F. C. Alcaraz, J. d'Albuquerque e Castro, S. L. A. de Queiroz, R. B. Muniz, L. E. Oliveira and J. J. Rodríguez-Núñez for discussions, Financial support from the Brazilian Agencies FINEP, CNPq and CAPES is also gratefully acknowledged. The authors are also grateful to Laboratório Nacional de Computação Científica (LNCC) for the use of their computational facilities.

References

[1] M. N. Baibich and R. B. Muniz, Braz. Journ. Phys. **22**, 253 (1992); B. Heinrich and J. F. Cochran, Adv. Phys. **42**, 523 (1993); K. B. Hathaway, in *Ultrathin Magnetic Structures II*, edited by B. Heinrich and J. A. C. Bland, (Springer, Berlin, 1994).

[2] S. S. Parkin, N. More, and K. P. Roche, **64**, 2304 (1990).

[3] P. Grünberg, S. Demokritov, A. Fuss, M. Vohl, and J. A. Wolf, J. Appl. Phys. **69**, 4789 (1991).

[4] M. N. Baibich, J. M. Broto, A. Fert, F. Nguyen Van Dau, and F. Petroff, **61**, 2472 (1988).

[5] See the special issue of Physics Today on Magnetoelectronics, April 1995.

[6] D. M. Edwards, J. Mathon, R. B. Muniz, and M. S. Phan, **67**, 493 (1991); J. Phys.: Cond. Matter **3**, 4941 (1991).

[7] P. Bruno and C. Chappert, **67**, 1602 (1991); **46**, 261 (1992).

[8] J. d'Albuquerque e Castro, M. S. Ferreira, and R. B. Muniz, **49**, 16062 (1994).

[9] J. E. Hirsch, **31**, 4403 (1985)

[10] J. E. Hirsch and S. Tang, **62**, 591 (1989).

[11] S. T. Ruggiero, T. W. Barbee, Jr., M. R. Beasley, **26**, 4894 (1982).

[12] J.-M. Triscone *et al.*, **64**, 804 (1990).

[13] Q. Li, X. X. Xi, X. D. Wu, A. Inam, S. Vadlamannati, W. L. McLean, T. Venkate-san, R. Ramesh, D. M. Hwang, J. A. Martinez, and L. Nazar, **64**, 3086 (1990).

[14] I. Banerjee, Q. S. Yang, C. M. Falco, and I. K. Schuller Solid State Commun., **41**, 805 (1982); C. S. L. Chun, G.-G. Zheng, J. L. Vicent, and I. K. Schuller **29**, 4915 (1984).

[15] J. S. Jiang, D. Davidović, D. H. Reich, and C. L. Chien **74**, 314 (1995).

[16] L. V. Mercaldo, C. Attanasio, C. Coccorese, L. Maritato, S. L. Prischepa, and M. Salvato, **53**, 14 040 (1996).

[17] Th. Mühge et al., **77**, 1857 (1996).

[18] G. Deutscher and P. G. de Gennes, in *Superconductivity*, edited by R. D. Parks, (Marcel Dekker, N.Y., 1969).

[19] Z. Radović, M. Ledvij, L. Dobrasavljević-Grujić, A. I. Buzdin, and J. R. Clem, **44**, 759 (1991).

[20] K. Usadel, **25**, 507 (1970).

[21] T. Paiva and R. R. dos Santos, **76**, 1126 (1996).

[22] T. Paiva and R. R. dos Santos, (unpublished).

[23] H. H. Roomany, H. W. Wyld and L. E. Holloway, **21**, 1557 (1980).

[24] E. Gagliano, E. Dagotto, A. Moreo, and F. C. Alcaraz, **34**, 1677 (1986); (E) **35**, 5297 (1987).

[25] E. Dagotto, **66**, 763 (1994).

[26] R. Micnas, J. Ranninger, and S. Robaszkiewicz, **62**, 113 (1990).

[27] A. Moreo and D. J. Scalapino, **66**, 946 (1991).

[28] M. Randeria, N. Trivedi, A. Moreo, and R. T. Scalettar, **69**, 2001 (1992).

[29] R. R. dos Santos, **50**, 635 (1994).

[30] E. Lieb and F. Y. Wu, **20**, 1445 (1968).

[31] F. Mila and X. Zotos, Europhys. Lett. **24**, 133 (1993); K. Penc and F. Mila, **49**, 9670 (1994).

[32] D. M. Luz and R. R. dos Santos, (unpublished).

[33] T. Paiva and R. R. dos Santos, (unpublished).

[34] T. Paiva and R. R. dos Santos, (unpublished).

[35] J. J. Rodríguez-Núñez, J. d'Albuquerque e Castro, R. R. dos Santos, and R. B. Muniz, (unpublished).

[36] J. E. Hirsch and M. Grabowski, **52**, 1713 (1984); D. K. Campbell *et al., ibid.* **52**, 1717 (1984).

QUANTUM PARTICLES IN MICROPOROUS CHANNELS

S. D. Mahanti and Viktor Cerovski
Department of Physics and Astronomy
Michigan State University, East Lansing, MI 48823, U. S. A.

ABSTRACT

Physical properties of quantum particles such as electrons, ^{3}He, and ^{4}He are dramatically modified when they are confined to move in microporous channels with diameter less than a nanometer. Examples of such systems are electrons in electrides, and He atoms in zeolites. Effects of confinement and the connectivity of the confining media on electronic properties (in electrides) and on the ground and low-lying excitations of He atoms (inside zeolites) will be briefly reviewed. In particular, in He systems consisting of a small number of atoms (≤ 8) moving in a *ring* geometry along the internal surface of tubular zeolitic channels, we will focus on the effects of statistics (bosons vs. fermions) and disorder on their low-lying excitations. We also briefly discuss the ground state and low-lying excitations of a system of ^{4}He atoms moving inside a narrow one dimensional channel using a quantum lattice gas model of attracting hard core bosons.

1. INTRODUCTION

Physical properties of quantum particles such as electrons and He atoms are dramatically modified from their bulk counterparts when these particles are confined to move inside microporous channels of diameter (d) less than one nanometer (10Å). If we define a length scale by the thermal wavelength of a particle $\lambda_{th} = h/(2\pi m k_B T)^{0.5}$, where m is the mass and T is the temperature, then $\lambda_{th} = 17.5$Å for m = 1amu and T=1K, and $\lambda_{th} = 43.3$Å for m = 1 electron mass and T = 300K. Thus one expects to see strong quantum effects for *electrons* inside electrides (d ≈ 3-5Å) at room temperature and for *He atoms* inside zeolites (d ≈ 5-12Å) at 1K or below. Electrides and zeolites are extensively studied microporous media which can be characterized by cages of different sizes connected by channels of different diameters and lengths [1,2].

Electrides are crystalline solids, first discovered by Dye [3], with the positive charges on complexed alkali metal cations balanced by an equal number of trapped electrons. The basic geometrical structure of these systems consists of one cavity per trapped electron with diameters from 3.7-4.8Å, produced by close-packing of the large (8-10 Å diameter) complexed cations. These cavities are connected by channels whose geometrical characteristics depend on the size and packing of the complexed cations. The trapped electrons move from one cage to another through the connecting channels. The cages and channels form a periodic microporous medium through which electrons move.

Zeolites, sometimes also denoted as molecular sieves, are another class of periodic microporous solids which provide a confining media for probing the structure, excitations and molecular transport of confined particles [4]. The composition of zeolites is usually $(SiO_2)_n(AlO_2)^- M^+$, where n denotes the silicon-to-aluminum ratio of the zeolite framework $(SiO_2)_n(AlO_2)^-$. The silicon and aluminum atoms are tetrahedrally surrounded by four oxygen atoms and combined in such a way that systems of (mostly) intersecting channels or of cavities interconnected by windows are formed. Probe particles such as He atoms move through these interconnected periodic microporous network structure.

In sec.II, we briefly discuss the main characteristic features of electrides such as their cavity and channel structure, and their electronic properties. In sec. III we discuss in some detail our work on He atoms trapped inside K-L zeolite. Finally we give a short summary.

2. ELECTRONS IN ELECTRIDES

As mentioned above, electrides are crystalline salts formed from complexed alkali-metal cations. They exhibit a broad range of electronic properties, some of them behave like metals and some as insulators. In particular, the crown ether electrides can be thought of as a novel class of Mott insulators (where the insulating property arises from strong electron electron interaction as in many transition metal oxides) in that the localized electrons are not located near a parent magnetic ions (transition metal ion), but rather exist in atom free region. This latter region is most likely the cavities described above. In fact, one of the fundamental questions that has been of interest in these systems is the nature of the valence electron in the complexed alkali atom itself and how this valence electron rearranges in the solid.

Kaplan and collaborators [5] have made extensive studies on the nature of the valence electron in *isolated molecular units* of these complexed alkali atoms. For example, using extensive Hartree Fock and limited CI calculations, they have found that an alkali dicrown, $Li(9-crown-3)_2$, a typical complexed alkali atom, has an unusual electronic structure. It is like a large distorted alkali atom, the "core" (radius r_c ≈ 4Å) consisting of many (≈ 50 to 100) bonded oxygen, carbon, and hydrogen atoms plus a trapped positive alkali ion, the valence electron distribution having a radius r_o ≈ 5 to 6 Å with extremely small density within the core. In other words, the isolated molecule behaves like a giant Rydberg atom. As these molecules are brought close to each other, the valence electron distribution gets dramatically modified (partly due to the large polarizability of the weakly bound valence electrons) as these electrons are forced into cavities formed by the hard cores.

The only serious calculation of electronic structure in a *solid* has been carried out by Singh *et al.* [6] using a local density functional approximation on $Cs(15\text{-crown-5})_2$. They concluded that indeed most of the valence electron density is in the cavity, and that it extends into major channel that connects adjacent cavities, providing a mechanism for inter-cavity tunneling or charge transfer. Quite interestingly, the charge isosurfaces computed by Singh *et al.* reflect, rather faithfully, the geometry of the cavities and channels which can be independently obtained by simply putting hard-sphere radii and van der Waals radii for the atoms (Cs^+, C, O, H etc.) [1]. This suggests a geometrical confinement picture to study the system of interacting valence electrons in electrides. Although extensive experimental studies of magnetic, optical, and transport properties of electrides have been made by Dye and collaborators [3], very little theoretical work on the electronic properties of these systems has been done. The reason for this is the difficulty in treating electron electron interaction properly in the presence of complicated geometrical confinement effects.

3. HE ATOMS INSIDE K-L ZEOLITE

We will now discuss the system of He atoms confined to move inside zeolitic microporous channels. Kato *et al.* [7] have made extensive measurements of the heat capacity (C) of He atoms inside K-L zeolite to investigate the low temperature (T ≤ 1.5K) statistical mechanics of interacting bosons and fermions in 1-dimension. This particular Zeolite consists of nonintersecting one dimensional channels consisting of cages of about 13Å diameter and 7.5Å long, interconnected through apertures of diameter about 7.4 Å. The silicate framework and the K^+ ions situated on the cage wall exert an attractive potential on the He atoms. As suggested by Kato *et al.* [7] and also found by our model calculations [8], this attractive potential gives rise to localized states for the He atoms arranged along the inner circumference of the tubular cage in a ring geometry. The number of such *cage states* is n_c/cage. Because of strong mutual repulsion, two or more He atoms cannot occupy the same *cage state* even if they are bosons. Thus the maximum number of He atoms that can be accommodated in these *cage states* is n_c/cage. Once the He concentration exceeds n_c/cage, the additional He atoms ($n\text{-}n_c$ per cage) move along the channel, occupying *channel states*. We will discuss the physics of the He atoms occupying the *cage states* and the *channel states* separately. Konishi *et al.* [9] have carried out similar low-T heat capacity measurements for He atoms inside ZSM-23 which has 1-dimensional channels of uniform diameter about 5.5Å. This system should show behavior similar to K-L zeolite when only the *channel states* are involved in the latter.

3.1. Physics of the cage states (intracage excitations)

The excitations of the He system in the manifold of *cage states* in K-L zeolite can be described rather well by Bose-Hubbard and Mott-Hubbard models for ^{4}He and ^{3}He systems respectively [10]. These models take into account both the quantum tunnelling of He atoms from one bound site to another and inter-particle interactions [7]. The Bose-Hubbard Hamiltonian is given by

$$H_{boson} = -t\sum_{i=1}^{n_c} (b_i b_{i+1} + hc) + \frac{U}{2}\sum_{i=1}^{n_c} n_i(n_{i-1}) - V\sum_{i=1}^{n_c} n_i n_{i+1} \tag{1}$$

where n_c is the number of binding (localized) sites, b_i (b_i^+) destroy (create) a boson (B) at the ith localized site. In eq.1, t is the hopping energy, U is the repulsive energy between two He atoms occupying the same site, and V is the strength of the attractive energy between atoms occupying neighboring sites. The boson operators satisfy the usual commutation rules and the number operator n_i has eigenvalues 0,1,2, etc. Since He-He repulsion is very strong when two atoms occupy the same *cage state*, we let $U \to \infty$ (hardcore bosons) and $n_i = 0$ or 1.

The ^{3}He system is similarly described by a Mott-Hubbard model Hamiltonian

$$H_{fermion} = -t\sum_{i,\sigma}^{n_c} (f_{i\sigma} f_{i+1\sigma} + hc) + \frac{U}{2}\sum_i^{n_c} n_{i\uparrow} n_{i\downarrow} - V\sum_i^{n_c} n_i n_{i+1} \tag{2}$$

Here the fermion destruction ($f_{i\sigma}$) and creation ($f_{i\sigma}^+$) operators associated with state i and spin σ satisfy usual anti commutation rules and $n_{i\sigma}$ has eigenvalues 0 or 1. The total number of fermions at site i is $n_i = n_{i\uparrow} + n_{i\downarrow}$. We again take $U \to \infty$ and for simplicity ignore the spin of the ^{3}He atoms and treat them as spinless fermions (SF). In this case, $n_i = 0$ or 1. Both hard core bosons (B) and SF systems can be described by a single Hamiltonian.

$$H_{B(SF)} = -t\sum_{i=1}^{n_c} (c_i c_{i+1} + h.c) - V\sum_{i=1}^{n_c} (n_i n_{i+1}), \tag{3}$$

where $c_i = b_i$ (f_i) and $n_i = c_i^+ c_i = 0$ and 1. The effect of spin will be discussed later.

The differences between B and SF lie in the commutation properties of the operators and different transfer energies t for ^{3}He and ^{4}He due to their mass difference. For same t and V, the spectra of B and SF are identical for an open chain. The same is true for periodic chains (the ring geometry in our case) when n ($= \Sigma n_i$) is *odd*. However, for *even* n, the spectra are *different*. Thus, in addition to the mass difference, the different statistics will give rise to differences in the heat capacities of the ^{3}He and ^{4}He systems in this ring geometry for small system sizes. Of course in the limit of infinite system SF and hard core bosons have identical spectrum (if they have same t and V). We have found that the difference in the statistics can lead to a larger heat capacity for the spinless fermions, whereas the mass difference, through smaller excitation energy gap (resulting form smaller t for ^{4}He) can lead to a larger low-T heat capacity for the bosons [10]. This observation is in accord with the experiments of Kato et al. [7] who find that the at about 1.5K, for same average number of particles/cage $<n>$, C for ^{3}He is larger than that for ^{4}He by about 25%.

The effect of incorporating spin of ^{3}He is much more difficult because of the expected lack of thermal equilibrium between states differing in total spin quantum number as in the classic ortho and para hydrogen molecule problem [11]. In fact, if we treat ^{3}He atoms as SF, we find that the right order of magnitude of C at T > 0.5K can be obtained by using reasonable values of the parameters t and V ($t \approx$ 10-15K and $V \approx$ 20-25K). We do not have any quantitative information about the value of t because of the complicated nature of the single-particle potential inside K-L zeolite. However the value of V looks reasonable if one takes into account both the direct and framework mediated attractive interaction between two He atoms occupying neighboring localized *cage states* [12].

Although our calculations gave reasonable values of C for T > 0.5K, they did very poorly at lower temperatures, the theoretical C values being much smaller than the experiment. We have explored the effect of disorder by treating the tunnelling parameter t as random but assuming the single site energies same for all cage states [13]. As expected from the earlier works (effect of disorder on the low-T heat capacity in solids by Anderson, Halperin, and Varma [14], by Phillips [15], and by Kaplan, Mahanti, and Hartman [16]) we find a disorder-induced enhancement in the low-T values of C for the spinless fermions. A similar enhancement is also expected for the hardcore bosons, these results are consistent with the experimental observations [7].

3.2. Physics of the *channel states* (intercage excitations)

In K-L zeolite, once all the cage states are filled up, additional He atoms go into *channel states*. In this state, the He atoms can be thought of as moving in one dimensional tubes of diameter < 7Å [7]. If we take into account the short range repulsion between the walls of the framework and the He atoms occupying the cage states, the effective diameter of the tube may even be smaller. Similar situation occurs in ZSM-23 where the tubes are uniform with diameter $\approx$ 5Å [9]. The energy of a single He atom in these tubular channels can be written as $E_{nk} = E_n + h^2k^2/2m^*$, where E_n is associated with the transverse (perpendicular to the tube axis) excitations and k is the wave vector associated with the motion along the channel. For simplicity we will assume that only the lowest band (n=1) is occupied. Thus we have a problem of quantum particles moving in one dimension (with an effective mass m^*) and interacting with each other via a Lennard Jones type potential. We have obtained the ground state and low-lying excitations of this systems as a function of density and interparticle interaction using a quantum lattice gas (QLG) model first proposed by Matsubara and Matsuda [18] for the superfluid transition problem in bulk ^{4}He.

In the QLG model, the He atoms are assumed to occupy lattice sites separated by a distance a. The hopping parameter (t) between neighboring sites is related to the kinetic energy of the particle and is given by $t = h^2/2m^*k_Ba^2$. If we assume that a = $\sigma_{He,He}$ = diameter of the He atoms, then the repulsive energy between two He atoms occupying the same site is essentially ∞ and the interaction between nearest neighbor atoms is -10K, and the next nearest neighbor atoms is $\approx$ 1K. The system can be approximated very well by 1-dimensional hardcore bosons with attractive nearest neighbor interaction (-V), similar to the one used for the *cage states*. (Eq.1), but in this case $n_c \rightarrow \infty$. We do not know the precise value of m^* and furthermore the effective tunneling parameter (t) depends on the interparticle interaction (as in the Hubbard model for electrons). We have therefore treated t as a variable and explored the nature of the ground and low-lying excited states as a function of the density $\rho = N_p/N$ (where N_p is the number of particles and N is the number of lattice sites) and V/t, in finite chains up to size N=16. We can compare our exact finite chain results to those for the ground state energy (equalities for certain range of parameters and inequalities for others) obtained by Yang and Yang [19] for the Ising-Heisenberg model.

In Fig.1 we give the ground state energy per particle $E_0/N_pt \equiv \epsilon(N,N_p)$ for the (N,N_p) system as a function of V/t for ρ = 1/4 and 1/2, for different values of N. As can be clearly seen, the scaling with N is dramatically different for $V/t < 2$ and $V/t > 2$. This is related to a quantum phase transition (in the limit of $N, N_p \to \infty$ with ρ finite and < 1) from a superfluid (Luttinger liquid [20]) to a phase separated solid. Our results for $V/t > 2$ also satisfy the inequality $\epsilon(N, N_p) \geq V/t$. This is consistent with the exact results [19]. In the thermodynamic limit, $\epsilon = -V/t$, independent of ρ. However, for finite size systems there are significant difference, the difference is due to both breaking and tunneling of N_p particle clusters. In fact, in the large V limit, we [17] find that for finite systems the low-lying excitations are associated with quantum tunnelling of N_p-particle clusters, the width of the tunneling band scaling as $(t/V)^{N_p-1}$.

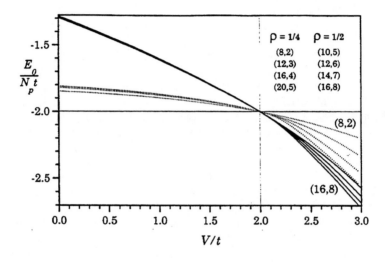

Figure 1. Ground state energy per particle as a function of the strength of the attraction V for densities1/4 and ½ for different system sizes. All the energies are expressed in units of the intersite hopping energy t.

4. SUMMARY

In summary, we have discussed briefly different types of systems whose physical properties can be studied using a model consisting of interacting quantum particles moving in tubular channels. The channel geometry dictates the nature of the single particle states such as *cage* states and *channel* states in K-L Zeolite, and consequently the nature of the ground and excited states of the interacting system. He atoms moving inside channels of rather uniform diameter (< 5Å) are excellent systems to probe the properties of Luttinger liquids (bosons or spinless fermions and fermions), separation of spin density and mass density fluctuations (in ^{3}He), and other non fermi-liquid behaviors. Direct application of known results of non-fermi liquid theories to these systems however will require a better understanding of the microscopic origin of the tunneling parameter t and the nature and the strength of the interparticle interaction.

ACKNOWLEDGEMENTS

We thank Dr. A. K. Rajagopal and Dr. Arghya Taraphder for helpful discussions. This work was partially supported by NSF grant CHE-9633798 and by the Center for Fundamental Research at Michigan State University. SDM would like to convey his sincere thanks to Professors B. K. Rao, P. Jena, and S. N. Behera for inviting him to participate in this exciting symposium on novel materials.

References

[1] J. L. Dye, M. J. Wagner, G. Overny, R. H. Huang, T. F. Nagy, and D. Tomanek, Jour. Am. Chem. Soc. **118**, 7329 (1996).

[2] T. F. Nagy, S. D. Mahanti, and J. L. Dye, Zeolites, 1997

[3] J. L. Dye, Science **247**, 663 (1990)

[4] D. W. Breck, *Zeolite Molecular Sieves*, Wiley, New York. 1974; R. M. Barrer, *Zeolites and Clay Minerals as Adsorbents and Catalysts*, Academic Press, London 1978.

[5] R. Rencsok, T. A. Kaplan, J. F. Harrison, J. Chem. Phys. **93**, 5875 (1990); ibid. **98**, 9758 (1993).

[6] D. Singh, H. Krakauer. C. Haas, W. Pickett, Nature **365**, 39 (1993).

[7] H. Kato, K. Ishioh, N. Wada, T. Ito, and T. Watanabe, J. Low Temp. Phys. **68**, 321 (1987); H. Kato, N. wada, T. Ito, S. Takanayagi, and T. Watanabe, J. Phys. Soc. Jpn. **55**, 246 (1986).

[8] W. Hammond, B. Y. Chen, S. D. Mahanti (unpublished); B. Y. Chen, S. D. Mahanti, and M. Yussouff, Phys. Rev. B**51**, 5800 (1995).

[9] K. Konishi, H. Deguchi, and K. Takeda, J. Phys. Condens. Matter **5**, 1619 (1993).

[10] B. Y. Chen, S. D. Mahanti, and M. Yussouff, Phys. Rev. Letters **75**, 473 (1995).

[11] D. M. Dennison, Proc. Royal Soc. London, Sec. A **115**, 483 (1927); R. K. Pathria, *Statistical Mechanics* (Pergamon Press, New York, 1978), p. 167.

[12] B. Y. Chen, Ph.D thesis, Michigan State University (1996).

[13] B. Y. Chen, S. D. Mahanti, and M. Yussouff, Phys. Rev B**54**, 11895 (1996).

[14] P. W. Anderson, B. I. Haplerin, and C. M. Varma, Philos. Mag. **25**, 1 (1971).

S.D. Mahanti and Viktor Cerovski

[15] W. A. Phillips, J. Low Temp. Phys. 7, 351 (1972).

[16] T. A. Kaplan, S. D. Mahanti, and W. M. Hartmann, Phys. Rev. Letters 27, 1796 (1971).

[17] V. Cerovski, S. D. Mahanti, and A. K. Rajagopal (submitted).

[18] T. Matsubara and H. Matsuda: Prog. Theor. Phys. 16, 569 (1956).

[19] C. N. Yang and C. P. Yang, Phys. Rev. 147, 303 (1966).

[20] F. D. M. Haldane, Phys. Rev. Letters 45, 1358 (1980).

OPTICAL AND FERMI-EDGE SINGULARITIES IN ONE DIMENSIONAL SEMICONDUCTOR QUANTUM WIRES

K.P. Jain[1], H.H. von Gruenberg[2], and R.J. Elliott[2]
[1] *Department of Physics, Indian Institute of Technology*
New Delhi - 110 016, India
[2] *Department of Theoretical Physics, University of Oxford*
Oxford, U.K.

ABSTRACT

We present a simple many-body treatment of the Fermi edge singularities **FES** in quasi one dimensional quantum wire associated with absorption and photoemission spectrum using Fermi golden rule for the transition probabilities with appropriate many-body wave functions. The functions are expressed in terms of Hartree-Fock determinants so that the problem of computing the transition probablities reduces to calculating these determinants. The edge singularity exponents are related to the phase shifts of the scattering states at the Fermi surface which depend substantially on the electron density. The essential result of this work is that it is possible to infer the edge singularity exponents of the infinite system from the size dependence of the many-body determinants.

1. INTRODUCTION

The fabrication of semiconductor nanostructures in one or two dimensions has led to the observation of new phenomena both in optics and in transport. The properties of doped semiconductor nanostructures, where electrons are confined in reduced dimensions, are radically different from those in the bulk. This is because elementary excitations - both single particle as well as collective and phase space filling exhibit profound changes on account of reduced dimensionality. For instance, the dynamical behaviour of an electron gas constrained to one dimensional (1D) motion exhibits novel properties since the electron or hole can only scatter forwards or backwards. Of special significance is the appearance of sharp peaks at the Fermi level in the optical spectra associated with **FES** and the behaviour of 1D plasmons. The FES have been recently observed at low temperatures in photoluminescence (PL) and photoluminescence-excitation (PLE) experiments on doped $GaAs$ and $In_xGa_{1-x}As$ quantum wires [1, 2], with electron densities in the 1D quantum limit, where only the lowest 1D conduction subband is occupied. These singularities have also been seen in X-ray absorption in metals and in two-dimensional semiconductor quantum well structures [4]. However, these effects are expected to be much more pronounced in 1D as compared to 2D since for a 2D electron gas, the hole recoil process have much greater phase space for indirect transitions which smear out the Fermi distribution with sufficiently high electron concentration. In 1D, on the other hand, no indirect transitions are possible since there are only two electronic states $\pm k$ for each $E(k)$ and the hole recoil which cannot compensate the electron momentum, is likely to be much less significant. The suppression of valence hole recoil processes in 1D is expected to enhance FES. With these qualitative remarks it is nevertheless clear that the effect of a finite mass hole is an important factor and must be taken into account in a quantitative description of the FES. Another factor is the role played by disorder, impurity scattering and hole localization in the robustness of the FES against low-energy single-particle or collective excitations of the Fermi sea [5, 6].

The FES, originally proposed by Mahan [7] for X-ray absorption in metals and later refined by Nozières and Dominicis [8] is due to many-body absorption transitions in which the Fermi sea responds collectively and dynamically to the sudden switching on of the localized hole potential. The response of the Fermi sea depends upon two competing effects: Anderson's orthogonality catastrophe [9] and final state electron-hole interaction (vertex corrections or excitonic effects). The former effect, which corresponds to a direct transition of the electron above the Fermi sea and the concomittant re-arrangement of the Fermi sea electrons to new states tends to suppress the absorption near the edge. The latter is due to transitions mediated by electrons inside the Fermi sea which enhance it, so that the power-law divergence is a resultant of the two. In the rigid Fermi sea approximation where the dynamical response of the Fermi sea electrons is neglected, the interaction between the hole and rigid Fermi sea leads to a bound state (Mahan exciton) below the Fermi energy. The dynamic response of the Fermi sea to the hole potential softens the Mahan exciton, leading to a power-law singularity.

Here we present a simple many-body treatment of the edge singularities in a quasi-one dimensional quantum wire associated with absorption and photoemission spectrum using the Fermi golden rule for the transition probabilities with appropriate many-body wave functions. These functions are expressed in terms of Hartree-Fock determinants so that the problem of calculating the transition probabilities reduces to numerically calculating a sufficient number of determinants. A random-phase-approximation (RPA) screened Coulomb interaction is used for the electron-hole potential, which determines the single particle properties such as the bound state and the phase shifts of the scattering states. Both these depend substantially on the electron density. The hole is taken to be localized at a point and has a infinite mass. The spectra can now be calculated as indicated above, where the electron-hole interaction plays the crucial role, as expected. In the presence of a bound state, a second threshold corresponding to the final state in which bound state is not occupied is resolved, as originally surmised by Nozières and Combescot.

One of the most important aspects of this work is that it is possible to explicitly determine the exponent of the singularities for $N \to \infty$ from a finite N calculation by studying the N dependence of the projection determinants. We thereby proceed in the same way as Anderson in his classic paper [9] who predicted the orthogonality catastrophe by investigating N dependence of a very similar determinant. The singularity exponents have been determined as a function of the electron density [10].

Clearly it is of some importance to understand the dependence of the singularity exponents for primary and secondary thresholds on the electron density. Our results are in quantitative agreement with Hopfield's thumb rule [11]: the exponents are given by $1 - n_f^2$, where n_f is the excess localized charge near the excited core in the ground state and is related to the scattering phase shifts by the Friedel sum rule. The primary threshold of the FES becomes sharper as the electron density decreases, finally going over into the exciton profile in the insulating limit. The secondary threshold, on the other hand, also evolves continuously towards the continuum absorption as the electron density is reduced.

2. THEORETICAL BACKGROUND

2.1. Many-body aspects

The first task is to define many-electron wave functions for processes in which there is a transition of an electron from the valence band of a semiconductor in the presence of a Fermi sea in the conduction band. The many-body wavefunction of the ground state $|\Phi_I\rangle$ is constructed from the N_f free single-particle states with a k-vector lower than the Fermi wave vector k_f so that

$$|\Phi_I\rangle = \prod_{k<k_f} c_k^\dagger |0\rangle \ , \qquad (1)$$

where $c_k^\dagger$ (c_k) denotes the creation (annihilation) operator of a conduction electron with energy $(m/m^*)(ka_0)^2 Ry$ (with $Ry = e^2/2a_0$ and m^* being the efffective mass of the conduction electrons) and where, furthermore, $|0\rangle$ denotes the vacuum state consisting of a filled valence band and an empty conduction band. The final state may be written in terms of a new set of creation operators, labelled by the quantum number λ,

$$c_\lambda^\dagger = \sum_{k=1}^N \psi_{k,\lambda} c_k^\dagger \ , \qquad (2)$$

which takes into account the effect of the hole in the valence band on the conduction electrons. Here N is the number of states. Assuming that the hole is fixed at the lattice site 0, taken as the origin, we can calculate the coefficients $\psi_{k,\lambda}$ by introducing a perturbing electron-hole potential $V(k)$ into the single particle Schrödinger equation

K.P. Jain, H.H. von Gruenberg and R.J. Elliott

$$(ka_0)^2\psi_{k,\lambda} + \frac{1}{Na_l}\sum_{k\prime}\frac{m^*}{m}V(k-k\prime)\psi_{k\prime,\lambda} = \frac{m^*}{m}E_\lambda\psi_{k,\lambda} \ . \tag{3}$$

Here a_l is the lattice constant. Thus the wave function of the final state reads

$$|\Phi'_F\rangle = c_v(0)|\Phi_F\rangle = c_v(0)\prod_{\lambda(F)}c_\lambda^\dagger|0\rangle \ , \tag{4}$$

where $c_v(0)$ is the annihiliation operator of an electron in the valence band at the lattice site 0. The absorption-(ABS) and photoemission (PES) spectrum can be calculated by Fermi's golden rule, so that for PES one gets

$$\rho(\omega)\Delta\omega = 2\pi|\mu|^2\sum_{F(\Delta\omega)}|\langle\Phi'_F|c_v(0)|\Phi_I\rangle|^2 \ , \tag{5}$$

and for ABS

$$\alpha(\omega)\Delta\omega = 2\pi|\mu|^2\sum_{F(\Delta\omega)}\left|\langle\Phi'_F|c_c^\dagger(0)c_v(0)|\Phi_I\rangle\right|^2 \ , \tag{6}$$

where μ is the dipole transition matrix element. The sum $\sum_{F(\Delta\omega)}$ runs over all final states with energy $E_F - E_I$ lying in the interval ω and $\omega + \Delta\omega$.

Our approach is based on rewriting the matrix elements between the many-body states in Eq.(5) and Eq.(6) as determinantal quantities, which can be done easily by using Eqs.(1) and (4) and standard commutation relations between the creation and destruction operators. Thus we obtain for the PES

$$\langle\Phi'_F|c_v(0)|\Phi_I\rangle = \langle\Phi_F|\Phi_I\rangle = Det\left\|\begin{matrix}\psi_{k_1,\lambda_1^F} & \cdots & \cdots & \cdots & \psi_{k_1,\lambda_{N_f}^F}\\ \cdots & \cdots & \cdots & \cdots & \cdots \\ \cdots & \cdots & \cdots & \cdots & \cdots \\ \cdots & \cdots & \cdots & \cdots & \cdots \\ \psi_{k_f,\lambda_1^F} & \cdots & \cdots & \cdots & \psi_{k_f,\lambda_{N_f}^F}\end{matrix}\right\| \ , \tag{7}$$

and for the ABS

$$\langle\Phi'_F|c_c^\dagger(0)c_v(0)|\Phi_I\rangle = \langle\Phi_F|c_c^\dagger(0)|\Phi_I\rangle = Det\left\|\begin{matrix}\psi_{\lambda_1^F}(0) & \cdots & \cdots & \cdots & \psi_{\lambda_{N_{f+1}}^F}(0)\\ \psi_{k_1,\lambda_1^F} & \cdots & \cdots & \cdots & \psi_{k_1,\lambda_{N_{f+1}}^F}\\ \cdots & \cdots & \cdots & \cdots & \cdots \\ \cdots & \cdots & \cdots & \cdots & \cdots \\ \psi_{k_f,\lambda_1^F} & \cdots & \cdots & \cdots & \psi_{k_f,\lambda_{N_{f+1}}^F}\end{matrix}\right\| \ , \tag{8}$$

where

$$\psi_\lambda(0) = \frac{1}{\sqrt{N}}\sum_{k=1}^N\psi_{k,\lambda} \ . \tag{9}$$

Note that the determinant in Eq.(7) is a minor of the determinant in Eq.(8). The latter possesses one more column than the former due to the additional electron from the valence band and one more row which is furnished by the electron probability amplitude $\psi_\lambda(0)$ at the hole site. The ABS and PES can now be determined from the Eqs.(5) and (6) by numerically calculating a sufficient number of determinants, Eq.(7) and Eq.(8) whose elements $\psi_{k,\lambda}$ are provided by the numerical solution of Eq.(3). The latter equation marks the point where the basic assumptions of our physical model enter, and shall be discussed in more detail now.

2.2. Single-particle aspects

It is now necessary to choose an appropriate electron-hole interaction potential which describes the essential physics of the 1D quantum wire, which we assumed to have a finite width. A harmonic confinement model [12] is used for the Q1D wire where the electrons in a zero thickness xy-plane are confined in the y-direction by a harmonic potential while being free to move in x-direction. The actual 1D electron-hole potential is then obtained by calculating the matrix elements of the 2D potential between the wave functions of the harmonic confinement potential. For the one-subband system considered here, only the ground state wave functions $\sim exp(-y^2/2b^2)$ with b being the wire width are necessary. Hence the unscreened potential $V^{us}(q)$ of a finite width wire is

$$V^{us}(q) = -\frac{e^2}{\epsilon_s}B(bq) \ , \tag{10}$$

where

$$B(bq) = e^{b^2q^2/4}K_0(b^2q^2/4) \ . \tag{11}$$

K_0 is the modified Bessel function of the second kind and ϵ_s the effective dielectric constant of the system.

We now introduce screening by taking the electronic polarizability $\chi(q)$ in the random-phase approximation (RPA) so that

$$\chi(q) = \frac{m^*}{\pi\hbar^2q}L(q) = \frac{2m^*}{\pi\hbar^2q}log\left|\frac{q+2k_f}{q-2k_f}\right| \ . \tag{12}$$

We are thus led to the dielectric screening function [13]

$$\epsilon(q) = 1 + \chi(q)V^{us}(q) = 1 + \frac{1}{a_Bq\pi}B(bq)L(q) \ , \tag{13}$$

where $a_B = \epsilon_s\hbar^2/m^*e^2$ is the radius of the first Wannier exciton and the electron-electron interaction is assumed to be equal to $-V^{us}$. The screened electron-hole potential in Eq.(3) becomes $V(q) = \frac{V^{us}(q)}{\epsilon(q)}$. The sum over k in Eq.(3) is carried out by introducing a cut-off so that $\langle 0|k| \leq \widetilde{K}(a_l = \pi/\widetilde{K})$ and a step width in k-space given by $\Delta k = 2\widetilde{K}/N$.

To actually calculate the single-particle energy spectrum from Eq.(3) a choice for the parameters a_B and b has to be made. Bearing in mind $GaAs$ quantum wires we

K.P. Jain, H.H. von Gruenberg and R.J. Elliott

take $b = 800a_0$ and $a_B = 270a_0$. We have found that a cutoff of k at $\widetilde{K} = 0.03\pi/a_0$ and a number of $N = 1000$ yielded a good convergence for the lowest eigenvalues and eigenstates of Eq.(3). There are two types of solutions of Eq.(3) corresponding to both bound and scattering states which will be considered in turn. Fig. 1 shows the bound state energy as a function of the Fermi energy obtained by numerically solving Eq.(3). The range considered is typical for doped quantum wires in the extreme quantum limit in which the only lowest 1D subband is occupied by electrons ($k_f a_B \approx 0.2$ in [1]). The energies $m/m^* E_\lambda$ are given in units of $m^*/m E_{ex} = (a_0/a_B)^2 Ry$, the binding energy of the first Wannier exciton of the system.

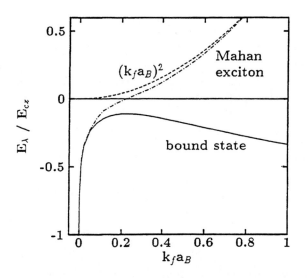

Figure 1. Mahan exciton and bound state of the single-particle energy spectrum, Eq. (3), as a function of the Fermi energy in a range typical for doped 1D $GaAs$ quantum wires. Energies and K_f are given in units of the binding energy E_{ex} and the Bohr radius a_B, respectively, of the first Wannier exciton.

In the 1D case there always exists a bound state, irrespective of the form of the potential, as opposed to the 3D systems where its existence depends on the strength of the potential. We see from Fig. 1 that it shows a pronounced dependence on the carrier density. In the region $k_f a_B \langle 0.2$, it rapidly approaches to the bottom of the conduction band. The efficiency of Fermi-sea electrons in screening the electron-hole interaction can once more be seen from the fact that the higher Wannier states are completely screened out in the range considered and can only be observed in the very low-density regime $k_f a_B \langle 0.001$.

The binding energy of the bound state reveals a somewhat unexpected behaviour for $k_f a_B \rangle 0.4$ where it increases again when one would expect a monotonic decrease (or atleast a saturation behaviour) with increasing k_f because it is not obvious why a higher carrier density should lead to less efficient screening. However, this effect can be traced to the complicated behaviour of the polarization function $\chi(q)$ in Eq.(12) as k_f is varied. We presume that the increase of the binding energy as k_f is increases as shown in Fig. 1 is an artefact due to breakdown of the RPA.

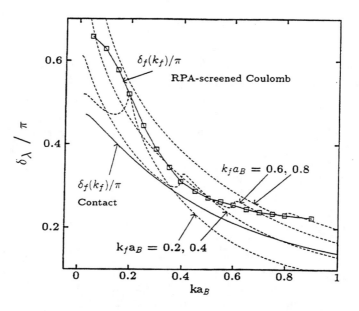

Figure 2. Phase shifts for the scattering states of Eq. (3) at four electron densities corresponding to $k_f a_B = 0.2, 0.4, 0.6, 0.8$ (dashed curves). 2) Phase shifts δ_f at the Fermi surface as a function of k_f (calculated values marked by empty boxes and connected with solid line). Each dashed curve intersects the δ_f-curve at it respective k_f value. 3) Phase shifts δ_f for an on-site potential (simple solid line).

Besides the bound state, the energy spectrum of Eq.(3) possesses $N - 1$ scattering states with $E_\lambda = 0$. In Fig. 2 we have plotted the phase shifts δ_λ as a function of $k a_B$ for four different Fermi energies. From these four curves, the phase shift δ_f at $k = k_f$ is of particular interest for the following. We therefore added another graph which connects these points and shows δ_f as a function of $k_f a_B$. To provide a reference curve for the latter graph we have performed another calculation using a simple contact potential (setting $m^*/mV(q) = const. = U$) instead of the screened Coulomb potential. The potential parameter U was chosen to give a comparative bound state energy of $E_\lambda/E_{ex} = -0.2$.

Comparing the different $\delta_{\lambda(k)}$ curves, one observes that, for a fixed k, $\delta_{\lambda(k)}$ tends to increase with growing k_f. This is due to the fact that the screening of the electron-hole potential for very low k values becomes less efficient. That the δ_f curve is falling in

spite of this tendency, originates in the simple fact that with greater k_f the distance
to the low k values, being mostly affected by the perturbation of the hole, increases.
It is interesting to note that althougt the curves of the phase shifts for a particular k_f
have some structure, we obtain a smooth and structureless δ_f- curve being not very
different from that obtained for the simple contact potential, though we failed to find
a value for the parameter U resulting in better agreement of both curves than that of
Fig. 2.

3. RESULTS

3.1. Fermi-edge singularity exponent

We are now in a position to calculate the ABS and PES from Eqs.(5) and (6) with
the determinants (7) and (8). The results are given in Fig. 3 which displays both
spectra for $k_f a_B = 0.6$ and $N = 500$ plotting against $k \sim \omega^{1/2}$ rather than ω to reveal
FES with clarity.

Consider the non-interacting limit when the electron-hole interaction is switched
off. We get a series of equidistant peaks of the height $\frac{1}{N}$ for the ABS starting at
$ka_B = 0.6$, reminiscent of a step function in the $N \to \infty$ limit. For PES one gets
a single peak of height one at $ka_B = 0$, since the only non-vanishing determinant
describes the transition to the state $|\Phi_F\rangle = |\Phi_I\rangle$.

Switching on the electron-hole interaction in Eq.(3) has a dramatic effect on the
spectra. In absorption, one observes a considerable increase of oscillator strength at k_f.
Obviously, transitions to the states near the Fermi surface become more favourable,
at the expense of transitions to the states $ka_B \rangle 0.8$. For the limit $N \to \infty$, this edge
structure evolves into the FES. In systems having bound state, there exists a second
threshold corresponding to final states in which the bound state is not occupied. This
was first pointed out by Combescot and Nozières [3]. For absorption this second threshold has a very low transition probability and is consequently not resolved in Fig. 3.
However, it can be clearly seen in the PES. The energy difference between both thresholds is the sum of Fermi energy and the binding energy of the bound state, i.e. just
the energy needed to take one electron from the bound state to the first level above
the Fermi surface.

To satisfy the sum rules it is necessary to calculate the determinants corresponding
to all different final states clearly an impossible task. Fortunately, not every excitation
is equally important. There are three sorts of excitations to be distinguished. In
absorption these are, first, states of the form $|\Phi_F\rangle = c^\dagger_{\lambda_v}|\Phi^0_F\rangle$, where only the valence
electron is excited and the remaining N_f electrons are in the ground state $|\Phi^0_F\rangle$ of the
system with hole. second, excitations leaving just one hole in the Fermi sea (Auger-
like process), and, finally, all the rest, i.e. excitations with more than one hole in the
Fermi sea (we call simply multiple particle excitations). For the spectra of Fig. 3, we
only took the valence electron and the Auger-like processes into account. The higher
excitations die out very near the edge so that the spectrum is exclusively governed

by valence electron excitations. That means that although the overall accuracy of the spectrum certainly suffers from neglect of the higher excitations, the very first part of the spectrum near the edge is exact even if one considers valence electron excitations only.

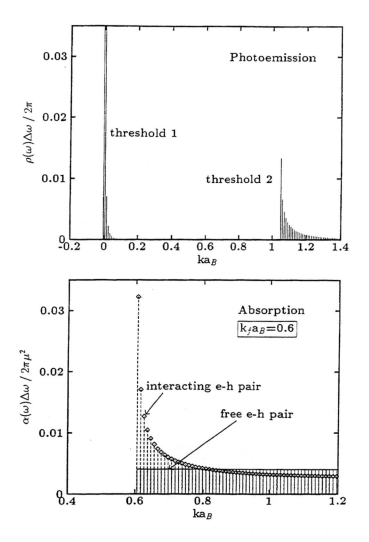

Figure 3. Photoemission and absorption spectrum for a $Q1D$ $GaAs$-like system with an electron density corresponding to $k_f a_B = 0.6$, calculated by evaluating determinants of the form of Eqs. (7) and (8), for a system of $N = 500$ single-particle states with properties depicted in Figs. 1 and 2. Peak structure near the Fermi edge (at $k_f a_B = 0.6$ in absorption) evolves into FES for $N \to \infty$.

At this point we focus on the main question addressed to in this paper - a calculation of the FES exponents. The idea is that if for any given N the first transition in a spectrum is accurately calculated as far as the higher excitations of the Fermi sea are concerned, it should also provide a reasonable approximation for the $(N \to \infty)$-spectrum near k_f which consequently should be obtainable from the first transitions plotted as a function of N. This is done in Fig. 4 for different N between $N = 300$ and $N = 1200$. Since $\Delta\omega \sim \omega^{1/2}\Delta k \sim \omega^{1/2}/N$, we have made the spectra comparable to each other by multiplying Eq.(6) by N. We see that due to the decrease of the energy quantum with expanding system size, the onset of the spectra are continuously shifted towards the edge. Yet, on approaching the edge, the probability of the first transition of each spectrum is growing and lies on a common straight line in a log-log plot. In other words, $\alpha(\omega)$ displays singular behaviour for $k \to k_f$. The exponent of this singularity can be determined from the slope of this line and is 0.47 for $k_f a_B = 0.6$. We thus see that by studying the N-dependence of the determinant one can infer the FES exponent for the infinite system from a finite size calculation.

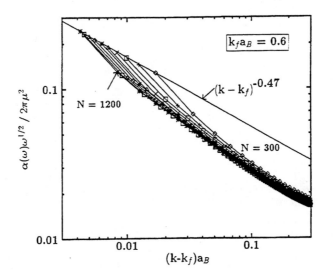

Figure 4. Study of the N-dependence of finite size absorption spectra ($N = 300, 400, \ldots, 1200$), for determining singularity exponent of the FES of the infinite system. N-dependence of the first transition of each spectrum results in a FES exponent of -0.47 for $k_f a_B = 0.6$.

3.2. Electron-density dependence of the singularity exponents

In Fig. 5 are given the first and second threshold exponents as a function of the electron density. In his classic paper [9], Anderson showed that the overlap of the ground states $|\Phi_F^0\rangle$ and $|\Phi_I\rangle$ of the system with and without hole, respectively vanishes as $N \to \infty$, leading to the orthogonality catastrophe. This result emerged from investigating a determinant of the form (7) furnished with the cofficients of the N_f lowest single-particle states. The N-dependence of this determinant was shown to be given by $N^{-\delta_f^2/2\pi^2}$ where δ_f is the phase shift at the Fermi surface. The exponent of the first singularity of the PES in Fig.5 has been calculated by means of the determinant $\langle\Phi_F^0|\Phi_I\rangle$, which is just the determinant that Anderson studied. And indeed, as evident from Fig. 5 the calculated exponents agree well with the Anderson result $1 - \delta_f^2/\pi^2$, where the phase shifts are those of Fig. 2.

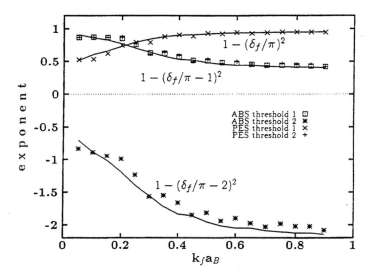

Figure 5. Exponents for the FES and the second threshold edge features in absorption and photoemission as a function of k_f, determined from finite size spectra in the way depicted in Fig. 5 (symbols) and compared with the exponents derived from Hopfield's rule with the $\delta_f(k_f)$ from Fig. 2 (continuous lines). (ABS=absorption spectrum, PES=photoemission spectrum).

With respect to the other cases of Fig. 5, our approach can be viewed as a natural extension of Anderson's approach, in the sense that the edge features of both ABS and PES and their thresholds can be described by a class of determinants similar to that of Anderson. According to Hopfield's rule [11], the exponents are given by $1 - n_f^2$ where n_f is the excess localized charge near the excited core in the ground state $\langle\Phi_F^0|$. This rule was found by Combescot and Nozières [3] to be generally correct. According to the Friedel sum rule [14], the hole potential gives rise to an excess charge n_f of δ_f/π so that with Hopfield the exponent for the first threshold of the PES is just $1 - (\delta_f/\pi)^2$. However, for the second PES threshold the bound state electron is absent, and so the net charge n_f becomes $\delta_f/\pi - 1$ and consequently the exponent is $1 - (\delta_f/\pi - 1)^2$. In the absorption process on the other hand, the electron from the valence band has to be taken into account so that the exponents are $1 - (\delta_f/\pi - 1)^2$ and $1 - (\delta_f/\pi - 2)^2$ for the first and the second threshold respectively. All of these exponents are plotted in Fig. 5 and show good agreement with our numerically determined exponents. The phase shifts δ_f are those due to the screened Coulomb potential.

4. CONCLUSIONS

We have studied FES in quasi-one dimensional quantum wires within a simple framework in which the transition probabilities are calculated by considering Hartree-Fock determinants appropriate to many-body transitions. A screened Coulomb interaction in the RPA has been used for the electron-hole interaction to determine the single-particle properties both with respect to the bound and scattering states of the system which constitutes the starting-off point of our many-body calculations.

The essential feature of this paper is that it is possible to infer the edge singularity exponents of the finite system from the size dependence of the many-body determinants. This procedure is in the spirit of the method used by Anderson to predict the orthogonality catastrophe. These exponents depend on the scattering phase shifts at the Fermi surface, and, via the k_f dependence of these phase shifts, on the electron density. We have been able to check this connection for 1D wires by calculating both the phase shifts as well as exponents. To our knowledge, this is the first explicit calculation of the singularity exponents of the FES in quantum wires. What is most interesting is that this result emerges from an extrapolation of a finite size calculation. The FES intensities and exponents are sensitive to electron correlation and density.

At this point we inject a caveat and ask a crucial question: what is the effect of the hole recoil on the singularities and how are they modified by this. Our treatment has been limited to an infinite hole mass case. Drawing attention to his earlier work [15], Nozières [16] has recently focussed attention on the effect of recoil on the edge singularities. In this he showed that for dimensions $d \geq 2$ the singularities vanish if recoil allows the hole to diffuse to infinity. However, when $d = 1$, or for $d \rangle 1$ if localization is included, the FES exponents are reduced due to recoil, by a factor that involves an angular average over the Fermi surface. For $1d$, the FES persists but the exponent is half as a result of recoil. Obviously more work is needed to elucidate the effect of hole recoil on FES.

Another question of interest is the robustness of FES against low-energy collective excitations of the Fermi-sea in quasi-1D systems, since one knows that these excitations are more significant here than single particle electron-hole transitions.

ACKNOWLEDGMENT

One of us (KPJ) would like to thank Philippe Nozières for a discussion on the nature of FES in reduced dimensions.

References

[1] J.M. Calleja et al., Surface Science **263**, 346 (1992).

[2] M. Fritze, A.V. Nurmikko and P. Hawrylak, Surface Science **305** 580 (1994).

[3] M. Combescot and P. Nozières, J.Phys. (Paris) **32**, 913 (1971).

[4] S. Schmitt-Rink, D.S. Chemla and D.A.B. Miller, Adv. Phys. **38**,89 (1989).

[5] J.F. Mueller, A.E. Ruckenstein and S. Schmitt-Rink, Phys. Rev. B**45**, 8902 (1992).

[6] P. Hawrylak, Solid State Commun. **81**, 525 (1992).

[7] G.D. Mahan, *Many-Particle Systems*, (Plenum, New York, 1981) 2nd ed.

[8] P. Nozières and C.T. De Dominicis, Phys. Rev. **178**, 1097 (1969).

[9] P.W. Anderson, Phys. Rev. Lett. **18**, 1049 (1967).

[10] H.H. von Gruenberg, K.P. Jain and R.J. Elliott, Phys. Rev. B**54**, 1987(1996).

[11] J.J. Hopfield, Comments on Solid State Physics, **2**, 40 (1969).

[12] G.Y. Hu and R.F. O'Connell, Phys. Rev. B**42**, 1290 (1990).

[13] G.Y. Hu and R.F. O'Connell, J.Phys. Cond. Matter. **2**, 9381 (1990).

[14] J. Friedel, Comments on Solid State Physics, **2**, 21 (1969).

[15] J. Gavoret, P. Nozières, B. Roulet and M. Combescot, Journal de Physique, **30**, 987 (1969).

[16] P. Nozières, J. Phys. I France **4**, 1275 (1994).

MANIPULATION OF NANOMETER OBJECTS: FRICTION, MECHANICAL PROPERTIES AND DEVICES

Richard Superfine,[1] Michael R. Falvo,[1] Scott Paulson,[1] Sean Washburn[1]
Russell M. Taylor III,[2] G. J. Clary,[2] Vernon Chi,[2] and Frederick P. Brooks, Jr.[2]
*[1] Department of Physics and Astronomy,
University of North Carolina, Chapel Hill NC 27599-3255
[2] Computer Science Department
University of North Carolina, Chapel Hill, NC 27599-3270*

ABSTRACT

We have developed and applied an advanced interface for scanning probe microscopy which provides an intuitive environment for the real-time control of the instrument and for presentation of the acquired data. This interface has been used for a variety of experiments which apply AFM manipulation to understand the properties of surface-bound objects, object/object and object/substrate interactions. Intricate manipulations have successfully created three-dimensional structures and begun the assembly of nanometer-scale electromechanical devices.

1. INTRODUCTION: NOVEL MATERIALS AND THE STUDY OF NANOMETER-SCALE OBJECTS

The study of novel materials produces many challenges in the areas of synthesis, modeling and characterization. For the latter, one would like to be able to determine mechanical and electrical properties, and correlate them with structure. In the case where the properties of the novel material come from individual structures, it is preferable to be able to study single units, rather than being limited to averaging over collections in macroscopic samples. The atomic force microscope (AFM) provides a wide range of characterization capabilities (electrical, mechanical, chemical, etc.) on the nanometer scale, while correlating these with structure in the form of detailed topography [1,2]. Impressive as these capabilities are, the AFM can also be used as a nanometer-scale manipulation tool [3-5]. The ability to manipulate objects efficiently on surfaces makes available a wide variety of experiments on the interactions between the sample and substrate [6,7], on the physical properties of individual objects and on the creation of unusual devices incorporating the nanometer objects.

We have studied nanometer-scale materials, including colloids, viruses and nanotubes, through the use and development of an advanced interface for scanning probe microscopy (SPM). For example, the simple pushing of an object with the AFM tip, with the measurement of the applied lateral force, measures the surface adhesion and friction [7]. While manipulating an extended object, such as a tubular virus or nanotube, the object often deforms. This behavior can be understood simply in terms of competition between the sample elastic properties and substrate interactions [8]. Finally, manipulation allows us to begin the study of unusual device structures incorporating nanometer samples which cannot be fabricated from traditional processing steps. Making these experiments possible is the nanoManipulator, an interface for SPM's [9,10]. The interface allows for real time control of the AFM tip while simultaneously recording topography, lateral forces, as well as device features such as conductivity. In addition, the data are rendered as three-dimensional, directionally illuminated surfaces providing immediate, intuitive interpretation of the SPM data.

2. THE NANOMANIPULATOR: AN ADVANCED INTERFACE FOR SPM

The nanoManipulator (nM) is an advanced interface for SPM's which allows intuitive control of the microscope operation and 3-D rendering of the collected data. It is an assembly of two personal computers, an HP 5000 workstation, a Silicon Graphics Inc. Onyx with an Infinite Reality Engine, a Sensable Devices PHANToM haptic interface and a scanning probe microscope (Topometrix Discoverer or Explorer Microscopes can be driven). The power of the nM is in the placement of the human operator directly into the "feedback loop" that controls surface manipulations. In practice, the user holds a stylus in hand. By moving the stylus laterally, the user directs the movement of the SPM tip across the sample. The haptic interface enables the user to "feel" the surface by forcing the stylus to move up-and-down in response to the surface topography. In this way the user understands the immediate location of the tip on the sample and can quickly and precisely maneuver nanometer scale objects. The nM interface is also a real-time logger of all SPM and experimental data so that sample and device modifications and the application of stress can be strictly correlated with their effect on electrical properties. A picture of the facility appears in Fig. 1 which shows a user holding the haptic device through which he controls the SPM. The experiment rendered on the monitor involves the manipulation of an 18nm diameter TMV particle across the surface in order to study its mechanical properties.

3. NOVEL MATERIALS I: VIRUSES

Viruses are assemblies consisting of an outer casing composed of proteins, enclosing genetic material [11]. Understanding the details of virus function is essential not only for the traditional goal of the blocking viral infection, but also for commandeering the operation of viruses to carry therapeutic genetic material into defective cells. This is the goal of gene therapy. Natural or designed viruses serve as one class of gene carriers, or vectors. Understanding the manner in which viruses bind at surfaces, including cells, invade cells and open to release the enclosed genes [12] is necessary for further progress in the field of gene therapy. The AFM can play several roles in such studies, including the measurement of the attachment of the virus to model substrates and cell membranes, and for measuring the mechanical properties of viruses. Beyond biological function, viruses are of great interest as macromolecular assemblies. TMV is one example of viruses that will self-assemble from individual proteins in solution [13]. Understanding self-assembly processes and the properties of the resulting structures at a wide range of length scales is at the forefront of modern materials science.

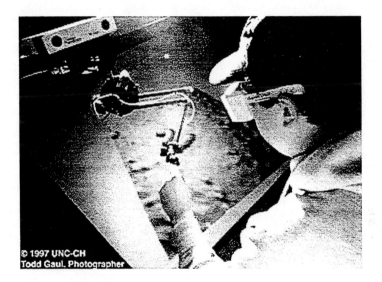

Figure 1. A user holds the haptic stylus in his left hand while using the nM system for reviewing an AFM manipulation of TMV. The stylus is connected through a mechanical linkage to motors which supply the force-feedback for feeling the surface topography. The glasses he is wearing provide 3-D visualization and head-tracking for realistic rendering of the surface topography.

One of the most well-studied viruses, tobacco mosaic virus (TMV), has served as a first subject. The TMV geometry is a thick-shelled rod 300nm long with 18nm outer diameter and 4nm inner diameter [14]. The shell of the virus consists of 2130 protein units wrapped helically around a single strand of RNA (three base pairs per protein). Inter-protein and protein-RNA interactions are non-covalent and are known to be critically dependent on osmolarity and pH [14]. Our methodology involves dispersing the viral particles on a substrate, imaging , then performing manipulation [8]. Under the forces of manipulation, the extended TMV are observed to deform. By modeling the applied forces and the virus response in analogy to a center-supported beam in a gravitational field, we can derive the relationship between the resultant virus shape and the relevant physical parameters: the substrate/virus friction and the virus Young's modulus. Fig. 2 shows the fit of a manipulated TMV particle (on a graphite substrate) to the derived equation

$$y' = R \left[(1/4) \, x'^2 - (1/6) \, x'^3 + (1/24) \, x'^4 \right]$$

where y' and x' are, respectively, the deflection and lateral coordinate normalized to the TMV half-length, L. The prefactor $R = f L^3/EI$ contains all of the physical parameters of the system: f is the frictional force per unit length acting along the particle, I is the moment of inertia of the cross section which is a function of the geometry of the beam and E is the Young's modulus of the beam material. The product EI, known as the flexural rigidity, is the parameter that describes the beam's resistance to deformation under bending. Using this model to describe the TMV system, the bent shape can be used to determine the prefactor, and hence measure the ratio of the substrate-TMV friction and the TMV flexural rigidity. The fit of the virus shape is consistent with the virus/substrate interaction expected on a low-friction surface, and the Young's modulus of 1.7 GPa is comparable to that measured for microtubules and polymers. In addition, by measuring the lateral force applied during manipulation, one can determine f and EI independently. Having established this protocol, we will pursue these studies on a range of substrates and solution conditions with TMV and other viruses.

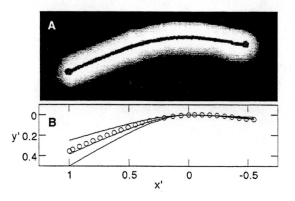

Figure 2. A. The AFM manipulation of a TMV particle has resulted in a bent shape. The line through the TMV is a medial line as determined by the cores method, software designed at UNC. B. A fit of the medial line to the beam equation results in the determination of a single parameter: the ratio of the TMV/substrate friction to the flexural rigidity.

4. NOVEL MATERIALS II: NANOTUBES

If one wanted to design a material for its mechanical properties, a good starting place would be the strongest known bonding network: the sp^2 bond of carbon-carbon such as is found in graphite sheets [15]. For use as a basic element of a nanostructure, such as a beam, one would roll this sheet into a cylinder. It is most striking that it is this precise structure which arises from the ashes of a violent electric arc or impinging laser pulse [16]. The properties of such carbon structures have long been appreciated and their use in composite materials has been pursued for decades. The current status of nanotubes, available as concentric, multiwall tubes and as bundles of single-walled tubes in various compositions of B, C and N [17], offer a wide range of systems for applications in fields such as composite materials and nanometer-scale electromechanical devices. Much work needs to be done to understand the basic physics at the nanometer length scale which governs the behavior of the nanotubes and their interactions with their surroundings. Of interest are the electrical [18] and mechanical properties [19,20] of individual nanotubes, substrate/nanotube interactions and the interactions between nanotubes and other objects on the nanometer scale.

4.1. Mechanical Properties

Important quantities characterizing the mechanical properties of a material include the small-strain elastic modulii, the material strength and the ductility. The basal plane elastic modulus of graphite, the largest of any known material, confers to the carbon nanotube (CNT) its predicted extraordinary stiffness. These expectations are consistent with experimental reports of measurements taken in the small strain limit [19,21]. The material strength characterizes the maximum stress for linear behavior, while the ductility quantifies the ability to sustain large strains without failure. The latter are thought to distinguish graphitic carbon from diamond, which has the next highest elastic modulus, but is brittle and consequently has a lower strength and ductility. The AFM can be applied to provide unique insight into these mechanical properties.

4.1.a. Manipulation-Induced Deformation

The analysis applied to the manipulated virus particles is valid for any extended structure. In Fig. 3 is shown a multiwall carbon nanotube after manipulation in ambient on a mica substrate. The topography data has been fit to the beam equation in order to extract the prefactor R appropriate for the nanotube elastic modulus and CNT/mica interaction. This prefactor is consistent with the expected interface friction and a Young's modulus of 1TPa, as expected from the relevant modulus of graphite [15]. Controlling this substrate-induced deformation will be critical to the operation of nanotube-based devices which rely on actuated nanotubes.

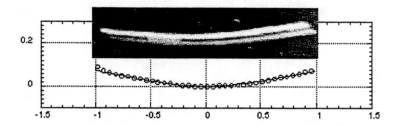

Figure 3. Inset: Carbon multiwall nanotube after manipulation, as bent by tip and surface forces. (The double image is a tip artifact is irrelevant to the analysis.) The fit of the beam equation to the medial line is shown with a sparse set of the data points. The axes are in units of distance normalized by twice the nanotube length.

4.1.b. High Strain Behavior

It is of great interest to understand the behavior of nanotubes under large strain. Theoretical predictions have indicated extraordinary flexibility [22] while TEM images of deposited samples show highly distorted tubes with unknown bending history [23,20]. Using the nanoManipulator system, we can take nanotubes through intricate, repeated manipulations in order to test their flexibility.

Fig. 4 shows a straight CNT lying amongst codeposited gold colloids on a mica substrate. In manipulating the CNT against these colloids, it was found that they acted as point constraints. The CNT was observed to increasingly bend as it was pushed between the colloids. Eventually, the end of the CNT passed the near-side colloid and, upon re-imaging, relaxed (Fig. 4) to an almost straight orientation subsequently constrained by another colloid. Fig. 4 shows the tube reorienting under imaging forces. This relaxation, where the tube's rigidity overcomes the substrate friction, indicates that the CNT retains its elastic properties even after severe bending.

The nM system can be used to apply the most severe bending strains. Fig. 5 shows an intermediate step in a long series of manipulations on a 20 nm diameter multiwall CNT. Even though the tube has undergone a 180° bend, it did not fracture apart. In fact, we have never observed the separation of a nanotube under repeated bending strains. Fig. 5 also shows the manipulation of a nanotube so that it wraps upon itself, taking advantage of the flexibility of the tube kinks. These experiments indicate a material of extraordinary ductility, and highlights its application in composite materials.

4.2. Nanotube/Colloid Interactions

Manipulation of surface-bound materials provides a way to measure the relative forces of adhesion and friction between materials and between materials and the substrate [6,7]. As an example, Fig. 6 shows the manipulation on mica of an adsorbed gold colloid by an AFM-manipulated carbon nanotube. In this case (as distinguished from above) the gold colloid was moved by the nanotube (a-c). The separation of the nanotube and colloid after the nanotube was moved away shows that the colloid/surface friction is larger than the colloid/nanotube adhesion. Experiments of this nature promise to elucidate the contact mechanics and surface forces of nanometer scale objects of a wide variety of shapes and compositions.

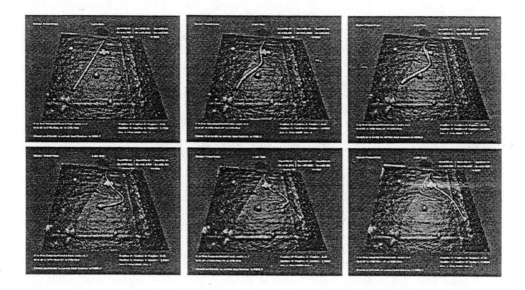

Figure 4. A carbon multiwall nanotube is manipulated amongst codeposited 20nm gold colloidal particles. The particles remain fixed on the substrate, acting as constraints to the moving nanotube. The sequence proceeds from top-left to top-right, bottom-left to bottom-right. The tube bends as it is slid through the colloids, until it slips free from the center particle. The tube is then unstable to imaging until it relaxes to hit a particle on the far right.

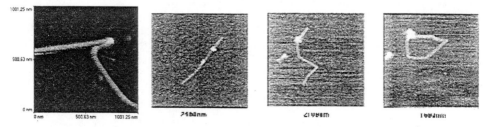

Figure 5. Left: Nanotube bent through 180° without separation. Images of later manipulation show that the tube remains intact even under repeated strains of this type. Right three images: Manipulation sequence wrapping a carbon nanotube upon itself.

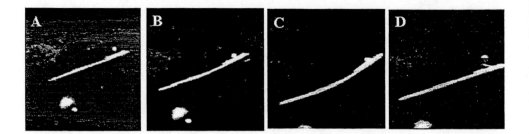

Figure 6. A carbon nanotube is manipulated against a gold colloid in order to assess the relative values of colloid/substrate friction and colloid/carbon nanotube adhesion. After making contact and pushing (a-c), the nanotube is manipulated by the AFM tip away from the colloid. The colloid remains in place, indicating that the colloid/substrate friction is greater than the colloid/nanotube adhesion.

5. MANIPULATION AND ELECTRO-MECHANICAL SWITCH

We have shown various ways in which the AFM can be applied fruitfully for the study of the mechanical properties of individual nanometer-scale objects, and to probe object/substrate and object/object friction and adhesion forces. AFM manipulation also provides the ability to incorporate objects into arrangements and devices which measure or exploit their electrical and mechanical properties. Fig. 7 (left) shows a nanotube which has been manipulated to lie across two other tubes, themselves having been manipulated into position. The top nanotube is truly off of the substrate in the regions between and to the left of the two underlying tubes. This arrangement is ideal for the study of tube/tube friction, testing the use of nanotube bearings in a nanometer-scale mechanical sliding device. Most interesting will be the study of the effects of humidity on this interaction, thus probing capillary condensation on the nanometer scale. The manifestation of adsorbed water in the adhesion of components is known as *stiction* [24], and is recognized as a serious problem in micron-scale electromechanical devices. It will be essential to control stiction for the realization of nanotube-based actuated devices. An example of a first step in the fabrication of such a device is shown in Fig. 7 (right) where we have manipulated nanotubes into position for an electrically-actuated mechanical switch. This is an image before metallization of two tubes positioned end-end for studies of tube-tube contact as initiated by the electric field introduced at the laterally placed tube. Each tube would be contacted separately by subsequently patterned wires. This is one of several examples of geometries that we are testing for a successful switch.

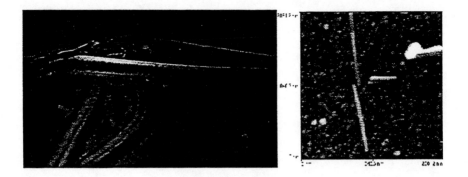

Figure 7. Manipulation of carbon nanotubes for studies for friction and device applications. Left: Carbon nanotube manipulated over two other tubes to form a bridge. This structure, which will be used for studies of tube-tube friction and contact mechanics, represents one of the first three dimensional structures formed from nanometer-scale objects using scanning probe microscopy. Right: Tubes manipulated to form adjustable end-end contact for studies of electron transport through tube junctions.

6. CONCLUSIONS

The development of scanning probe microscopy over the past decade has made possible the nanometer-scale correlation of a wide range of sample material properties including topography, chemistry, friction, elasticity, to name but a few. Through the development and application of an advanced interface for SPM's, we have explored a variety of experiments in which surface-bound objects can be studied to understand object/object and object/substrate interactions, in addition to probing the properties of the object itself. The deformation of an extended object under manipulation can be used to quantify the mechanical properties and substrate interaction of individual objects. Complex stresses can be applied to objects in order to explore their response under large, repeated strains. Objects can be manipulated against each other to elucidate the competition between object/object and object/substrate interactions. Finally, we have shown that intricate arrangements can be achieved which will lead to the study and realization of nanometer-scale actuated electromechanical devices. Understanding and controlling the nanometer world is increasingly an interdisciplinary endeavor. The experiments discussed here have been made possible by a collaboration of computer scientists and physical scientists which have brought the wonders of the atomic world to our fingertips.

ACKNOWLEDGMENTS

The authors would like to gratefully acknowledge contributions of Otto Zhou and the full cast of the nanoManipulator development team. support of NSF (ASC-9527192, DMR-9512431, CDA-9504293), the NIH (5-P41-RR02170), and the corporate support for instrumentation development by Topometrix, Inc. and Silicon Graphics, Inc.

222 R. Superfine, M. R. Falvo, S. Paulson, S. Washburn, R.M. Taylor III, et al.

References

[1] S. N. Magonov, M.-H. Whangbo, *Surface Analysis with STM and AFM : Experimental and Theoretical Aspects of Image Analysis* (VCH Publishers Inc., New York, 1996).

[2] H.-J. Guntherodt, D. Anselmetti, E. Meyer, Eds., *Forces in Scanning Probe Methods* (Kluwer Academic Publishers, Dordrecht, The Netherlands, 1995).

[3] T. Junno, K. Deppert, L. Montelius, L. Samuelson, Appl. Phys. Lett. **66** 3627 - 3629 (1995).

[4] T. A. Jung, R. R. Schlitter, J. K. Gimzewski, H. Tang, C. Joachim, Science **271** 181-184 (1995) .

[5] A. L. Weisenhorn, *et al.*, Science **247** 1330 - 1333 (1990) .

6. P. E. Sheehan, C. M. Lieber, Science **272** 1158 - 1161 (1996) .

[7] R. Luthi, et al., Science **266** 1979 - 1981 (1994) .

[8] M. R. Falvo, et al., Biophys. J. **72** 1396-1403 (1997) .

[9] R. M. Taylor, *A Virtual-Reality Interface for a Scanning Tunneling Microscope*, SIGGRAPH '93 (ACM SIGGRAPH , New York, 1993).

[10] M. Finch, *et al.*, *Surface Modification Tools in a Virtual Environment Interface to a Scanning Probe Microscope*, ACM Symposium on Interactive 3D Graphics (ACM SIGGRAPH, Monterey, CA, 1995).

[11 S. Casjens, in *Virus Structure and Assembly* S. Casjens, Eds. (Jones and Bartlett, Boston, 1985).

[12] U. F. Gerber, M. Willetts, P. Webster, A. Helenius, Cell **75** 477-486 (1993) .

[13] D. L. D. Caspar, A. Klug, *Physical Principles in the Construction of Regular Viruses, Cold Spring Harbor Symposium on Quantitative Biology* (The Biological Laboratory, Cold Spring Harbor, L.I., New York, 1962), vol. 27, pp. 1-24.

[14] J. W. Davies, Eds., *Molecular Plant Viruses, Volume 1* (CRC, Boca Raton, 1985).

[15] M. S. Dresselhaus, G. Dresselhaus, P. C. Eklund, *Science of Fullerenes and Carbon Nanotubes* (Academic Press, San Diego, 1996).

[16] T. W. Ebbesen, P. M. Ajayan, Nature **358** 16 (1992) .

[17] A. Loiseau, F. Willaime, N. Demoncy, G. Hug, H. Pascard, Phys. Rev. Lett. **76** 4737-4740 (1996) .

[18] S. J. Tans, M. H. Devoret, C. Dekker, Nature **386** 474 (1997) .

[19] M. M. J. Treacy, T. W. Ebbesen, J. M. Gibson, Nature **381** 678-680 (1996) .

[20] R. S. Ruoff, K. M. Kadish, *Fullerenes: Recent Advances in the Chemistry and Physics of Fullerenes and Related Materials* (The Electrochemical Society, Pennington, NJ, 1995), vol. 2.

[21 C. Lieber, Bull. Am. Phys. Soc. **42** 591 (1997) .

[22] B. I. Yakobson, C. J. Brabec, J. Bernholc, Phys. Rev . Lett. **76** 2511-2514 (1996) .

[23] R. S. Ruoff, D. C. Lorents, Carbon **33** 925-929 (1995) .

[24] R. Legtenberg, H. A. C. Tilmans, J. Elders, M. Elwenspoek, Sensors and Actuators A **43** 230-238 (1994) .

LOCALIZED IMPURITY IN NANOCRYSTALS - PHYSICS AND APPLICATIONS

Rameshwar N. Bhargava
Nanocrystals Technology
P. O. Box 820, Briarcliff Manor, NY 10510, U. S. A.

ABSTRACT

The role of Transition and Rare-Earth impurities in nanocrystals is discussed. The changes in the luminescent properties of an localized atom in a quantum dot are explained on the basis of the interaction between the excited states of the atom and the quantum dot. This size dependent interaction, which depends on the size of the quantum dot and also on the nature of the atomic impurity, make these materials very attractive for various optoelectronic applications.

1. INTRODUCTION

In the past decade, the properties of nanocrystals have been extensively studied [1,2]. In particular, the optical properties and their size dependence based upon quantum confinement are well understood. In spite of the fact that the properties of these nanocrystals show enhanced efficiency , their practical usage is still lacking. Two critical limitations have prevented many of the applications of these nanocrystalline materials.

The first of the limitations being that the size of the nanocrystals are still difficult to control. This size variation leads to variation in the optical properties of the nanocrystals. This variation of the size dependent properties could be eliminated by an incorporation of a localized impurity in a nanocrystal. In this doped nanocrystalline materials, for example, the optical emission is controlled by the properties of the localized impurity and not by the size of the host nanocrystal. This has been basis of the development of Doped Nanocrystals

(DNC). One of the additional advantages of these doped nanocrystals is that we could vary
the properties of the nanocrystals by doping them with different atoms. For example, we
could have red, green and blue emitting phosphors from yttrium oxide nanocrystals by
incorporating europium, terbium and thulium ions, respectively. By incorporating other
transition and rare earth ions we could develop optical, magnetic and magneto-optical
materials. The design of a new class of materials by incorporating localized atomic
impurities in nanocrystalline size host is expected to dominate the next generation of
artificially designed materials. These doped nanocrystalline materials could lead to devices
with very high densities, ultrafast speeds and high efficiencies.

The second limitation for the lack of usage of these nanocrystalline materials relates
to the reduced luminescent efficiency of free standing, non-passivated nanocrystalline
materials. The practical photonics applications of these nanocrystals [NC] are still lacking
due to the fact that the surface related non-radiative recombination dominates in the strong
confinement limit [3,5]. The precipitous decrease of the phosphor luminescence efficiency
due to the surface-related nonradiative processes in the region between 1 to 0.01 μm is
shown in Fig. 1. This reduction in luminescence efficiency has prevented the use of smaller
particles in the past.

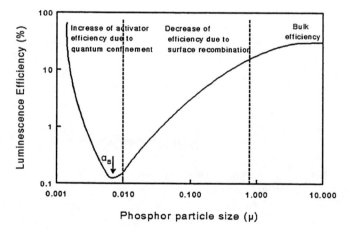

Figure 1. The size dependent efficiency of doped nanocrystalline phosphors.

To eliminate the non-radiative contribution from the surface states, one needs to
passivate the free standing quantum dots. The passivation reduces the non-radiative
contribution of the surface and thereby enhances the luminescent efficiency. These coated
or matrix based encapsulated NC so far have found limited applications due to the inherent
difficulty in the passivation of the individual nanocrystalline surface.

To further decrease the contribution of the surface related nonradiative recombination,
an impurity in a quantum dot was introduced for the first time and reported by Bhargava *et*

al. in 1994 [6]. By incorporating an impurity in a quantum confined structure, the dominant recombination route can be transferred from the surface states to the impurity states. This can occur when the size of the particle is smaller than the excitonic Bohr radius of the electron and hole (strong confinement limit). If the impurity induced transition can be localized as in case of the transition metals or the Rare-Earth elements, the radiative efficiency of the impurity induced emission increases significantly. The choice of the localized impurity and the size of the host quantum dot are the two key factors determining the efficiency of the impurity. The more the confinement, the better is the efficiency.

The choice of host also helps to reduce the non-radiative contribution of the surfaces. For example, insulators such as oxides surfaces contribute non-radiatively much less than the semiconducting sulfides. This is discussed in the next section.

A major advantage of placing a transition or rare-earth luminescent activator in a quantum dot is that the rate of transfer of carriers from the host to the impurity can be modified without changing internal electronic levels. This allows us to improve the efficiency without any significant changes in the chromaticity of the phosphors.

2. EXPERIMENTAL

We have studied two doped nanocrystalline systems: Zinc Sulfide doped with Manganese ($ZnS:Mn^{2+}$) [6-8] and Yttrium Oxide doped with Terbium ($Y_2O_3:Tb^{3+}$) [8,9]. The corresponding bulk phosphors of these two systems are well known for their applications in the electroluminescent devices and CRT displays, respectively. Both phosphors were prepared by room temperature chemical synthesis [7,8]. The process utilizes a modified sol-gel technique using organometallic synthesis.

When a localized impurity is introduced in a Zero-Dimensional (0-D) quantum dot, the optical properties of the impurity changed drastically [5,6]. In case of $ZnS:Mn^{2+}$ nanocrystals, the photoluminescent efficiency was shown to be better than that of bulk $ZnS:Mn^{2+}$ materials while the luminescent decay time was shortened by five orders of magnitude. This was explained by the hybridization of the d-states of the Mn^{2+} atom with the s-p states of the host ZnS nanocrystal. This interaction under quantum confinement renders the parity forbidden d-d transition of the Mn-atom to an allowed transition changing the luminescent decay time from 2 msec. to 4 nsec. This change occurs in the particle size range of 20 to 50 Å.

As discussed earlier [5], in DNC phosphors the hybridization among the s-p like excited electrons in the host (ZnS) and the d-electrons of the activator (Mn^{2+}) increases the rate of transfer of carriers from the host to the activator. This transfer rate should be strongly dependent on the quantum confinement i.e. on the size of the nanocrystal. An important consequence of this size-dependent transfer rate is that quantum efficiency increases as the particle size decreases. This size-dependent quantum efficiency is observed in two different DNC phosphor systems. The variation of the efficiency as a function of the size of the nanocrystals for $ZnS:Mn^{2+}$ and $Y_2O_3:Tb^{3+}$ are shown in Figs. 2(a) and 2(b), respectively. In both cases, the efficiency increases approximately as $1/D^2$ where D is the diameter of the particle [9].

If the surfaces play a dominant non-radiative role, the efficiency of the activator luminescence is expected to decrease with the decrease of the particle size. This is the opposite of what we observe in DNC phosphors. It should be noted that the efficiencies observed in DNC phosphors are comparable to that observed in the bulk samples [6].

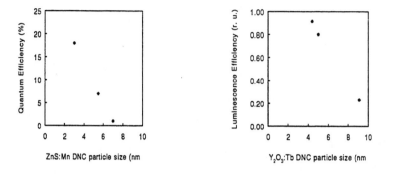

Figure 2. Variation of the external luminescent efficiency as a function of the size of the nanocrystals (a) ZnS:Mn^{2+} and (b) Y_2O_3:Tb^{3+}.

3. Tb ATOM IN YTTRIUM OXIDE QUANTUM DOT

To understand further the modification of properties of a 'localized impurity' in a 'quantum-confined host' we have recently studied terbium (Tb) atom in a nanocrystal of yttrium oxide (Y_2O_3). In this system, the atomic states of the Tb^{3+} atom derived from the 4f core states are observed in the luminescent spectra. These states hardly shift under quantum confinement. However, the excited states interact well with the host so as to provide insight to the modulation of the atomic properties. We propose the model for the size dependent modulation of the luminescent efficiency of a localized impurity. The model is based on the modulation of the excited states of the localized atom imposed by the quantum confinement by the host. After presenting the latest data, we will briefly discuss the model for the 'quantum confined atoms' (QCA). The implications of engineering materials at this ultrafine scale for various applications is also discussed.

In case of oxide hosts, the band gaps are larger than the 254 nm exciting radiation. In these cases the absorption of the radiation normally occurs either via internal atomic transitions (f-d) or via charge transfer states (CTS) between the host lattice atoms and the impurity. For the same activator ion, these CTS absorption spectra are different for different hosts. The variation of the CTS spectra due to microstructural changes are well known [9]. In the case of Tb^{3+}, the $4f^8$ - $4f^7$ 5d transition dominates the absorption in the region 200 nm to 300 nm.

These absorption bands associated with the f-d transitions in Y_2O_3:Tb^{3+} DNC phosphors show variation which may be associated with the varying degree of quantum confinement in different samples. The variation of the absorption (PLE) and the activator luminescent efficiency respectively, are shown in Figs. 3(a) and 3(b) for two different DNC

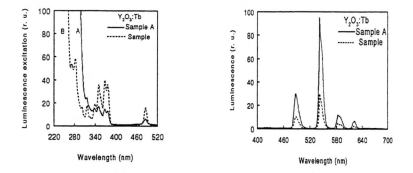

Figure 3. Variation of the PLE (a) and PL (b) in two different samples of Y_2O_3:Tb^{3+}.

samples of Y_2O_3:Tb^{3+}. In Fig. 3(a) the photoluminescent excitation (PLE) spectra for two different samples of Y_2O_3:Tb^{3+} are shown. In the region below 300 nm, the absorption in sample A is shifted from that of sample B. We associate this red shift as a result of the quantum confinement. This red shift also results in a increased luminescent efficiency due to higher absorption at 254 nm as in the case of sample A. This is shown in Fig. 3(b) where the relative PL intensity in the sample A is about 3.3x higher than the sample B. As a result of quantum confinement, the modulation of the transfer rates and $4f^8$ - $4f^7$ 5d transition is observed in doped nanocrystalline phosphors. This is discussed below.

4. MODEL OF THE QUANTUM CONFINED ATOM

The ionized Tb^{3+} in Y_2O_3 has highly localized $4f^8$ - electrons and their electronic transitions are hardly effected by the host lattice. These electronic states for Tb^{3+} in Y_2O_3 are given in Fig. 4. However the quantum confinement of such a localized atom in a nanocrystal, does effect the transition between $4f^8$ and $4f^7$ - 5d absorption bands [12]. In fact, because of the proximity of the boundary of the host quantum dot and the excited states of terbium atom, these absorption band become much broader than in the bulk materials. This is depicted in Fig. 4.

The absorption bands are broad enough to the point that the individual levels of the Tb^{3+} ion are over shadowed by these broad absorption bands. The additional observation is that the absorption spectra shifts significantly to the low energy. This enables the transfer of energy without going through the $4f^8$ discrete levels of the Tb^{3+}. The transition via discrete steps slow down the transfer of electrons due to multi-phonon emission. The continuous broad absorption band are created by the interaction of the quantum confinement and the $4f^7$ - 5d, $4f^7$ - 6s and other higher excited states. This modulation of the excited states results in broad absorption bands as shown in Fig. 4 [15]. These bands can lead to

R.N. Bhargava

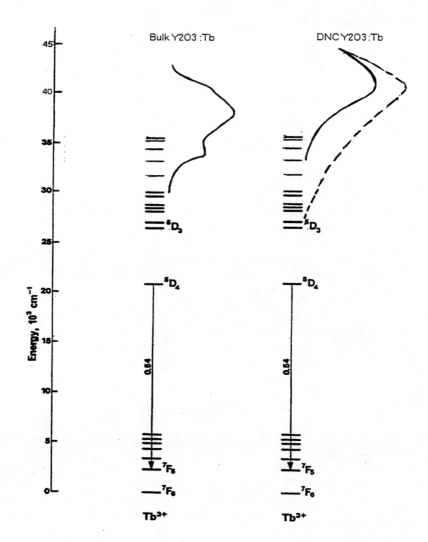

Figure 4. The photoluminescent excitation (absorption bands) observed in bulk and DNC Y_2O_3:Tb^{3+} phosphor. Aborption in the bulk corresponds to $4f^7$ - 5d bands. In DNC, the PLE is broadened and shifted to high energy.

large absorption which then results in an ultrafast transfer of carriers to $5D_3$ and $5D_4$ states of the Tb^{3+}. This rapid transfer of carriers would allow us to develop new devices including lasers from localized impurities.

The model of the localized atom could be visualized as a quantum mechanical funnel (QMF) [15] where large energy is collected (absorbed) and transferred to a single atom. This is schematically depicted in Fig. 5.

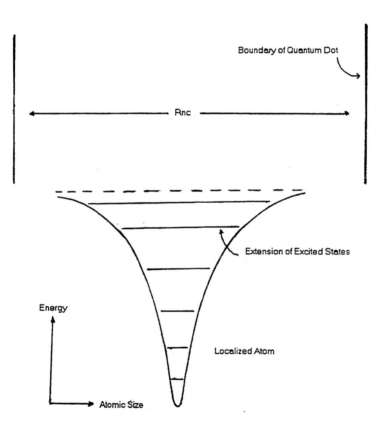

Figure 5. The model of the quantum confined localized atom could be viewed as a quantum mechanical funnel (QMF).

This model of the QMF explains the large absorption cross section observed for the quantum confined localized atom. In QCA, the energy transfer from the host to the atom is extremely fast making the atom a superactive atom. The effective absorption cross section for such an atom is much larger since its absorption is modulated by the quantum confinement of its excited states. This single localized QCA borrows the density of states from the host. The size of this QMF would be dependent on the size of the quantum dot. The effectiveness of the funnel will depend on the quantum confinement as well as the choice of the atom. Higher the degree of localization (e.g. rare-earth and transition metals) larger would be the modulation factor in QCA.

5. CONCLUDING REMARKS AND APPLICATIONS

As discussed above, the impurity in a quantum dot allows one to modify the properties of the nanocrystal. In particular, the contribution of the surface related non-radiative recombination is significantly reduced by the introduction of the impurity. This establishes the usefulness of the doped nanocrystalline materials in the luminescent field. The modification of the carrier lifetime via introduction of an impurity can be beneficial in designing new class of quantum dot sensors for the X-ray and UV region. In such a sensor the minority carrier lifetime should decrease with decreasing size of the particle.

Besides modification of the surface properties for the efficient impurity-induced extrinsic luminescence, we also have the advantage of modulating the properties of the impurity itself. The lifetime shortening of the impurity luminescence is one of such results. The 0-D quantum confinement imposed by the host, particularly in the strong confinement limit (16), results in mixing of the electronic states of the host and the impurity. This results not only in higher efficiencies but provides a mechanism of converting slow and inefficient luminescent center to a fast and efficient recombination center. This confinement of a localized impurity-atom in a quantum dot leads to interaction among the excited states of the atom and the host. This size dependent interaction, which depends on the size of the quantum dot and also on the nature of the atomic impurity offers multitudes of applications. The science of confined atoms in nanocrystalline solids is expected to revolutionize the science of artificially created materials.

At present several applications of these DNC materials are being developed by Nanocrystals Technology. In brief, we exploit the following key properties of DNC for various technological applications a) high absorption, b) high efficiency, c) ultrafast speed and d) ultra-high density. Some of the applications are :

Displays The rare-earth and transition metals yield efficient phosphors. In active displays such as CRT's, we need high intensity without light-saturation for next generation High definition TV (HDTV). Since DNC phosphors are expected to be superlinear phosphors due to its ultrafast luminescent recombination speed, saturation for HDTV can be eliminated. In case of a flat-slim display such as the Field Emission Display (FED), the DNC phosphors can be excited at much lower voltages (~ 1kV), which results in a reduced TV-depth from 18" to 1/2". Other potential applications would be in Plasma and Electroluminescent displays.

Sensors The DNC particles can yield ultrafast sensors since the transfer rate from the host to quantum dots is < 20 psec. These ultrafast sensors could be also used for X-ray, (γ-ray detection and as a fast scintillator phosphor.   Future DNC materials may be used as photoelectric (solar-cells) sensors.

Lasers The impurity in the DNC materials retains its own atomic levels within the band gap of the host DNC material.   Since the carrier trapping to these impurity states is fast compared to recombination times in DNCs, these states can be considered as intrinsic part of the DNC host states.   This combined host and impurity energy levels scheme, as shown in Fig. 4, is very similar to a four-level energy level scheme of an efficient gas laser.   An additional advantage these doped nanocrystalline materials possess over their bulk counterpart is that available high density of states of the host, prevents the excited state absorption in the rare-earth atoms, as shown in Fig. 4.   This absorption within the hosts provides a large excitation cross-section and simultaneous rapid transfer to the impurity.   We believe that a new class of lasers will be fabricated from these doped quantum dots.

There are many other applications which involve implementation of these DNC materials directly in systems.   The key issue being the handling, incorporation in matrices and fabrication of device-structures.   To achieve these in a cost-effective manner, new processes will have to be developed to handle these ultra-small particles.

The potential applications of these doped nanocrystalline materials have yet to be exploited.   The 'caged atom' in a nanocrystal behaves like 'frozen gas'.   This suggests that the plasma like conditions can be achieved in DNC materials at room temperature.   It should be noted that the size of these particles are similar to the size of critical components of the biological systems.   Thus, by studying DNC materials, we may begin to unravel the physics of the biological systems (17), a major goal for the materials scientists in the 21st century.

ACKNOWLEDGMENTS

The work reported is a result of stimulating collaborative effort between many colleagues listed in the various publications, and specially to Alexander Efros and Alexei Ekimov for the model of the quantum confined atom.   I am indebted to them as well as to my coworkers Tim Goldburt, Rolande Hodel, Bharati Kulkarni, Victor Veliadis and Vishal Chhabra at Nanocrystals Technology, for their contributions which has led to the model of quantum confined atom.

References

[1]   See, for examples, the review articles in *Spectroscopy of Isolated and Assembled Semiconductor Nanocrystals*, Eds. L. E. Brus, A. L. Efros, and T. Itoh, J. Lum. **70**(1-6), 1-484 (1996).

[2]   A. D. Joffe, Advances in Phys. **42**, 173 (1992).

[3]   L. Brus, J. Phys. Chem. **90**, 2555 (1986) and references cited therein.

[4]   Y. Wang and N. Herron, J. Phys. Chem. **95**, 525 (1991) and references cited therein.

[5] P. Alivisatos, Materials Res. Bulletin p.23 Aug. 1995.

[6] R. N. Bhargava, D. Gallagher, X. Hong, and A. Nurmikko, Phys. Rev. Lett. **72**, 416 (1994).

[7] R. N. Bhargava, D. Gallagher, and T. Welker, J. Lum. **60**, 275 (1994).

[8] R. N.Bhargava, J. Lum. **70**, 85 (1996).

[9] D. Gallagher, W. E. Heady, J. M. Racz, and R. N. Bhargava, J. Mater. Res. **10**, 870 (1995).

[10] E. T. Goldburt and R. N. Bhargava, Electrochemical Proceedings, **95-25**, 368 (1995).

[11] E. T.Goldburt, B. Kulkarni, R. N. Bhargava, J. Taylor, and M. Libera, Mat. Res. Soc. Symp. **424**, 441 (1997).

[12] R. N. Bhargava, Proceedings of the 1996 Int. Conf. on Luminescence, Prague (to be published).

[13] *Luminescent Materials*, Eds. G. Blasse and B. C. Grabmaier (Springer-Verlag, Berlin, 1994), Ch. 2.

[14] A. Daud, M. Kitagawa, S. Tanaka, and H. Kobayashi, Tattori University Report, (Japan) **25**, 153 (1994).

[15] The model of QCA and QMF were evolved out of the discussions with A. L. Efros and A. Ekimov (private communication).

[16] Y. Kayanuma, Phys. Rev. B **38**, 9797 (1988).

[17] R. P. Feynman, IEEE Journal of Microelctromechnical Systems **1**, 60 (1992).

SIZE DEPENDANT PROPERTIES AND LIGHT EMISSION FROM NANOCRYSTALLINE SEMICONDUCTORS

S. N. Sahu, S. N. Sarangi, D. Sahoo, B. Patel, S. Mohanty,[1]
and K. K. Nanda
Institute of Physics
Sachivalaya Marg, Bhubaneswar-751005, India

ABSTRACT

Nanocrystalline semiconductors viz. CdS, CdSe and PbS grown respectively by chemical and electrochemical techniques are characterized using RBS, PIXE, AFM, optical micrograph and XRD analysis. The growth pattern of the CdSe nanocrystallites appear to follow a diffusion limited aggregation and associated with fractals. Size quantization effects have been observed through optical absorption and Raman scattering studies which gave blue shift and size dependant properties. Photoluminescence studies show red shift with increasing crystalline size. CdS/Au junction interface have been studied through capacitance-voltage measurements and the size dependant properties have been evaluated. CdS nanocrystallites exposed to UV light at room temperature emit red and green emission.

1. INTRODUCTION

Recent studies on optical, electrical and interfacial behavior of nanocrystalline semiconductors belonging to II-VI [1, 2] group show unusual size dependant properties. As one narrow down the crystalline size, the continuum electronic bands break into discrete states which shift to higher energies and results in widening of the band gap. When the size becomes comparable to exciton Bohr radius, their electronic wave functions experience three-dimensional quantum confinement and the effects can in general be observed in their optical absorption, Photoluminescence and Raman studies. Nanocrystalline semiconductors showing quantum size effect when appropriately doped by an impurity and exposed to photons of higher energy, the photo-generated carriers

---

[1]*Permanent address:* P. G. Department of Physics, Utkal University, Bhubaneswar-751004, India.

undergo an efficient radiative recombination due to their strong wave function overlap between the host and the impurity and emits photons in the visible range whose color can be monitored by the crystalline size and the impurity state [2].

2. EXPERIMENTAL DETAILS

Nanocrystalline semiconductors viz. CdS, CdSe and PbS are grown respectively by chemical [3, 4] and electrochemical [5] techniques. The physical properties are evaluated through Rutherford Back Scattering (RBS), Proton Induced X-ray Emission (PIXE), Atomic Force Microscope (AFM), optical micrograph and X-ray Diffraction (XRD) studies whereas the optical properties are evaluated through optical absorption, Photoluminescence (488 nm Ar^+ laser) and Resonant Raman (RR) (with different excitation by Ar^+ laser) scattering studies. The interfacial and some semiconductor parameters are evaluated from Mott-Schottky plots using variable frequency LCR meter. Finally the nanocrystalline samples at 300^0K are exposed to UV lamps to witness the visible light emission.

3. RESULTS AND DISCUSSIONS

3.1. Physical Properties

The crystalline size of the chemically prepared CdS were controlled by its thickness/ reaction time period. Higher thickness yield larger crystallites. For the electrodeposited samples the sizes are controlled by their electrolysis current density, pH and thickness. A batch of CdS samples prepared have thicknesses as 1.4 μm, 75 nm, 50 nm and 35 nm and are designated as S1, S2, S3, and S4 respectively. RBS analysis show stoichiometry with S1 whereas Cd-excess have been detected as one narrow down the thickness. RBS analysis also show the surface oxygen. PIXE analysis identified no impurity even at ppm level. XRD analysis show peak broadening with nanocrystalline samples and associated with mixed cubic and hexagonal phases of the deposit.

The AFM images of CdS samples S1 and S3 (Fig. 1) clearly show the non-spherical nature of the crystallites. In order to estimate the crystalline size, a random orientation of crystallites is being assumed. The linear dimension of crystallites along only one direction of the micrographs have been estimated. For every sample, such measurement was made for more than a hundred particles and the results were averaged. The average crystalline sizes obtained are 70 and 10 nm respectively, for samples S1 and S3.

(a)

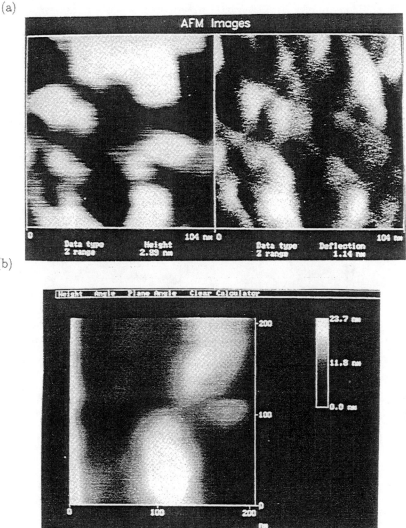

(b)

Figure 1. AFM images of CdS samples S3 (a) and S1 (b). The verage crystalline sizes 70 and 10 nm respectively.

The optical micrograph of CdSe and CdS samples are shown in Fig. 2. These micrographs reveal the fractal growth of nanocrystallites whereas the bulk show to consist of finer interconnected particles with a non-uniform surface topography. The growth of electrodeposited CdSe nanocrystalline samples follow a diffusion limited aggregation (DLA) growth pattern. The dimension of the fractal is found to be 1.65±.04.

Nanocrystallites samples giving a fractal growth show alone quantum size effect to be discussed in later section.

(a)

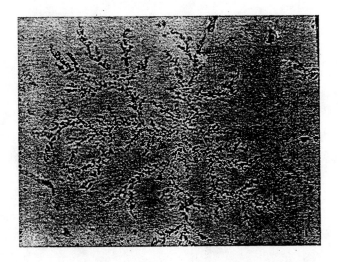

(b)

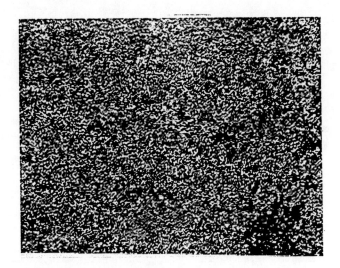

(c)

(d)

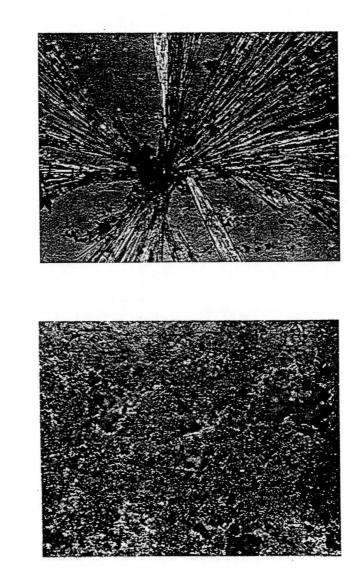

Figure 2. Optical micrograph of CdSe (a & b) and CdS (c & d). The fractals are associated with nanocrystalline samples. For CdSe nanocrystalline samples, the fractal dimension is 1.65 ±0.04.

3.2. Optical Absorption

Fig. 3(a) shows the room temperature absorption spectra of CdS samples, having different crystalline sizes. Usual interband transition is seen with bulk, S1. As the crystal size decreases, there is a blue shift of the absorption edge along with an oscillatory structure, which indicates the size quantization effect [6, 7]. It shifts from ~515 nm for bulk, S1 to ~417 nm for nanocrystalline sample, S4. An absorption peak at 395 nm for CdS sample, S4 agreeing well with Salata et al [8] who have synthesized their CdS nanoparticles in polymer matrix with average size ~ 45 A°. This peak is due to 1S–1S quantum particle transition as suggested by Brus [9]. Our nanocrystalline sample, S4 shows a shoulder (weak band) around 460 nm in good agreement with earlier reported results [10]. However, with increasing crystalline size the shoulder shifts to a relatively lower wavelength and disappears in bulk crystallites as seen from Fig. 3(a). In addition to the peak at 395 nm, another peak at 346 nm is also observed. The energy difference between these two peak is 0.47 eV and the difference depends on the crystalline size [11]. Each peak of the absorption spectrum correspond to transitions to different excited states of the conduction band [12, 13]. Our overall observations show the crystalline size of sample S4 to be in the range 45 to 50 A° estimated from the blue shift. Note that the crystalline sizes estimated from AFM and from blue shift do not agree well.

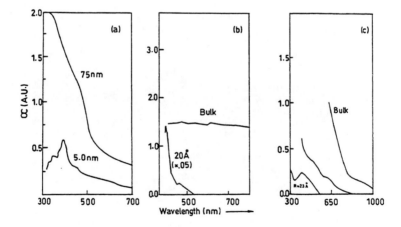

Figure 3. Optical absorption spectra of (a) CdS; (b) PbS and (c) CdSe.

The optical absorption spectra of PbS nanocrystalline thin films grown on ITO by electrodeposition technique is shown in Fig. 3(b). Note that the absorption threshold shifts to the lower wavelength with decreasing crystalline size achieved through control of electrolysis temperature and current density. We could get a band gap of 2.9 eV for PbS nanocrystallites prepared at 2 mA/cm^2 current density and electrolyte temperature 273 ^{0}K. 2.9 eV band gap corresponds to a crystalline size of ~2.0 nm indicating the electron-hole confinement with absent of excitonic structure in good agreement with earlier reported result [1]. The absence of excitonic structures may be attributed to the weak exciton binding energy due to strong Coulomb screening and broad size distributions. So far the report shows that excitonic structures have been observed only in large band gap semiconductors such as ZnS and CdS where exciton binding energy is large. In narrow gap semiconductors, the large exciton radius and the strong screening give small exciton binding energy and the absence of excitonic structures in the absorption spectrum.

Fig. 3(c) shows the optical absorption spectra of CdSe bulk and nanocrystalline film prepared on ITO conducting glass by electrodeposition technique. With decreasing crystalline size through a control of electrolysis current density, a shift in the absorption threshold towards lower wavelength along with oscillatory structure is clearly seen. We can change the band gap from 1.74 eV to 2.54 eV.

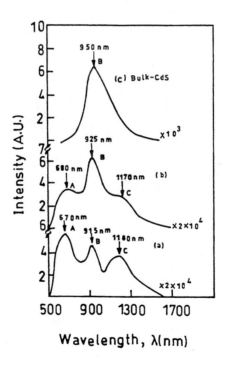

Figure 4. PL spectra of CdS having different crystalline sizes, 75 nm (a); 7.5 nm (b) and 5.0nm (c). The excitation wavelength is 488 nm Ar$^+$ laser.

3.3. Photoluminescence

The Photoluminescence (PL) spectra of CdS samples of different crystalline size are shown in Fig. 4. As shown in this figure, we observed three peaks at 670, 925 and 1160 nm respectively for sample S4 marked as A, B and C. Whereas, a single peak (B) is observed in case of bulk-CdS, S1. No green edge luminescence was observed for any of our CdS samples in contrast to others [14]. The absence of green edge PL may be due to very low PL intensity [15] or due to wavelength limitation with InGaAs detector used for PL detection. Note that with increasing crystalline size there is a clear red shift associated with peaks A and B attributed to the size and impurity effect where as the shift of the peak C contrary to A and B is attributed to the surface related band [16].

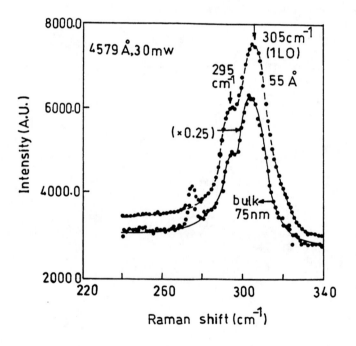

Figure 5. Resonant Raman spectra of nano- and bulk-CdS samples. The excitation wavelength is 457.9 nm.

3.4. Raman Scattering Studies

Fig. 5 shows the RR spectra of CdS bulk, S1 and nanocrystalline, S3 samples. Note that a low frequency wing supposed to be surface phonon (SP) contribution is clearly seen at 295 cm^{-1}. A similar SP contribution at 292 cm^{-1} also seen with polymer capped sample [17]. Further, one can clearly see an enhancement of 1LO peak intensity in case of nanocrystalline CdS compared to the bulk as shown in Fig. 5. However, upon increasing the excitation from 457.9 to 514.5 nm, the low frequency wing shifts to higher frequency ascribed to the contribution from surface related defects.

3.5. C-V Measurement

Capacitance - Voltage measurements with configuration ITO/CdS/Au were carried out in order to estimate semiconductor parameters and evaluate interfacial properties. Shown in Fig. 6 are the Mott-Schottky plots ($1/C^2$ Vs V) at 100 kHz for bulk-CdS, S2 and nanocrystalline CdS film, S4. From the slope and intercepts, the donor concentration, N_D, built-in potential, V_{bi} and depletion width, W are estimated and found to be : $N_D = 3.52 \times 10^{15}$ cm^{-3}, 0.18 V and 11.12 nm (S4) $N_D = 2.67 \times 10^{15}$ cm^{-3}, 0.75 V and 21.1 nm (S2) These data suggests an increase in barrier height and depletion width with increasing crystalline size, whereas, the carrier concentration follow an opposite trend.

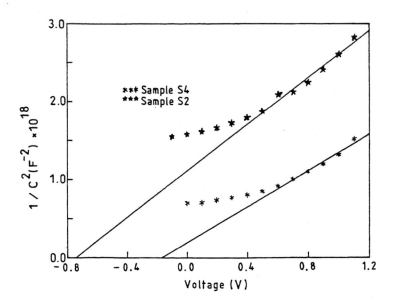

Figure 6. Mott-Schottky plot (100 kHz) of CdS/Au junction.

4. LIGHT EMISSION FROM NANOCRYSTALLITES

Bulk semiconductors when appropriately doped with impurities and excited with high energy photon, luminescent efficiently and has large luminescence life time. However, when the crystalline size is reduced to nanometer range and appropriately doped, yield efficient luminescence with shorter life time due to increased oscillator strength and good wave-function overlap between the host and the impurity.

Rutherford Back Scattering (RBS) analysis of our CdS samples show excess Cd in the nanocrystalline sample, S4 which in principle should act as an acceptor in CdS [19]. Thus, if these nanocrystallites are now excited with higher energy photons, one expects to see visible light emission whose color would depend upon the crystalline size and the trap levels. It is observed that the bulk-CdS powder do not emit any light and look black whereas the nanocrystalline powders yield red as well as green emission at 300 ^{0}K when exposed to UV light, suggesting a size distribution in the nanocrystalline CdS samples. CdS sample, S4 prepared on filter paper cooled to 80 ^{0}K exposed to UV light the red emission intensity is found to increase by many fold and appears as if a 25 W bulb is lighted-up. Our nanocrystalline sample, S4 when capped with PVA and arrested its growth, emits exclusively green light under UV.

ACKNOWLEDGMENTS

Helpful discussions with Prof. S. N. Behera and Prof. V. S. Ramamurthy is gratefully acknowledged.

References

[1] Y. Wang and N. Herron, J. Chem. Phys. **91**, 257 (1987).

[2] R. W. Seigel, Scientific American **275(6)**, 74 (1996).

[3] S. N. Sahu, J. Mat. Sc. Mater. Electron. **6**, 43 (1995).

[4] K. K. Nanda, S. N. Sarangi and S. N. Sahu, Current Science **72**, 110 (1997).

[5] R. K. Pandey, S. N. Sahu and S. Chandra, *Hand Book of Electrodeposition* (Marcel Dekkar, Inc., New York, 1996), p.205.

[6] L. E. Brus, J. Phys. Chem. **90**, 2555 (1986).

[7] A. Henglein, Chem. Rev. **89**, 1861 (1989).

[8] O. V. Salata, P. J. Dobson, P. J. Hull and J. L Hutchinson, Thin Solid Films **251**, 1 (1994).

[9] L. E. Brus, IEEE J. Quantum Electron. **22**, 1909 (1986).

[10] P. Hoyer, N. Baba and H. Masuda, Appl. Phys. Lett. **66**, 2700 (1995).

[11] Al. L. Efros and A. L. Efros, Sov. Phys. Semicond. **16**, 722 (1982).

[12] Ch. -H. Fischer, H. Weller, L. Katsikas and A. Henglein, Langmuir 5, 429 (1989).

[13] A. I. Ekimov, Al. L. Efros, A. A. Onushchenko, Solid State Commun. **56**, 921 (1985).

[14] T. Arai, T. Yoshida and T. Ogawa, J. Appl. Phys. **26**, 396 (1987).

[15] T. Arai, H. Fujumura, I. Umezu, T. Ogawa and A. Fujii, J. Appl. Phys. **28**, 484, (1989).

[16] M. Agata, H. Kurase, S. Hayashi and K. Yamamoto, Solid State Commun. **76**, 1061 (1990).

[17] S. K. Deb, Solid State Physics **36C**, 268 (1994).

[18] R. N. Bhargava et al., Phys. Rev. Lett. **72**, 416 (1994).

[19] S. M. Sze, *SEMICONDUCTOR DEVICES: Physics and Technology* (John Wiley and Sons, New York, 1985).

APPLICATIONS OF GaAs NANOWIRE CRYSTALS

T. Katsuyama, K. Hiruma, K. Haraguchi, M. Shirai, K. Hosomi,
and J. Shigeta
Central Research Laboratory, Hitachi Ltd.
Kokubunji Tokyo 185, Japan

ABSTRACT

Ultrathin GaAs wire crystals as thin as 15-40 nm have been grown by metal-organic vapor-phase epitaxy. The growth site on the substrate can be artificially controlled by positioning Au-alloy droplets. The lateral growth of these wire crystals was also demonstrated by using nanolithography techniques.

Furthermore, intense light emission from p-n junctions formed in GaAs nanowire crystals was observed, thus showing the possibility of the realization of ultra-small light emitting diodes. Current-voltage (I-V) characteristics of nominally non-doped GaAs nanowire crystals were also studied. Typical I-V curve is approximated with the polynomial form of $V^{1.5}$. This property suggests that a space-charge layer is formed in the wire crystal, thus demonstrating the possibility of new electronic devices.

1. INTRODUCTION

The development of semiconductor fabrication techniques facilitates the construction of one-dimensional structures that can confine electrons based on electric potentials. Various fabrication methods have been proposed to make quantum wires [1-5]. However, it is still difficult to fabricate ideal quantum wires because these methods introduce a lot of problems, such as the difficulty of controlling film thickness, the roughness of or damage to etched surfaces, and the inaccuracy of photolithographic techniques. Therefore, in an effort to solve those problems, we have proposed a novel nanostructure fabrication method. This method can fabricate very small needle-shaped wire crystals (whiskers) [6-13]. Since these wire crystals exhibit excellent two-dimensional quantum confinement effects of carriers, they are potentially applicable to some opto-electronic devices such as quantum wire transistors and quantum wire lasers.

Here, we demonstrate the fabrication method of these nanowire crystals. In particular, the controllability of the growth site and direction is discussed [10-12]. The electrical and optical properties of such nanowire crystals are also discussed in connection with the possibilities of the applications of these wire crystals to opto-electronic devices. GaAs nanowire crystals with p-n junctions are discussed together with nominally nondoped GaAs nanowire crystals with uniform compositions [7, 9].

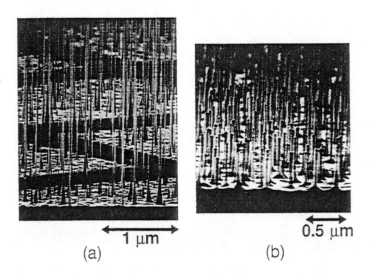

(a)   1 μm   (b)   0.5 μm

Figure 1. Typical nanowire crystals grown on a substrate.

2. FABRICATION METHOD

2.1. Mechanism

Typical nanowire crystals grown on a substrate are shown in Fig. 1. These wire crystals are composed of InAs. Fig. 1(a) shows relatively long wire crystals. On the other hand, Fig. 1(b) shows relatively short wire crystals. Nanowire crystals composed of GaAs can also be grown. The growth procedure of the wire crystals is shown schematically in Fig. 2. This procedure is similar to those reported by Wagner and Ellis for Si whiskers [14].

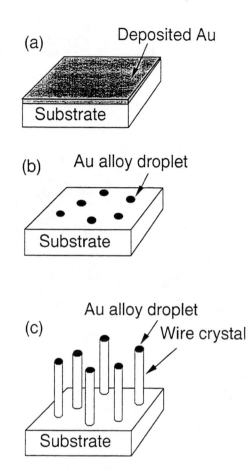

Figure 2. Growth procedure of wire crystals.

However, our wire crystal is extremely small compared to the Si whisker. In our procedure, Au was first deposited on a GaAs substrate surface by vacuum evaporation, thus forming Au clusters. The average thickness of the deposited Au was 0.1 nm. After Au deposition, the substrate was annealed at 500°C for about 10 min, and Au alloy droplets formed on the substrate. The wire crystals were grown by using a conventional reduced-pressure metal-organic vapor-phase epitaxy (MOVPE) method. Trimethylgallium (TMG) and arsine (AsH_3) were supplied as source materials and the growth temperatures ranged from 350 to 600°C. The wire crystal growth is based on the vapor-liquid-solid (VLS) growth model [14]. Wire crystal growth begins with gallium atoms from the Au-Ga-As alloy droplets, sticking to the surface arsenic atoms on the substrate. These droplets are in a saturated phase so they absorb source materials from the vapor during growth. This might be attributable to the difference in the vapor pressures of gallium and arsenic atoms. Actual alloy droplet can be seen on the top of the wire crystals, as shown in Fig. 3.

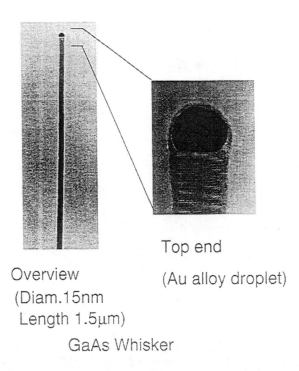

Top end
(Au alloy droplet)

Overview
(Diam.15nm
Length 1.5μm)

GaAs Whisker

Figure 3. Individual GaAs nanowire crystal.

The wire crystal grows in the same direction as that of the As dangling bond of the substrate surface [10]. Therefore, when a (111)B substrate is used, the wire crystal will only grow perpendicular to the substrate. The wire crystal shape and size relates to the growth conditions. The lateral size (diameter) increases as the deposited Au thickness, growth temperature, and growth time increase. The length of the wire crystal, however, is almost constant irrespective of the Au thickness. Furthermore, the growth temperature plays an important role in controlling the wire crystal shape. At higher temperatures (460-500°C) the wire crystal is needle-shaped, while, at lower temperatures (about 420°C) the wire crystal shape is cylindrical.

2.2. Growth Control

Here, we demonstrate artificial control of the growth position of the wire crystal [11]. Such control is essential for various device applications of nano wire crystals. The growth procedure for site-controlled wire crystals is shown schematically in the inlet of Fig. 4. We controlled the location by using a SiO_2 window mask formed by electron-beam lithography. After patterning of the window on the electron beam (EB) resist layer (PMMA), Au atoms with an average thickness of 1 nm were deposited by vacuum evaporation on both the window region and on the region masked with the resist layer. The EB resist layer was removed together with the Au clusters on the masked region. Next, the sample with the SiO_2 mask was annealed at 500°C for about 10 min. The annealing process formed a single alloy droplet inside the windows only. After removing the SiO_2 mask, the sample was placed in an MOVPE chamber to grow the nanowire crystals. Source gases were trimethylgallium and AsH_3. Fig. 4 shows a scanning electron microscope image of site-controlled GaAs wire crystals grown on a GaAs (111)B substrate. Wire crystals with diameters of about 80 nm are exactly placed at the four corners of the $2\mu m \times 2\mu m$ square. The growth direction of these wire crystals is [111]B only; this result coincides well with the results showing that nanowire crystals grow preferentially in the $<111>$ As dangling bond direction which is perpendicular to the (111)B substrate surface.

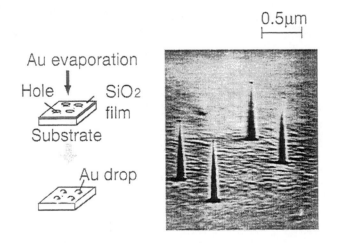

Figure 4. Site-controlled GaAs nanowire crystals.

Next, fabrication methods for making lateral nanowire crystals are described [12]. To grow the nanowire crystals laterally, Au-alloy droplets must be formed on the side wall of a ridge formed on the substrate. This was done by the following method. We selected a GaAs substrate whose [111] As direction was parallel to the surface and the SiO_2 pattern side edge is normal to the [111] As direction (Fig. 5(a)). GaAs (110) and (211) As substrates were used. The SiO_2 masking pattern was formed by conventional photolithography. Then the substrate was partially etched away and a step (about $5\mu m$ high) was formed. SiO_2 was deposited again on all of the substrate surface except for the GaAs step side. A small amount of Au (less than one monolayer) was evaporated on the substrate, which was then heat-treated at 500°C (Fig. 5(b)). Wire crystals were grown by MOVPE after the SiO_2 mask was removed (Fig. 5(c)). The grown wire crystals were about 30 - 60nm in diameter and about 2μm long. Fig. 5(d) shows a scanning electron microscope (SEM) image of the GaAs nanowire crystals, and a schematic configuration.

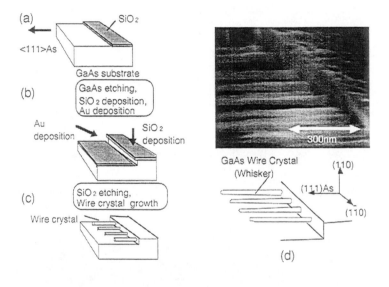

Figure 5. Lateral growth of GaAs nanowire crystals.

Bridge-type nanowire crystal arrays can be fabricated by using a technique similar to the one described above [12]. The substrate used for this experiment has a shallow 1μm-wide ditch. One side wall of the ditch has a (111) As face, and the other side wall has a (111) Ga face. Since the ditch is so narrow, the whole ditch is subject to Au diffusion. Since one side wall has a (111) Ga face, nanowire crystals will grow only on the side wall with the (111) As surface even if the Au atoms diffuse over the entire area. Fig. 6 shows the bridge-type GaAs nanowire crystal arrays we obtained. It took 2,000s for the top of a wire crystal to reach the opposite side. The time is one order of magnitude longer than that needed for conventional wire crystal growth. This is because only a small amount of source gases arrive at the side wall which has a small area relative to the flat surface. Thus we can control the position and direction of the wire crystal growth. These are very important for device applications of nanowire crystals.

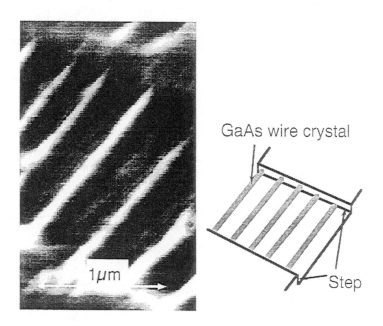

Figure 6. Bridge-type GaAs nanowire crystals.

3. APPLICATIONS

3.1. GaAs nanowire crystals with p-n junctions

The wire crystals were grown on n-type GaAs(111)B substrates by AsH_3 and TMG. The substrates were doped with Si of $1x10^{18}$ cm^{-3}. The growth occurred along the [111] direction, such that the wire crystal axes were just perpendicular to the substrates. The length of the wire crystals is about 2 μm and the diameter is about 100 nm. During the growth process, doping species were changed in order to obtain the p-n junction at the middle point of the wire crystal [7, 9]. During the first half of the growth, silicon doping was carried out using disilane (Si_2H_6), resulting in n-type doping of about of $1x10^{18}$ cm^{-3}. Carbon doping followed, performed naturally due to the small supplying rate of AsH_3/TMG during the latter half, thus leading to p-type doping of about $8x10^{18}$ cm^{-3}. Carbon doping is advantageous because it prevents the self diffusion of the dopant and reduces the diffusion of Zn from the Au/Zn electrode used for the ohmic contact.

Fig. 7 schematically illustrates the fabricated diode structure. There are considered to be 3.3×10^6 crystals involved in this device with 1x1 mm area. A conventional Au/Ge/Ni electrode was employed as the ohmic contact on the substrate side. Au/Zn electrodes were fabricated as ohmic contacts at the top of the p-type doped wire crystal. The wire crystals were buried with spin-on-glass (SOG).

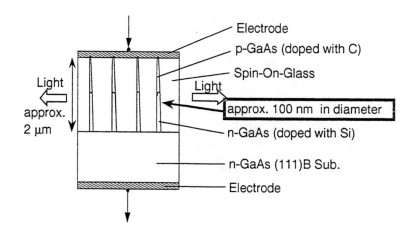

Figure 7. Diode structure made of GaAs nanowire crystals with p-n junctions.

We measured the electrical characteristics of the devices. Fig. 8 shows the current-voltage (I-V) characteristics of the fabricated diode shown in Fig. 7. A sudden rise in current was obtained at 0.6 V in the forward bias. Moreover, soft breakdown occurred at nearly -2.5 V in the reverse bias. Thus the device shows the non-linear I-V characteristics inherent to the p-n junction.

Such a p-n junction can be well characterized by electroluminescence measurements [7, 9]. Emission spectra were measured at 77 K, while current was being applied as shown in Fig. 9. It can be seen from the figure that emission intensity varies linearly with injection current density. The peak position for all three cases is 1.485 eV (835 nm). Full width at half maximum remained constant at 0.051 eV even with varying current. The peak position shows that the emitted light comes from the p-n junction, because the peak position in the case of emission from GaAs bulk is estimated to be at 1.483 eV (836 nm), taking the shallow doping level of carbon into account. The difference in the peak positions between the bulk and the wire might be considered to result from the quantum effect. Still more clear evidence of the quantum confinement effect is the polarization dependence of the electroluminescence. In Fig. 10, the electroluminescence spectra at 77 K are shown [15]. The solid line represents

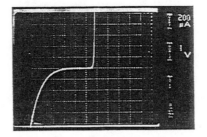

Figure 8. Current-voltage characteristics of fabricated diode.

the electroluminescence spectrum for the light whose electric field vector is parallel to the wire crystal axis, while the dashed line indicates the spectrum of the electric field vector perpendicular to the wire crystal. The intensity of *perpendicular* light is 70% smaller than that of *parallel* light at 77 K. The measurement at 4.2 K shows a similar ratio (76%). In the case of bulk samples with p-n junctions, we could not find any polarization anisotropy at either temperature. Therefore, these results suggest the existence of the two-dimensional quantum confinement effect of the carriers. However, it should be noted that the polarization anisotropy is much stronger than the measured ratio because of the waveguide structure of the sample having lossy metal electrodes. The polarization anisotropy for each individual wire crystal was estimated to be 20% based on the calculation considering sample geometry. This value is in good agreement with the value given by a cylindrical approximation theory proposed by Sercel *et al*. These results are the evidence of the quantum confinement effect of the carriers.

3.2. Nominally non-doped GaAs nanowire crystals

We measured the current-voltage characteristics of nominally non-doped GaAs nanowire crystals [16]. The fabrication procedure for the samples was as follows. We used an n-type GaAs (111)B substrate doped with 1×10^{18} cm^{-3} Si. First, circular windows (about 1μm in diameter) were formed in a SiO$_2$ masked area. A very minute quantity of gold (less than one monolayer) was then deposited on the substrate and the sample was annealed at 500°C. After that, wire crystals were grown in the < 111 > As direction on the GaAs substrate surface. The grown wire crystals are slightly tapered, so the diameter used here is defined as the diameter at the midpoint of a wire crystal. The diameter was about 100 nm. As the next step, the wire crystals were buried with transparent glass material (spin-on glass), and baked at 200°C. An electrodes (Au/Ge/Ni) was then formed at the tops of the wire crystals and another was formed on the substrate. Fig. 11 shows a fabricated sample used for electrical measurement as well as its schematic illustration. In the following, the positive current direction is defined as being from the top to the bottom of a wire crystal.

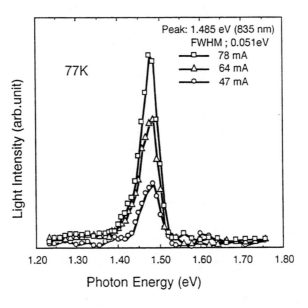

Figure 9. Emission spectra for GaAs nanowire crystals with p-n junction under forward biased conditions.

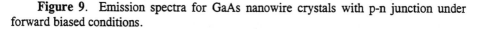

We measured the electrical properties of wire crystals at room temperature after an alloy treatment of the Au/Ge/Ni electrodes. In this case, the alloy treatment consisted of heating the sample at temperatures ranging from 300°C to 500°C in a H_2 atmosphere. When samples were not alloyed, no current was obtained. Schottky-like characteristics were obtained from samples alloyed at 390°C. These two results indicate that the current flows within the wire crystal itself, but does not flow at the surface of the wire crystal. We obtained Ohmic contact from the sample alloyed at 430°C.

Then, we measured the I-V characteristics at 77 K and at 4.2 K. Though the I-V curve at room temperature is almost linear, the I-V curve at 77 K shows nonlinear characteristics. The typical I-V curve (Fig. 12) is approximated with the polynomial form of $V^{1.5}$ rather than the exponential form appeared in the diffusive transportation of the carriers. This property suggests that a space-charge layer is formed in the wire crystal. This may be due to both the formation of the n-i-n structure along the wire crystal and the narrowness of the current path in the wire.

Furthermore, we observed a reproducible step-like current fluctuation at 77 K in some samples (Fig. 13). The step height is as high as 0.4 μA. The arrow along the curve in Fig. 13 indicates the bias direction. We believe the steps in the current are caused by changes in the current path. The dotted lines correspond to the I-V curve for each individual wire crystal given the current flow in this sample. Note that the typical current step (0.4 μA)

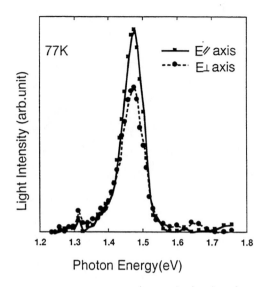

Figure 10. Polarization dependence of the emission spectra for GaAs nanowire crystals with p-n junctions.

corresponds to a change in the current density as large as 400 A/cm². This seems to be a result of the transition of the current path due to the effect of charges trapped around the surface of the wire crystal.

4. SUMMARY

Ultrathin GaAs wire crystals as thin as 15-40 nm have been grown by metal-organic vapor-phase epitaxy. The growth mechanism is based on the vapor-liquid-solid (VLS) growth model. Au-alloy droplets play an essential role for the crystal growth. The crystal growth site on the substrate can be artificially controlled by positioning the Au-alloy droplets. The lateral growth of these wire crystals have also demonstrated by using nanolithography techniques.

Furthermore, GaAs nanowire crystals with p-n junctions have been studied. Intense light emission from the p-n junction was obtained, thus showing the possibility of the realization of ultra-small light emitting diode. We also measured the current-voltage (I-V) characteristics of nominally non-doped GaAs nanowire crystals. Typical I-V curve is approximated with the polynomial form of $V^{1.5}$. This property suggests that a space-charge layer is formed in the wire crystal, thus demonstrating the possibility of new electronic devices.

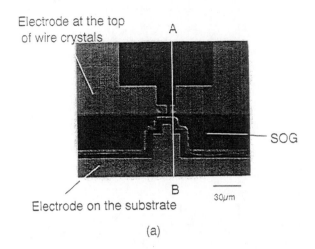

(a)

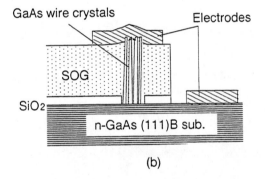

(b)

Figure 11. Fabricated sample with nominally nondoped GaAs nanowire crystals. (a) Top view. The line AB corresponds to the cross-section in Fig. 11(b). (b) Cross-sectional view.

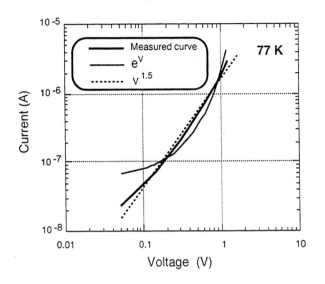

Figure 12. Current-voltage characteristics of GaAs nanowire crystals when electrons are injected from the top of the nanowire crystals. The thick solid line is a measured curve, the thin gray line is the approximation using the exponential of voltage, and the dotted line is the approximation using the polynomial.

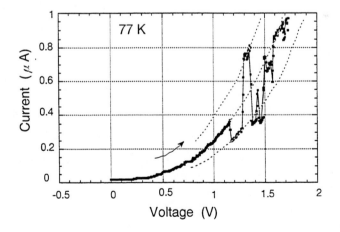

Figure 13. Current-voltage characteristics of GaAs nanowire crystals showing the reproducible current fluctuation. The arrow along the curve shows the bias direction.

ACKNOWLEDGMENT

This work was performed under the management of FED as a part of the MITI R&D Program (Quantum Functional Devices Project) supported by NEDO.

References

[1] M. Tanaka and H. Sasaki, Appl. Phys. Lett., 54, 1326 (1989).

[2] P. M. Petroff, A. C. Gossard, and W. Wiegmann, App. Phys. Lett. 45, 620 (1984).

[3] P. M. Petroff, A. C. Gossard, R. A. Lorgan, and W. Wiegmann, Appl. Phys. Lett. 41, 635 (1982).

[4] N. Randoll, M. A. Reed, and Y. C. Kao, J. Vac. Sci. Technol. B 8, 1348 (1990).

[5] Y. Nagamune, Y. Arakawa, S. Tsukamoto, and M. Nishioka, Phys. Rev. Lett. 69, 2963 (1992).

[6] K. Hiruma, T. Katsuyama, K. Ogawa, M. Koguchi, and H. Kakibayashi, Appl. Phys. Lett. 59, 431 (1991).

[7] K. Haraguchi, T. Katsuyama, K. Hiruma, and K. Ogawa, Appl. Phys. Lett. 60, 745 (1992).

[8] K. Hiruma, M. Yazawa, K. Ogawa, T. Katsuyama, M. Koguchi, and H. Kakibayashi, J. Appl. Phys. 74, 3162 (1993).

[9] K. Haraguchi, T. Katsuyama, and K. Hiruma, J. Appl. Phys. 75, 4220 (1994).

[10] K. Hiruma, M. Yazawa, T. Katsuyama, K. Ogawa, K. Haraguchi, M. Koguchi, and H. Kakibayashi, J. Appl. Phys. (Appl. Phys. Rev.) 77, 447 (1995).

[11] T. Sato, K. Hiruma, M. Shirai, K. Tominaga, K. Haraguchi, T. Katsuyama, and T. Shimada, Appl. Phys. Lett. 66, 159 (1995).

[12] K. Haraguchi, K. Hiruma, T. Katsuyama, K. Tominaga, M. Shirai, and T. Shimada, Appl. Phys. Lett. 69, 386 (1996).

[13] K. Hiruma, H. Murakoshi, M. Yazawa, and T. Katsuyama, J. Cryst. Growth, 163, 226 (1996).

[14] R. S. Wagner and W. C. Ellis, Appl. Phys. Lett. 4, 89 (1964).

[15] T. Katsuyama, K. Hiruma, T. Sato, K. Ogawa, M. Shirai, K. Haraguchi, and M. Yazawa, Topical Meeting of Quantum Optoelectronics, QThD3-1, Dana Point, California, 1995.

[16] K. Haraguchi, K. Hiruma, T. Katsuyama, M. Shirai, and T. Shimada, 2nd Int. Workshop Quantum Functional Devices, p.26, Matsue, Japan, 1995.

MATERIALS FOR MICROELECTROMECHANICAL SYSTEMS

David J. Nagel
Naval Research Laboratory
Washington, DC 20375-5345, U. S. A.

ABSTRACT

The production of microelectromechanical systems has developed into a significant industry in the recent past. These complex systems challenge materials science and technology in ways both old and new. Initially, materials and processes from the microelectronics industry were employed to make these devices. Then, materials and associated techniques, beyond those needed for integrated circuits, were brought to bear. Now, microsystems offer many opportunities to apply new materials in order to make more capable and less expensive devices, much like the situation for larger-scale engineering systems.

1. INTRODUCTION

Microelectromechanical Systems, commonly called MEMS, are structures, some part of which has micrometer-scale dimensions, that contain both electronic and mechanical functions [1]. They can be thought of as chips with moving parts. The flexible components react to the environment in *sensors*, or controllably influence the environment as *actuators* [2].

MEMS grew out of the microelectronics industry. Currently a $2B a year industry, it is projected to grow rapidly to about $10B by the end of this decade. Many of the same materials and processes employed to produce integrated circuits are used to make MEMS. Silicon is, in fact, a marvelous mechanical material [3]. While MEMS do not contain as many discrete components as integrated circuits, they can be more complex than ICs in their interaction with the environment, the variety of materials employed in their construction and their internal structure.

   A useful way to categorize both ICs and MEMS is to note their number of component parts. Fig. 1 plots T, the number of transistors, versus M, the number of moving parts [4]. It is noteworthy that the microelectronics and the information revolutions of the past few decades involved ever increasingly complex ICs with M=0. Adding the mechanical components (M>0) opens an entire new "space" with rich functionality. The various types of applications indicated by shaded regions in Fig. 1 can be viewed as falling into three groupings. Most fall near the line of equal T and M values. Two of those have to do with optical technologies, first optical switches and aligners, and second computer displays. Another with nearly equal numbers of electronic and mechanical parts involves micro-pumps and -valves. It is noteworthy that microactuators cannot exert large forces in general. However, neither the control of light nor moving fluids through microanalyzers requires strong actuators. The MEMS application in Fig. 1 labeled parts handling involves large numbers of actuators controlled by relatively few transistors in devices designed to move and assemble small and light components. The applications with high T/M ratios, the last category, require relatively more sophisticated electronic functions.

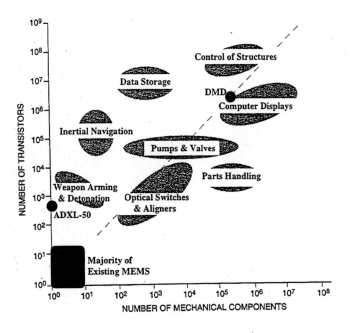

Figure 1. MEMS applications and examples, arrayed according to the number of transistors and moving parts [4].

The three darker shaded areas in the first figure contain noteworthy examples of current commercial MEMS. Many devices with small numbers of T and M are available. The commercial fluister microvalve from Redwood Microsystems Inc. [5], for example, has one electrical and one mechanical component, shown schematically in Fig. 2. Resistive heating expands a trapped fluid, moving a micron-scale silicon membrane to open or close the valve, which is housed in a transistor can. The fluister is built of deep etched silicon, bonded to pyrex. Note that it requires robust, leak-free bonding of these materials. Further, the fluister is necessarily connected to the outside world (the controlled fluid). Many other examples of MEMS with both T and M small are available as prototypes or products.

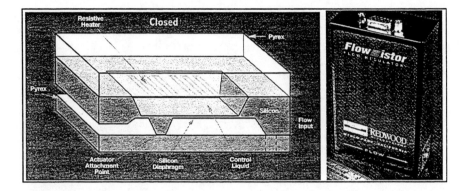

Figure 2. Schematic of the Redwood Microsystems MEMS fluistor, and a photograph of the Flow-istor regulator based on the fluistor.

The MEMS accelerometers made by Analog Devices Inc. [6] for automobile air bag triggers have several hundred transistors but only one moving part. The mass, shown in the center of the chip in Fig. 3, deflects relative to the rest of the structure upon impact. It is suspended by silicon arms, part of the basic wafer, which also serve as plates of an inter-digitated capacitor. The motion causes a capacitance change, which produces the output electrical signal. The ADXL accelerometers, while relatively simple mechanically, have very tight tolerances on the 1.27 μm spaces between the capacitor fingers. Their electronic functionality is relatively complex. The ADXL devices, made primarily by IC processes and sealed in transistor or surface mount packages, sell for under $10 each in large numbers. MEMS accelerometers with differing designs, complexity and materials are available from at least six companies.

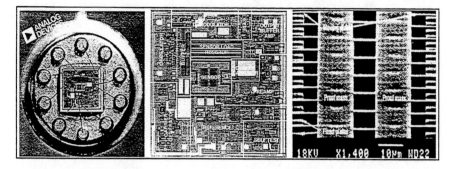

Figure 3. Photograph of the ADXL-50 (left), annotated micrograph of the 3 x 3 mm chip with the multi-function electronics and proof mass (center) and electron micrograph of the micromachined moving part with its supports (right). Note the 10 μm scale marker near the proof mass.

The commercial devices with the most transistors and moving parts are the digital mirror devices (DMD) sold by Texas Instruments, Inc. [7]. They are the heart of a wide range of display devices. DMD projectors for conference rooms are now sold by seven companies. In the future, they may displace cathode ray tubes in computer monitors and televisions, and be used for theater projectors. DMD have large numbers of electrostatically-activated 16 x 16 μm mirrors (pixels) which throw the light from the source either onto the screen or into a beam dump, as shown in Fig. 4.

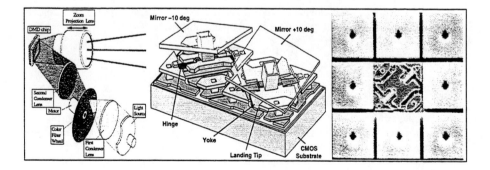

Figure 4. Schematics of the operation of the DMD (left) and of two of the mirrors (center). Electron micrograph of the 16 μm square mirrors; one is removed to show the support layer (right).

Color is obtained with a filter wheel rotating at 60 Hz between the light and the DMD. Brightness is determined for each pixel by the time its mirror places light into the screen for each color cycle. The mirror switching time is less than 20μ sec. Current DMDs have 508.8K (= 848 x 600) pixels, and devices with 1,310.7K (1280 x 1024) pixels planned for production. DMD are sealed in a package with a transparent, coated window and have materials which go well beyond conventional ICs. For example, an organic layer on the two landing pads in each pixel prevents stiction. The two torsional support hinges for each micromirror are about 5 μm long, 1.5 μm wide, and only 60 nm thick. Each of them (over 2 million on some devices) have to twist up to 10^{10} times during the useful lifetime of DMD device.

ELECTRONIC   MECHANICAL

MAGNETIC   THERMAL

FLUIDIC   CHEMICAL   BIOLOGICAL

OPTICAL   MICROWAVE
(ElectroMagnetic)

Figure 5. Physical mechanisms active in MEMS.

The examples just discussed illustrate the diversity of materials and structures in MEMS. They also show that the mechanisms operative in MEMS go well beyond the electronic and mechanical functions which form the core of the name. These additional functions are indicated schematically in Fig. 5. The number and interactions of these mechanisms give MEMS their versatility and complexity, and enable their widespread use.

2. MEMS LIFE CYCLE

The four primary steps for MEMS between concept and scrap are design, manufacture, testing and use. These and other steps are indicated in Fig. 6. This sequence applies to the engineering of any system. However, there are significant differences between ordinary civil, electrical, mechanical, and other macroscopic systems produced by conventional engineering disciplines, and the microsystems of interest here. For example, size effects can influence the flow of electrons and fluids when the scale of geometry is comparable to mean free paths and boundary layer thicknesses. Surface effects become more important, and bulk properties can vary significantly, as the scale of devices shrink. The distances over which materials diffuse at elevated and even ordinary temperatures can compare to the dimensions of MEMS components. The long term reliability of MEMS can

D.J. Nagel

be affected. Present MEMS generally do not face these limits, but they may become important for the reliability of some future microsystems.

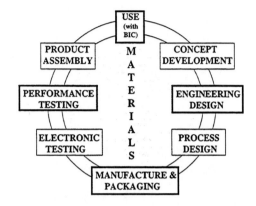

Figure 6. Schematic of the primary operations (heavy boxes) and secondary, but necessary, operations (lighter boxes) for the design, manufacture, testing and use of MEMS, arrayed clockwise. BIC stands for built-in calibration. *Materials are central to all aspects of MEMS.*

The design, manufacture, testing, and applications of MEMS are discussed in the following four sections. Then the four main aspects of materials, namely synthesis, processing, characterization, and determination of properties, are reviewed. Finally, the evolving situation regarding materials for MEMS is summarized.

3. DESIGN OF MEMS

The design of many engineering systems permits separate consideration of electronic, mechanical, thermal, fluidic, and other properties. Electro-mechanical systems, for example, motors and generators, and electro-mechanical-fluidic-thermal systems, such as HVAC systems, demand consideration of multiple engineering disciplines. However, their macroscopic size permits more separate consideration of the salient factors than in the case of MEMS.

The interaction between the multiple mechanisms active in complex MEMS requires employment of *integrated, iterative programs* for their design. For example, design of electronic and mechanical aspects of an electronically-actuated cantilever involves performing self-consistent electrostatic field and finite-element deformation calculations. Commercial computer-aided design (CAD) software packages for such electro-mechanical designs are available. The Defense Advanced Research Program Agency in the U.S. now has a

Composite CAD Program to expand the list of physical mechanisms which can be treated consistently in the design of MEMS, and to produce facile codes.

Two necessities for CAD of MEMS and other engineering systems are the algorithms (the codes) embodying the salient physical principles, and the parameters (properties) which describe the behavior of specific materials. Initially, it was hoped to include adequate tables of properties for MEMS materials in the CAD packages. But, this is difficult because of the influence of composition, microstructure and geometry on properties. Control of impurities in thin film processes is a challenge. Even if accomplished, materials of the same composition can exhibit different crystal and defect structures, with different properties, depending on deposition temperature, for example. Finally, the bulk properties of materials, even with controlled composition and crystallinity, also depend on size for nanometer dimensions. Molecular-dynamic simulations of quartz nano-beam resonators showed a size-dependent bulk modulus for beam cross sections on the order of nanometers.

The design of ICs requires both engineering or functional, and manufacturing or process, design steps. The same will be increasingly true for MEMS. At present, codes for functional CAD are more developed than for process CAD of MEMS.

The intrinsically complex and integrated character of mechanisms active in MEMS, and their variable material properties, essentially requires a paradigm shift in their design. Put another way, the design of MEMS requires an integration of conventional engineering disciplines. Because design is an essential engineering function, MEMS technology challenges traditional approaches to engineering education.

4. MANUFACTURE OF MEMS

Ordinary ICs seem to be planar, but they have complex geometries in the third dimension due to the two dozen or more sequential steps in their production. MEMS are even more intrinsically three dimensional. Also, with MEMS, there is the additional requirement to produce the mechanical part without damaging or degrading the electronic part, and vice versa. A vast literature regarding the several processes to produce MEMS is available from the past 15 years. In much of it, the cross sectional drawings of the results of processing steps look much like those used in IC production. The usual three types of processes are involved in various sequences during production of MEMS, namely patterning, material addition, and material removal.

Three distinct production processes commonly used to make MEMS are essentially based on IC procedures for silicon, namely surface micromachining, bulk micromachining, and reactive ion etching. In the surface micromachining, repeated patterning, deposition, and removal steps modify the top few microns of a wafer. Movable parts are freed from the substrate by dissolution of a sacrificial layer (usually by HF etching of SiO_2). Flexible, rotatable, and erectable structures attached to the substrate have been made by surface micromachining. The ADXL accelerometer (Fig. 3) and the DMD (Fig. 4) are both made by surface micromachining. In bulk micromachining, features are etched into silicon to depths as great as about 500 μm. Chemicals which attack some crystalline planes much faster than others, so-called orientation dependent anisotropic etchants, are employed. The liquid-filled cavity in the fluistor shown in Fig. 2 is made by this bulk micromachining process.

The third silicon-based process employed for the production of MEMS is reactive ion etching (RIE), a method used heavily for IC production because it is anisotropic. That is, RIE yields quite vertical sidewalls for trenches etched into wafers to depths typically on the order of 1 µm. The same plasma-based RIE process used for ICs can etch 1 µm wide trenches to depths exceeding 50 µm, that is, with an aspect ratio (depth/width) of 50. New commercial equipment can etch silicon to depths over 200 µm with aspect ratios near 20. To make more three-dimensional structures for MEMS, the single crystal reactive etching and metallization (SCREAM) process was developed. It employs RIE, followed by coating of the sidewalls of deep trenches by etch-resistant materials. Then additional isotropic etching *under* the protected structures results in high-aspect ratio structures released from the substrate over spans up to 5 mm. These structures have aspect ratios of about 50. If, instead of an isotropic etch to release the structures, another sequence of sidewall coating and RIE is done, then structures on the order of 1 µm width and 100 µm deep are obtained prior to isotropic etching for release. SCREAM is a process based on IC production developed to meet the needs of MEMS production.

RIE is the starting point for another high-aspect-ratio deposition technique developed for production of MEMS. In the HEXSIL process, the topography produced in a silicon wafer by RIE is first oxide coated and then filled with polysilicon. After polishing of the surface to restore planarity, and expose it, the oxide layer is etched away as in surface micromachining. The highly-three-dimensional polysilicon structure is then removed from the silicon crystal template, which can be reused.

Yet another process not used for IC production, which adds relatively thick material onto a surface, is now central to production of high-aspect-ratio structures for some MEMS. LIGA, a German acronym for sequential patterning (Lithographie), electrodeposition of metal (Galvanoformung), and micromoulding (Abformung), produces remarkable structures already found in commercial products. Highly parallel x-rays are used to expose thick layers of plastic, which, after development, are themselves often the final "product", more than 10 mm thick in some cases. Alternatively, a thick photoresist can serve as a guide for subsequent electrodeposition, after development to open spaces down to a conducting substrate. The electroplated metal (often nickel) is typically up to about 1 mm thick. It can be used as a mold for replication in plastic, either by injection or embossing.

Lasers play some roles in the microelectronics industry, but are increasingly important in the production of MEMS. Lasers can be used to etch features in polymers down to about 1 µm in width. Repeated etching, using high-precision X-Y translation tables under computer control, can produce structures a few 10s of microns deep. They might be parts of MEMS, or used (after flash evaporation of a metallic layer) as molds for electrodeposition, similar to LIGA. Lasers can also play a role in deposition of materials. Thin films of complex materials are made by pulsed laser deposition. And, lasers can decompose vapors and deposit patterns of materials on a surface, essentially by laser chemical vapor deposition. Finally, lasers under control of the output of a computer-aided design program can cross-link polymer precursor liquids to make solids with complex shapes. This last process, called rapid prototyping, is generally used to produce macroscopic objects. However, lasers in current commercial systems for rapid prototyping can focus to spots below 10 µm. It seems clear that structures with details on the scale of 1 µm could be made by extensions of the current equipment. Lasers, incidentally, are useful for the testing of MEMS. While not a part of manufacturing

MEMS, laser metrology can reveal the temporal and spatial behavior of a device once it is made.

Bonding of materials is basic to the production of many MEMS devices. Again, the fluistor serves as an example. Sealing together either similar or dissimilar materials over relatively large areas approaching 1 cm^2 is needed. Bonding is often challenging due to the need for very clean surfaces, in order to get essentially molecular contact to achieve leak-free and long-lived performance. Processes which bond silicon to silicon, glasses, polymers, and other materials are used much more for MEMS than for ICs. So, here is another instance where the demands of MEMS have pulled processing beyond what is normally found in the microelectronics industry.

The packaging of MEMS is part of the overall production of useful devices. It has all the same problems met in packaging ICs, plus special challenges due to the moving parts. In the case of sealed microsystems, such as accelerometers, the situation is more similar to that of ICs. However, maintenance of the proper internal atmosphere in an accelerometer package to avoid friction is, for example, a requirement not present for ICs. MEMS sensors and actuators, which have to communicate directly with ambient conditions, face much greater problems. Humidity, particles, and even the material to be analyzed can degrade devices. Single-use MEMS, which are economically feasible due to their mass production, are employed in some instances, especially clinical medicine. Packaging can account for much more than half of the cost of a finished MEMS component.

5. TESTING OF MEMS

Testing of MEMS has two major goals: (a) determination of performance (sensing or actuation) as a function of input and environment parameters, and (b) determination of the failure modes and associated probabilities, which limit the reliability and lifetime of a device. While design and production are distinctly separated in time, electronic testing of MEMS can sometimes be done during production at wafer or die levels. This avoids expensive packaging and later testing of bad devices.

Calibration of MEMS is a fundamental aspect of performance testing since the quality of sensing or activation is usually more critical than mere operation. That is, quantitative performance is required. Measurement of output signal voltage as a function of the desired input parameter (for sensors) or determination of the output force or displacement for the input drive signal (for actuators) is needed. Built-in calibration (BIC) is necessary for some MEMS. The ADXL air bag triggers have a self-test capability. MEMS in some future systems will be calibrated during their service life by two-way wireless radio-frequency links.

Full testing and calibration, including the effects of unwanted physical inputs, such as mechanical noise, over a range of environmental parameters, especially temperature, are complex and costly. Testing of manufactured items, such a ICs, for functionality and reliability is done on statistical bases. Test protocols for MEMS are undoubtedly well developed for mass produced parts such as the ADXL and the DMD. However, this aspect remains to be thoroughly developed for most MEMS devices. In particular, test structures such as are used in the IC industry to determine critical dimensions, switching speeds and other important parameters, will be of increasing use to the MEMS industry.

272 D.J. Nagel

While MEMS are themselves systems, they can be viewed as components for subsystems and systems. For example, a MEMS RF switch might be part of the antenna subsystem of a radar systems (in an airplane or automobile, which are part of transportation systems!). Assembly of MEMS into products is yet another distinct step. The point is that the ultimate performance and reliability of MEMS are influenced, if not determined, by the overall product containing the MEMS and the environment the product encounters during use.

6. APPLICATIONS OF MEMS

It might seem that the use of a MEMS component is entirely determined by its initial design, as realized during manufacture and verified by testing. However, one of the advantages of MEMS is that capable microsystems can find alternative uses in the hands of creative people. A good example of this is the range of uses for the ADXL accelerometers. Although they were developed for air bag deployment, they are not merely threshold detectors. They work over a wide range of accelerations (e. g., 0 to 50 g's, with sensitivity of milli-g, for the ADXL-50). Hence, they are being employed for measuring accelerations due to disparate causes, for example, machinery vibrations and roller coasters.

While the specific applications of MEMS are important, it is not possible to discuss them in any detail here. The major point is that, before long, the use of MEMS will be widespread. Twenty years ago, relatively few individuals owned integrated circuits, but now most people in developed countries have ICs in some of their possessions. Ten years ago, very few people owned lasers, but now many have them in compact disc players. Presently, few people own MEMS, but ten years from now, they will be quite commonplace, and twenty years from now they are likely to be widely distributed.

The expected dispersion of MEMS can be examined from either of two viewpoints, the industries they will impact, or their locations and functions. These categories, often with correlations between them, can be separately listed (Fig. 7). The lists, even though they are really not complete, are so long and comprehensive to appear quite useless. However, they serve as a systematic way to examine the penetration of MEMS into the world quite generally, much the same as happened with ICs. Certainly, some major industries have already been heavily influenced by MEMS, notably the automobile and healthcare industries. Others will be impacted in the foreseeable future. Of course, some will never be dramatically influenced by MEMS. It will be interesting and important to watch the applications of MEMS grow.

7. MATERIALS FOR MEMS

The discussion above already illustrates much about the materials used in the production of MEMS. Some materials are transient, in the sense they are not part of finished MEMS. Photoresists and sacrificial oxides are examples. Others become part of the devices themselves. Both are critical to successful production and performance of MEMS.

The four major steps in the lifecycle of MEMS have been discussed already. Material scientists have another sequence of four steps in the development and use of materials. The

first is synthesis, namely the production of the material of interest. This might be done in bulk, for later application in specific instances. Silicon wafers provide an example. On the other hand, the material might be synthesized in place, as in the production of a thin film by sputtering or pulsed-laser deposition.

INDUSTRIES	LOCATIONS & FUNCTIONS
CONSTRUCTION	AT HOME
VEHICLE or PLATFORM	AT WORK
Marine	Office
Terrestrial	Factory
Air Vehicles	Laboratory
Spacecraft	Other Buildings
EDUCATION	AT SCHOOL
FOOD & DRINK	GROCERY STORES & RESTAURANTS
CLOTHING	PERVASIVE
COMPUTERS & INFORMATION	PERVASIVE
TRANSPORTATION & TRAVEL	COMMUTING & RECREATION
R & D INSTRUMENTS	RESEARCH
MEDICAL	PERSONAL HEALTH
ENVIRONMENTAL	ENVIRONMENTAL HEALTH
ENERGY	EXPLOITATION & GENERATION
MANUFACTURING	PROCESS CONTROL
BROADCAST & TELEPHONE	COMMUNICATIONS
SPORTS	ATHLETICS
ENTERTAINMENT	RADIO, TV, MOVIES & GAMES
DEFENSE	MILITARY OPERATIONS

Figure 7. Uses of MEMS for various purposes at various locations.

The second step in the materials sequence is processing. In bulk materials, such as steel, this could be a thermal anneal or a mechanical process such as rolling. From one viewpoint, the processing of materials for MEMS is more limited, often being only annealing without the usual mechanical options being available. From another perspective, MEMS processing is also broader, because the forming steps used for ordinary materials, such as machining of steel, are done by wet or dry chemical processes during production of MEMS. The ability to perform forming steps by simultaneous processing of thousands of individual MEMS on a wafer is the basis for the low-cost potential of MEMS compared to parts which have to be handled individually. The requirements and costs of packaging generally erode, but do not destroy the advantage of mass processing.

The third step is characterization, that is, determination of the composition and structure of the material. The behavior of materials in service depend on their basic and low-level constituents and on the way they are arranged on atomic, mesoscopoic and macroscopic scales. Factors such as internal stresses, and the type and arrangement of point or extended defects, come into play here.

The fourth phase is the measurement of material properties, that is, the quantitative response of materials to the application of any outside "stress", such as heat, forces, voltages, etc. The properties of a material, initially and over the lifetime of the object, determine the use of the material. Note that these four aspects of material science are iterative. That is, if the characteristics of a material are not satisfactory, changes can be made in synthesis. If the properties are sub-standard, again aspects of synthesis are altered, and so on until appropriate performance in use is achieved. This situation is similar to the four main steps, design, production, testing, and use, for MEMS. Problems at any stage require redesign or changes in the manufacturing processes.

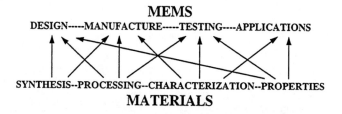

MEMS

DESIGN-----MANUFACTURE-----TESTING----APPLICATIONS

SYNTHESIS--PROCESSING--CHARACTERIZATION--PROPERTIES

MATERIALS

Figure 8. Steps involved in MEMS and materials, and the interactions between them.

The important point regarding materials for MEMS is that the two sequences described above interact greatly with each other. This is indicated schematically in the Fig. 8. A modification of some aspect of one material in a MEMS will propagate through both the materials and the MEMS sequences. The impact of the introduction of a new material will be correspondingly greater. That potential for achievement of better performance in use is, of course, a primary motivation for the introduction of new materials into MEMS.

There are three stages which can be identified in the use of materials, and the associated processes, for production of MEMS. They overlap the past, present, and future. The first period served as the basis for the emergence of a MEMS industry and is already past. It is characterized by the use of IC materials and processes with few additions and changes. In this stage, silicon, both crystalline wafers and polycrystalline deposits, and silicon compounds, especially oxides and nitrides, and metallic conductors, such as aluminum and gold, were dominant. While still centrally important to MEMS, they have been joined by many other materials in the next stages.

The second stage involved the employment of available materials, not normally used for ICs. This phase began early and overlapped the first stage because materials such as pyrex were used for some of the early MEMS. Polymers also came into additional use, for example, the thick photoresists employed in the LIGA process. So did other metallic materials such as electroplated nickel. Silicon carbide was introduced for its high temperature properties. The ability of make MEMS out of another refractory material, namely glassy carbon, has been demonstrated recently. Thin films of artificial diamond, with their remarkable mechanical, optical and thermal properties, are attracting attention in the MEMS community. Available conducting polymers may soon be found in MEMS. Self-assembled organic monolayers were also introduced into some MEMS in the past few years. The hallmark of this stage was the introduction of magnetic, piezoelectric and other functional materials which, while not new materials in themselves, were new to MEMS. This second stage continues and blends into the last phase.

The third stage is characterized by the incorporation of totally new materials, not yet realized, into MEMS. Traditionally, materials have been considered separately in four classes: metals and alloys; inorganic compounds, especially ceramics and glasses; polymers and organics, including most biomaterials; and semiconductors of many types. The functionality of each of these material classes continues to improve as modern experimental and theoretical tools are applied to their synthesis, processing, characterization and property determination. Examples abound, including magnetic materials, optoelectronic materials, conducting polymers, and materials for solar cells plus many other devices. However, material science is undergoing a major change in which *mixed materials* drawn from these traditional classes are made for specific reasons. Dispersions of metallic nanocrystals in oxides for their non-linear optical properties provide one example. Semiconductor quantum dot materials is another. The optical properties of such new materials will be of increasing interest. The same is true of the chemical aspects of emerging materials, as analyzers on chips come to market. Interactions of man-made and biological materials also promise to be increasingly important, especially if biomedical analysis applications of MEMS advance as expected. The general point for MEMS is that new materials are likely to be exploited in many cases in the future. Whether or not the desire for particular materials properties for use in MEMS will provide the primary drive for the development of new materials remains to be seen.

The bulk properties of materials are certainly not irrelevant to MEMS design, production, testing and use. However, the large surface-to-volume ratios in thin film materials, such as used in MEMS, makes the control and analyses of thin film and surface characteristics and properties critical for all aspects of MEMS. The ability to lay down and modify materials with the needed composition and structure is as important to the MEMS industry as it is to the IC industry. This is now true for performance reasons. As the market for MEMS increases, and competition with it, then the costs of material production and modification during all stages of MEMS engineering will become more important.

Materials are as central to MEMS as they are to any other engineering system. They determine what MEMS can be made and dominate their performance. Because most of the critical materials in MEMS are produced in place, attention to materials processing is also inescapable. A current study of materials for MEMS by the National Research Council in the U. S. embraces fabrication methods as well as diverse materials.

ACKNOWLEDGMENTS

Larry Hornbeck of Texas Instruments, Inc. provided the micromirror schematic in Fig. 4. His assistance, and that of Steve Walker in proofreading the manuscript, are appreciated. This study was supported by the Defense Advanced Research Projects Agency, with travel funds provided by the Indo-US Science and Technology Initiative.

References

[1]   Journal of Microelectromechanical Systems, published since 1992 by the IEEE.

[2]   Sensors and Actuators, a journal published by Elsevier Sequoia since 1981.

[3]   K. E. Petersen, Silicon as a Mechanical Material, Proc. IEEE, 70 (5), 420-457 (May 1982)

[4]   K. J. Gabriel, Engineering Microscopic Machines, Sci. Am., 150-153 (Sept 1995)

[5]   Redwood Microsystems, Inc., Menlo Park CA (http://www.redwoodmicro.com/)

[6]   Analog Devices, Inc., Norwood MA (http://www.analog.com/)

[7]   Texas Instruments, Inc., Dallas TX (http://www.ti.com/)

Bibliography and Webliography

[1]   W. S. Trimmer (editor), Micromachines and MEMS; Classical and Seminal Papers to 1990, IEEE Press, New York (1997). A useful compendum of important papers, including the 1987 report of a National Science Foundation Workshop on MicroElectroMechanical Systems Research.

[2]   http://mems.isi.edu, the web site maintained at the Institute for Scientific Information by the Defense Advanced Research Projects Agency. It is a useful gateway to most of the information on MEMS on the internet.

ACCELERATED MOLECULAR DYNAMICS SIMULATION OF MATERIALS DEFECTS WITH THE HYPER-MD METHOD

Arthur F. Voter

Theoretical Division, Los Alamos National Laboratory
Los Alamos, NM 87545, U. S. A.

ABSTRACT

Molecular dynamics (MD) is a powerful tool for investigating detailed atomic-scale behavior. However, the accessible time scales are limited to nanoseconds, so that many processes of interest are simply out of reach. Often, the dynamics of slower processes are characterized by infrequent-events, i.e., the system makes an occasional transition from one metastable state to another. Examples include surface or bulk diffusion, thermal dislocation dynamics, or fracture at low strain rates. When the nature of the transition is understood, transition state theory (TST) can be used to compute the exact rate constant without dynamics. However, in many cases, the reactive events that will occur are not known in advance, or the transition states are extremely complicated. We discuss a new method, "hyper-MD," for treating this type of case in the solid state. The method works in continuous space and makes no assumptions about the nature of the transition states. Substantial time scale enhancements appear possible. As an example, the diffusion of an Ag adatom on the Ag(100) surface is examined in a 10 μs simulation.

1. INTRODUCTION

The molecular dynamics (MD) method is a powerful tool for detailed studies of defects in materials. For a given interatomic potential, atom positions are evolved with no other assumptions than classical mechanics, so the dynamical behavior in a simulation is "truth" for the system under study. Direct and meaningful comparisons to experiment can be made, with discrepancies attributable solely to the inaccuracy of the potential. For many defects (e.g., surfaces, grain boundaries, vacancies, dislocation cores) useful results can be obtained from simulation of hundreds or thousands of atoms on a standalone workstation, making MD an increasingly popular approach. A major drawback of MD, however, is that it is limited to time scales of nanoseconds or less. Even massively parallel computers, which have extended the size scale to millions of atoms, have had little impact on the accessible time scale. due to the inherently sequential nature of the equations of motion.

For many systems, the dynamics on longer time scales can be characterized as "infrequent events" behavior. In infrequent-event systems, most of the atomic motion consists of simple vibrations within a particular potential basin. After a large number of these vibrational excursions, just the right conditions occur to cause a transition to an adjacent potential basin, after which the system vibrates about the minimum of the new potential basin.

For infreqent-event systems, the time scale limitations of MD can be overcome using transition state theory (TST) [1, 2, 3, 4], in which the rate constant for escape from one state (basin) to another is approximated as the flux through the dividing surface separating the two states. Conceptually, the TST rate for a two-state system can be viewed as being computed from a long MD trajectory that passes back and forth between the two states [5]. The number of crossings of the dividing surface per unit time is then the desired rate constant, while the vibrational motion is ignored. The appeal of TST is that one need not actually run any trajectory; the flux through the dividing surface is an equilibrium property of the system, depending only on the thermal partition function of the 3N-dimensional potential basin and the partition function of the 3N-1-dimensional dividing surface (N is the number of atoms). This can be evaluated using Monte Carlo [6], for example, or approximated harmonically, as discussed below. Thus, the TST rate constant can be computed directly for a given temperature, even if it corresponds to very long times between transitions.

The TST rate is not exact, because there can be correlations between successive crossings of the dividing surface. However, the exact rate constant can be recovered using dynamical corrections theory [4, 7, 8, 9]. This is accomplished by examining the behavior of short-duration MD trajectories that begin at the dividing surface; qualitatively, trajectories that recross the dividing surface before thermalizing in one state or the other contribute to a correction term that lowers the rate constant. For material defects, the dynamical corrections effects are typically small (a few percent or less in many cases), so that TST is an excellent approximation to the true rate.

Most defect properties of interest to materials scientists fall into this infrequent-event class, if one is not near the melting point. Moreover, the harmonic approximation to TST [10] is usually a very good approximation to exact TST (although less so at very high temperatures). In the harmonic approximation, the rate depends only on the properties (energy and normal mode frequencies) at the minimum and the saddle point, and yields the well-known Arrhenius form, in which the rate depends exponentially on the barrier height divided by the temperature. Thus, TST provides the formal underpinning for the usual conceptual view of long-time dynamics in the solid state: evolution from state to state consists of activated barrier-crossing events. TST is also the basis of lattice-mapped kinetic Monte Carlo simulations, in the system is evolved from state to state by choosing from a list of possible transition events.

As elegant and accurate as TST is, however, there is a catch: to use it requires knowledge of the dividing surfaces, or at least requires knowing where the saddle points are. If the transition states are complicated, or if the states to which a system might evolve are not known in advance, TST is hard or impossible to apply. Many such situations occur in materials science. Examples include the complicated healing events after radiation damage or ion implantation, the dynamics of a dislocation core in the presence of other defects, or the advance of a crack front. Even the motion of relatively simple defects can turn out to be more complicated than expected [11, 12, 13], calling into question the validity of kinetic Monte Carlo simulations based on simple event rules.

In this paper we discuss a recently developed alternative approach [14, 15]. Beginning from the TST approximation, a method is derived for extending the time scale of MD simulations without any advanced knowledge of either the location of the dividing surfaces or the states through which the system may evolve. A bias potential (ΔV_b) is designed to raise the energy of the system in regions other than at the TST dividing surfaces. A simulation on the biased potential gives accelerated evolution from state to state, while the elapsed time becomes a statistical property of the system. The instantaneous gain (or "boost") in the rate at which time advances (relative to direct MD) depends exponentially on the bias potential. In essence, this new method ("hyper-MD") captures the advantages of both MD and TST into a single simulation approach.

2. ACCELERATED TIME SCALE ON THE BIASED POTENTIAL SURFACE

We first derive the expression for the accelerated time scale when dynamics are performed on the biased potential. Although the hyper-MD method is designed to treat many-dimensional, many-state systems with arbitrarily complicated TST dividing surfaces, the essence of the derivation can be captured in a one dimensional system with a single escape path (although a many-state system will be considered to make certain points).

Consider a one-dimensional system characterized at time t by position $x(t)$ and potential energy $V(x(t))$, with a TST boundary at $x = q$. The system is in state A when $x < q$ and escape has occurred if $x > q$. Throughout, this system is assumed to obey TST exactly; any crossing of the dividing surface leads to thermalization in the newly-entered state. The TST rate constant for escape from state A is given by the flux through the TST boundary,

$$k_{A\to}^{\text{TST}} = \langle |dx/dt|\, \delta(x - q)\rangle_A . \tag{1}$$

Here $\delta()$ is a Dirac delta function and $\langle \cdots \rangle_A$ indicates a classical, canonical-ensemble average restricted to the configuration space of state A.

Now consider adding to $V(x)$ a nonnegative bias potential $\Delta V_b(x)$ which, by construction, is zero at the TST boundary. (We discuss below how this can be accomplished even when the position of the dividing surface is unknown.) Assume that the biased system still obeys TST, and that the bias potential does not introduce significant local minima that disrupt the rapid ergodic sampling within the state. If dynamics are run on this biased potential $(V + \Delta V_b)$, the escape rate increases, because the biased well is not as deep as the original well. Also, if there are multiple states adjacent to state A, dynamical evolution on the biased potential preserves the probability of escape to a given adjacent state (B, say), relative to other adjacent states [14]. The consequence of these two properties is that a trajectory on the biased potential will evolve from state to state in a correct fashion, but at an accelerated rate. We now derive the corrected time scale for this accelerated evolution.

Equation (1) can be manipulated [14] using importance-sampling techniques [16] to give averages over the biased-potential state (A_b),

$$k_{A\rightarrow}^{TST} = \frac{\langle |dx/dt| \delta(x - q) \rangle_{A_b}}{\langle e^{\beta \Delta V_b(x)} \rangle_{A_b}}. \tag{2}$$

Here $\beta = 1/k_B T$, where k_B is the Boltzmann constant and T is the temperature. An important step in deriving Eq. (2) utilizes the fact that $\Delta V_b = 0$ where $\delta(x - q) \neq 0$, simplifying the numerator from $\langle |dx/dt| \delta(x - q) exp(\beta \Delta V_b(x)) \rangle_{A_b}$. Consequently, the numerator is exactly the TST escape rate from the biased state.

Now consider a thought experiment in which the averages in Eq. (2) are evaluated using a long trajectory, confined to state A_b by reflecting barriers along the TST surface(s). This trajectory is coupled to a heat bath to obtain proper sampling of the canonical phase space. By making the trajectory extremely long, the averages can be computed to arbitrarily high accuracy. From the trajectory results, the average *escape time* for state A (inverse of $k_{A\rightarrow}^{TST}$) for state A can be expressed as

$$\begin{aligned} \tau_{esc}^A &= \frac{< e^{\beta \Delta V_b(x)} >_{A_b}}{< |dx/dt| \delta(x - q) >_{A_b}} \\ &= \frac{\frac{1}{n_{tot}} \sum_{i=1}^{n_{tot}} e^{\beta \Delta V_b(x(t_i))}}{n_{esc}/(n_{tot} \Delta t_{MD})} \\ &= \frac{1}{n_{esc}} \sum_{i=1}^{n_{tot}} \Delta t_{MD} e^{\beta \Delta V_b(x(t_i))}, \end{aligned} \tag{3}$$

where Δt_{MD} is the integration time step, n_{tot} is the total number of MD steps, t_i indicates the time at the ith MD step, and n_{esc} is the number of attempted escapes (reflections off the barrier). Utilizing the equivalence between an ensemble average and a time average, the numerator in the first line of Eq. (3) has been evaluated using the n_{tot} equal-time snapshots. The denominator, which is the TST escape rate from A_b, has been expressed as the number of TST surface crossing attempts divided by the total trajectory time. Inspection of the last line of Eq. (3) suggests a simple definition for the time evolved per MD step (Δt_{b_i}) on the biased potential. By requiring that the total time evolved, divided by the number of attempted escapes, equal the average escape time, we arrive at

$$\Delta t_{b_i} = \Delta t_{MD} e^{\beta \Delta V_b(x(t_i))}, \tag{4}$$

where the total ("boosted") time that the system has evolved is accumulated as

$$t_b = \sum_i^{n_{tot}} \Delta t_{b_i}. \tag{5}$$

Equations (4) and (5) are the heart of the hyper-MD method. At each integration step, the t_b clock is advanced by Δt_{b_i}, which depends on the instantaneous strength of the bias potential. (Note that where $\Delta V_b(x_i) = 0$, $\Delta t_{b_i} = \Delta t_{MD}$; the method is equivalent to direct MD where there is no bias potential.) While individual values of Δt_{b_i} have little meaning, at long times, by construction, t_b converges on the correct result. What is a "long" time depends on the statistical properties of the time-dependent boost factor, $e^{\beta \Delta V_b(x(t))}$. If escape from a state requires so many steps that the average boost factor ($\langle e^{\beta \Delta V_b(x)} \rangle_{A_b}$) is well approximated, individual escape times will be meaningful. In contrast, an aggressive choice for ΔV_b may stimulate escape in so few steps that predicted escape times are very noisy. Then the time scale becomes meaningful only after many transitions. In either case, the escape-time estimates are unbiased and their errors are uncorrelated, so, from the central limit theorem, the relative error in t_b decreases as $t^{-\frac{1}{2}}$, even if every state is different.

3. DEFINING THE BIAS POTENTIAL

Implementing the hyper-MD method requires a definition for ΔV_b that does not assume advanced knowledge of the transition states. Formally, ΔV_b must be zero along all the dividing surfaces, must not block rapid ergodic sampling within the current state, and must not introduce TST-violating correlations among the transitions. In practice, however, these requirements need only be met to a good approximation. For example, we know that the saddle points are the most important regions, so the bias potential should be designed to at least behave properly there. Candidate forms for ΔV_b can be tested and refined on benchmark systems. In the following, we discuss an approach for constructing ΔV_b that has been successful so far in applications to metal systems. The form is certainly not unique, so variations on this approach may lead to improvements in the future.

For a system with N atoms, a position in configuration space is defined by the $3N$-dimensional vector $\mathbf{r}$. The form for ΔV_b will be based on local properties of the potential via the $\mathbf{r}$-dependent gradient vector, $\mathbf{g}$ ($g_i \equiv \frac{\partial V(\mathbf{r})}{\partial x_i}$, where x_i are components of $\mathbf{r}$), and Hessian matrix, $\mathbf{H}$ ($H_{ij} \equiv \frac{\partial^2 V(\mathbf{r})}{\partial x_i \partial x_j}$).

The key is to design a function that can recognize regions of configuration space that might be part of a dividing surface. Although the true definition of a TST dividing surface is intrinsically nonlocal (requiring steepest-descent trajectories), Sevick, Bell, and Theodorou [17] have proposed a local definition that senses the ridge line between two valleys in the potential. They approximate the dividing surface by the set of points in $\mathbf{r}$-space for which the lowest eigenvalue of $\mathbf{H}$ is negative and where the projection of $\mathbf{g}$ onto the lowest eigenvector of $\mathbf{H}$ is zero. I.e.,

$$\epsilon_1 < 0 \quad \text{and} \quad g_{1p} = 0, \tag{6}$$

where ϵ_1 is the lowest Hessian eigenvalue ($\epsilon_1 = \mathbf{C}_1^\dagger \mathbf{H} \mathbf{C}_1$) and g_{1p} is the projected gradient ($g_{1p} = \mathbf{C}_1^\dagger \mathbf{g}$). Near a saddle point this is an excellent approximation to the true, nonlocal TST surface.

Using these same quantities ($\epsilon_1(\mathbf{r})$ and $g_{1p}(\mathbf{r})$), an expression for ΔV_b can be constructed that goes smoothly to zero in regions where Eq. (6) holds. One way to accomplish this is to consider a sine wave with barrier height h and period $2\pi d$,

$$V^{\cos}(x) = (h/2)\cos(x/d), \tag{7}$$

as a representative potential for an activated process. Because of its simple form, $V^{\cos}$ can be expressed as a function of its first and second derivatives, without explicitly specifying x, if the period and barrier height are known; i.e.,

$$V^{\cos}(x) = V^{\cos}(\epsilon_1(x), g_{1p}(x)) = \frac{-\epsilon_1 h/2}{[\epsilon_1^2 + g_{1p}^2/d^2]^{\frac{1}{2}}}, \tag{8}$$

where we have recognized that $g_{1p} = dV^{\cos}/dx$ and $\epsilon_1 = d^2 V^{\cos}/dx^2$. Thus, we can define a bias potential, $\Delta V_b^{\cos}$, as that function of ϵ_1 and g_{1p} that exactly "cancels" this potential,

$$\Delta V_b^{\cos}(\epsilon_1, g_{1p}) = \frac{h}{2}\left[1 + \frac{\epsilon_1}{[\epsilon_1^2 + g_{1p}^2/d^2]^{\frac{1}{2}}}\right]. \tag{9}$$

I.e., $V^{\cos}(x) + \Delta V_b^{\cos}(x) = h/2$ for all x; the energy everywhere is raised to that of the transition state, as shown in Fig. 1. In a sense, $\Delta V_b^{\cos}$ represents the "perfect" bias potential for $V^{\cos}$. However, dynamics on this biased potential would lead to TST-violating multiple-jump events, which we want to avoid. A slightly weaker bias term is a better choice (see Fig. 1), as it leaves some residual depth in the well, allowing the trajectory to thermalize between transitions.

Because ϵ_1 and g_{1p} are scalars, Eq. (9) is readily applied to the $3N$-dimensional system. An arbitrary potential will not have basins that are exact sine waves, of course, but we can nonetheless use Eq. (9) as a definition for the bias potential. The idea is to set $2\pi d$ to a physically reasonable transition length (e.g., a nearest-neighbor distance), and take h to be somewhat smaller than the lowest anticipated barrier in the system.

To improve the overall quality of the bias potential, $\Delta V_b^{\cos}$ is augmented by a second term ($\Delta V_b^{\Delta\epsilon}$), which provides a smoothly increasing repulsion between the two lowest eigenvalues of $\mathbf{H}$ when they differ by less than $\Delta\epsilon_c$. Defining $q = (\epsilon_1 - \epsilon_2)/\Delta\epsilon_c$,

$$\Delta V_b^{\Delta\epsilon} = \begin{cases} a[1 - 3q^2 + 2q^3], & \text{if } \epsilon_2 - \epsilon_1 \le \Delta\epsilon_c \\ 0, & \text{if } \epsilon_2 - \epsilon_1 > \Delta\epsilon_c. \end{cases} \tag{10}$$

This term is helpful because in regions where the two lowest eigenvalues cross (or nearly cross), the projected gradient often swings through zero as the eigenvector changes direction, driving $\Delta V_b^{\cos}(\mathbf{r})$ to zero if ϵ_1 is negative. The contribution from $\Delta V_b^{\Delta\epsilon}$ fills in these $3N - 1$–dimensional "crevasses" in $\Delta V_b^{\cos}(x)$, maintaining a higher overall boost. Provided that $\Delta\epsilon_c$ is chosen to be much smaller than $\epsilon_2 - \epsilon_1$ for a typical saddle point in the system, the total bias potential will remain zero along the TST surface in the crucial saddle point region, as desired. The fact that $\Delta V_b^{\Delta \epsilon}$ keeps ϵ_1 and ϵ_2 apart also improves the numerical stability of the algorithm for computing ϵ_1, ϵ_2 and g_{1p}, which is described elsewhere [15].

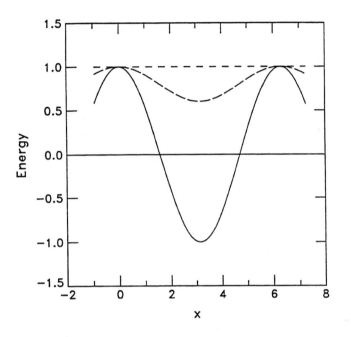

Figure 1. Model sine-wave potential used to define part of the bias potential, as discussed in text. The solid line is $V^{cos}(x)$ (with $h = 2$ and $d = 1$), the dotted line is $V^{cos}(x) + \Delta V_b^{cos}(x)$, and the dashed line is $V^{cos}(x) + 0.8\Delta V_b^{cos}(x)$, which might be a reasonable choice for the bias potential strength in this case.

To summarize, implementation of hyper-MD proceeds as follows: Using some knowledge of the system being studied, values are chosen for the bias potential parameters h, d, $\Delta \epsilon_c$, and a, as discussed above. At each hyper-MD integration step, ϵ_1, ϵ_2 and g_{1p} (and their derivatives with respect to $\mathbf{r}$) are computed. at the present position of the trajectory, The total bias potential is then calculated as

$$\Delta V_b(\mathbf{r}) = \Delta V_b^{cos}(\mathbf{r}) + \Delta V_b^{\Delta \epsilon}(\mathbf{r}) \tag{11}$$

using Eqns. (9) and (10). The boosted-time clock is advanced according to Eq. (4), and the equations of motion are advanced using the sum of the forces from the original potential $(-dV/dr = -\mathbf{g})$ and the bias potential $(-d\Delta V_b/dr)$.

4. SAMPLE APPLICATION: Ag/Ag(100) SURFACE DIFFUSION

As a demonstration of the hyper-MD method, we present a simulation of adatom self diffusion on the Ag(100) surface at $T = 400$ K, a temperature where direct MD is unfeasible. Although the strength of the hyper-MD approach is that mechanisms need not be known in advance, in this example we make use of the known mechanisms and barrier heights to show that the method works correctly.

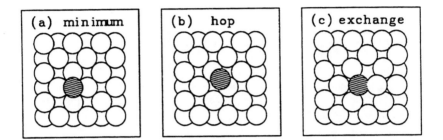

Figure 2. Surface-normal (z) view of the Ag/Ag(100) system. The adatom is shaded for clarity. (a) minimum-energy geometry; (b) saddle point for hop event; (c) saddle point for substrate-exchange event, in which a substrate atom becomes the new adatom.

The simulation cell, shown in Fig. 2(a), is periodic in the x and y directions, with a free surface normal to z. The lattice was expanded to the quasiharmonic lattice constant for $T = 400$ K. The atoms in the top three layers were allowed to move during the simulation, for a total of 55 moving atoms including the adatom. Deeper layers were held fixed in the proper fcc geometry.

The interatomic forces were modeled using an embedded atom method (EAM) interatomic potential. The EAM potential is a semiempirical form [18] that augments a pair potential with a local, density-dependent term, offering a good description of transition metals [19, 20], at a computational cost that is only a few times that of a pair potential. The Ag potential employed here was fit to the bulk lattice constant, cohesive energy, elastic constants, and unrelaxed vacancy formation energy, and the bond length and bond energy of the gas-phase diatomic molecule [21]. Because the hyper-MD forces depend on up to third derivatives of the interatomic potential, the potential was modified slightly to achieve a continuous third derivative at the cutoff distance for the pairwise terms.

In this system, there are two predominant diffusion mechanisms. The adatom can execute a conventional hop to a nearest-neighbor site (Fig. 2(b)), or can exchange with a substrate atom (Fig. 2(c)) to make what is effectively a second-nearest-neighbor jump. This fcc(100) exchange mechanism, which was unknown before 1990 [11, 12, 22], offers a simple example of the type of event that might be neglected if we rely purely on our intuition to determine how a system will evolve. Using molecular statics, the barriers for these two mechanisms were found to be 0.542 eV (hop) and 0.555 eV (exchange). At the saddle points, the Hessian eigenvalues are $\epsilon_1 = -0.87$ eV/A^2, $\epsilon_2 = 0.39$ eV/A^2 (hop) and $\epsilon_1 = -0.54$ eV/A^2, $\epsilon_2 = 0.44$ eV/A^2 (exchange).

The bias potential parameters were chosen as $h=0.3$ eV, $d=0.46$ A$^\circ$, $\Delta\epsilon_c=0.4$ eV/A$^\circ$2, and $a=0.6$ eV. Here we made use of the known barrier heights and saddle properties, although in general situations this should not be necessary. For example, a lower bound on the lowest barrier in the system can be estimated from the length of a direct MD simulation that shows no events. h can be chosen to be less than this bound, and a conservative (i.e., small) value for ϵ_c can be specified. The value for d depends on the lattice constant and setting a to a few tenths of an eV has worked in test cases so far.

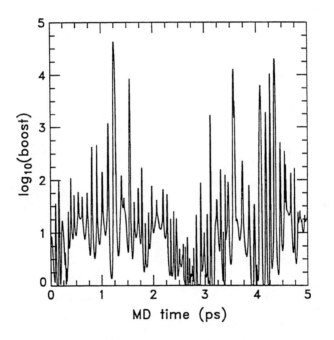

Figure 3. Instantaneous boost factor versus MD time for a small segment of the Ag/Ag(100) hyper-MD simulation at $T = 400$ K.

To control the maximum boost during the simulation, the bias potential was limited to less than ΔV_{max}, by using

$$\Delta V_b^{limited} = \frac{\Delta V_b}{1 + \Delta V_b/\Delta V_{max}} \qquad (12)$$

with $\Delta V_{max} = 0.8$ eV. Also, to decrease the chance that the bias potential inadvertently blocked parts of the true TST surface, an offset was applied to the projected gradient; $g_{1p} = 0$ was employed in Eq. (9) when g_{1p} was less than 0.03 eV/ A$^\circ$.

A.F. Voter

To ensure that at long times the trajectory sampled properly from the canonical ensemble at T=400K, a Langevin thermostat with a coupling rate of 2×10^{12} s^{-1} was applied. The integration was executed using a Langevin-Verlet procedure described by Allen and Tildesley [23], with an MD time step of $\Delta t_{MD} = 2 \times 10^{-15}$ s. The total simulation discussed here consisted of 3.7×10^6 steps.

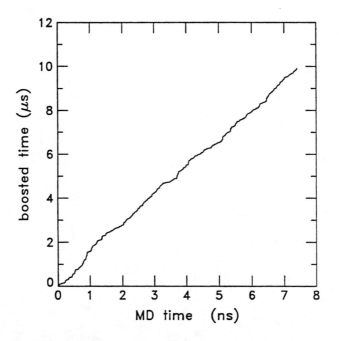

Figure 4. Evolution of the cumulative boosted time (t_b) during the hyper-MD simulation. The average boost factor is 1356.

Fig. 3 shows the instantaneous boost ratio for a small segment of the total simulation and Fig. 4 shows the evolution of the cumulative boosted time, t_b. The instantaneous boost fluctuates between 1 and $\sim 10^5$, with an average of 1356, so that the MD time of 7.4 ns has been converted to a boosted time of $t_b = 9.89 \pm 0.5$ μs. In this system, each hyper-MD time step requires $\sim$30 times more work than a normal MD step, so the true computational boost is about $\sim$45. The slope of the cumulative boosted time evolution does not vary dramatically during the simulation (Fig. 4). As discussed above, a more aggressive choice for the bias potential could lead to greater uncertainty in t_b if it is observed over a short simulation interval. Also, a more complicated system can exhibit changes in the local slope of t_b as the simulation evolves through energetically inequivalent states. The present example has only one unique state (although at much higher temperatures, other states can be accessed, such as the formation of adatom-surface-vacancy pairs).

Fig. 5 indicates the diffusive events taking place during the simulation. The atom number of the current adatom (defined as the atom with the highest z value) is plotted, along with the x position of this atom. The time axis is MD time, which can be converted to boosted time using Fig. 4. Each exchange event changes the adatom number (i), and may or may not change x_i, depending on whether the exchange occurs in the x direction (as in Fig. 2(c)) or the y direction. Every hop event changes x_i. Between these events the vibrational motion of the adatom can be seen.

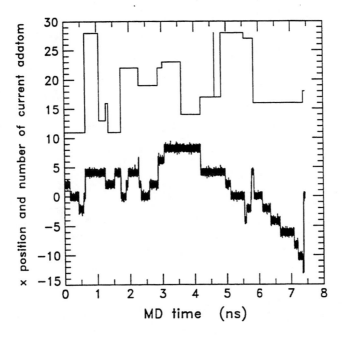

Figure 5. Adatom hopping and exchange events during the hyper-MD simulation. The bottom curve is the x position (A°) of the adatom, while the top line indicates which atom number is the current adatom (shifted by 10 for clarity). Each exchange event changes the adatom number.

Over the whole simulation, 23 hop events and 16 exchange events were observed, corresponding to rate constants of $2.3\pm0.5\times10^6$ s^{-1} (hop) and $1.6\pm0.4\times10^6$ s^{-1} (exchange). These are in agreement with the full harmonic rate constants [2.84×10^6 s^{-1} (hop) and 1.84×10^6 s^{-1} (exchange)] calculated using the Vineyard expression [10],

$$k^{HTST} = n_{path} \frac{\prod\limits_{i}^{3N} \nu_i^{min}}{\prod\limits_{i}^{3N-1} \nu_i^{sad}} \exp(-\beta E_a). \tag{13}$$

Here E_a is the static barrier height, $\{\nu_i^{min}\}$ are the $3N$ normal mode frequencies at the minimum, $\{\nu_i^{sad}\}$ are the $3N - 1$ nonimaginary normal mode frequencies at the saddle point, and n_{path} is the number of escape paths (4 for either mechanism here).

For the hop mechanism, the exact TST rate constant was also computed, using the displacement-vector importance-sampling method [6]. The TST surface was defined as the saddle plane (the plane orthogonal to the imaginary-frequency eigenvector at the saddle point). This yielded $3.0\pm0.2\times10^6$ s^{-1}. Although dynamical corrections were not computed, past experience indicates they will only have a few-percent effect (at most) on the rate constant in this system. Inspection of Fig. 5 shows that a few of the hyper-MD crossing events are closely spaced in MD time, indicating that perhaps the bias potential has induced some correlation in the TST surface crossings.

The characteristics of a hyper-MD run depend intimately on the definition of the bias potential and its interaction with the system under study, and appear hard to predict *a priori*, although this may change as we gather experience with this new atomistic method. Properties of interest include the average boost factor, the change in boost factor as the system evolves through different states, and the possible inaccuracies caused by partial blocking of some pathways (i.e., if $\Delta V_b \neq 0$ along part of a TST surface) or the introduction of correlated crossing events.

A dependence that is easier to quantify is the effect of temperature. As expected from its exponential form, the average boost factor climbs rapidly as the temperature is decreased. The effective simulation rate (events per cpu time) scales as the true rate constant multiplied by the average boost factor, divided by the extra work of hyper-MD relative to normal MD per time step. In test runs on the present system, we find that the boost factor increases faster than exponentially with $1/T$ (because at lower T, the system spends more of its time in regions where the bias potential is high). Because rate constants decrease exponentially with $1/T$, this offers the tantalizing possibility that below a certain temperature, the effective simulation rate actually levels off, or even *increases* as the temperature is lowered further. This temperature range should be accessible on computers available a few years from now.

5. CONCLUSIONS

The hyper-MD method presented here offers a new approach for simulating defect dynamics in situations where direct MD is too slow and TST is inapplicable due to the complexity of the transition states or incomplete knowledge of the possible escape paths. By performing dynamics on a biased version of the original potential, there is accelerated evolution from state to state. Time becomes a dependent variable with statistical error bars, accumulated easily from the instantaneous bias strength as the trajectory runs. A bias potential with the required features, primarily that it be zero at the TST boundaries, can be constructed from the Hessian and gradient, even without advanced knowledge of the dividing surfaces. This bias potential can be evaluated efficiently [15]. For a valid bias potential, the resulting long-time dynamics are exact if TST holds. For the Ag/Ag(100) example presented here, the boost factor was more than 10^3, with a true computational boost of 45. In other systems, true computational boosts of 10^3 and more have been realized. Although much remains to be learned about the optimal design of bias potentials and the full range of applicability, the hyper-MD method already looks very promising.

References

[1] R. Marcelin, Ann. Phys. **3**, 120 (1915).

[2] S. Glasstone, K.J. Laidler, and H. Eyring, *The Theory of Rate Processes*, (McGraw Hill, New York, 1941).

[3] P. Hanggi, P. Talkner, and M. Borkovec, Rev. Mod. Phys. **62**, 251 (1990).

[4] J.B. Anderson, Adv. Chem. Phys. **91**, 381 (1995).

[5] Restricted to these two states for the purposes of this conceptual calculation.

[6] A.F. Voter, J. Chem. Phys. 82, 1890 (1985).

[7] C.H. Bennett, in *Algorithms for Chemical Computation*, edited by R.E. Christofferson (American Chemical Society, Washington, DC, 1977), p. 63.

[8] D. Chandler, J. Chem. Phys., **68**, 2959 (1978); J.A. Montgomery, Jr., D. Chandler, and B.J. Berne, J. Chem. Phys. **70**, 4056 (1979).

[9] A.F. Voter and J.D. Doll, J. Chem. Phys. **82**, 80 (1985).

[10] G.H. Vineyard, J. Phys. Chem. Solids **3**, 121 (1957).

[11] P.J. Feibelman, Phys. Rev. Lett **65**, 729 (1990).

[12] G.L. Kellogg and P.J. Feibelman, Phys. Rev. Lett. **64**, 3143 (1990). Pt(100) of exchange mechanism

[13] J.C. Hamilton, M.S. Daw, and S.M. Foiles, Phys. Rev. Lett. **74**, 2760 (1995).

[14] A.F. Voter, J. Chem. Phys. in press, 1997 (expected March 15).

[15] A.F. Voter, submitted for publication.

[16] J.P.Valleau and S.G. Whittington, in *Modern Theoretical Chemistry*, edited by B.J. Berne (Plenum, New York, 1977), Vol. 5, p 137.

[17] E.M. Sevick, A.T. Bell, and D.N. Theodorou, J. Chem. Phys. **98**, 3196 (1993).

[18] M.S. Daw and M.I. Baskes, Phys. Rev. B **29**, 6443 (1984).

[19] A.F. Voter, in *Intermetallic Compounds: Principles and Practice*, edited by J.H. Westbrook and R.L. Fleischer, John Wiley and Sons, Ltd, 1995), Vol. 1, p. 77.

[20] M.S. Daw, S.M. Foiles, and M.I. Baskes, Mater. Sci. Reports **9**, 251 (1993).

[21] A.F. Voter, in *Modeling of Optical Thin Films*, M.R. Jacobson, Ed., Proc. SPIE, **821**, 214 (1987).

[22] C. Chen and T.T. Tsong, Phys. Rev. Lett.,**64**, 3147 (1990).

[23] M.P. Allen and D.J. Tildesley, *Computer Simulation of Liquids* (Oxford, New York, 1987), p. 263, Eq. (9.24).

DIRECT ATOMISTIC SIMULATION OF QUARTZ CRYSTAL OSCILLATORS. BULK PROPERTIES AND NANOSCALE DEVICES

Jeremy Q. Broughton,[1] Christopher A. Meli,[1] Priya Vashishta,[2] and Rajiv K. Kalia[2]

[1] *Complex Systems Theory Branch, Naval Research Laboratory*
Washington DC 20375, U. S. A.
[2] *Physics Department, Louisiana State University*
Baton Rouge, LA 70803, U. S. A.

ABSTRACT

Current experimental research aims to reduce the size of quartz crystal oscillators into the sub-micron range. Devices then comprise multimillion atoms and operating frequencies will be in the gigahertz regime. Such characteristics make direct atomic scale simulation feasible using large scale parallel computing. Here, we describe Molecular Dynamics simulations on bulk and nanoscale device systems focussing on elastic constants and flexural frequencies. Here we find: (a) in order to achieve elastic constants within 1% of those of the bulk requires approx. one million atoms; precisely the experimental regime of interest; (b) differences from continuum mechanical frequency predictions are observable for 17 nanometer devices; (c) devices with 1% defects exhibit dramatic anharmonicity. Future work will involve the direct atomistic simulation of operating characteristics of a micron scale device.

1. INTRODUCTION

Quartz crystal oscillators (QCO) are used in a multitude of applications [1, 2]. They are used as timing devices, as accelerometers, as mass balances and as viscometers. Such devices can be excited in many different ways, but the one to be considered in this series of papers will be the "flexural plate"; that is, one which oscillates by virtue of being plucked in the direction normal to the longitudinal axis. Conceptually they are the simplest of mechanical machines; they function like a violin string except that instead of catgut, they are made of quartz (SiO_2). The observed resonant frequencies are functions of the length of the device, its cross-sectional area and its elastic constants (which, for an anisotropic material like α-quartz, vary with orientation). The latter, of course, are functions of temperature and degree of compression of the device.

Simple continuum elasticity theory [3] tells us that the frequencies (ω) should scale as:

$$\omega \propto \frac{d}{L^2}\sqrt{\frac{Y}{\rho}} \qquad (1)$$

where Y is the Young's modulus of the material along the longitudinal direction. ρ is the density of the plate, L is its length and d its thickness. Young's modulus gives the rate of change with elongation of longitudinal pressure in the rod when the two other directions are allowed to move freely in the hydrostatic bath. For our purposes, the hydrostatic pressure (**P**) will be one bar, which is very close to one atmosphere. (Note that we boldface the symbol for pressure, because in general it will be a tensor). As regards the precision of Molecular Dynamics (MD), this pressure is essentially zero. The constant of proportionality in Eqn. 1, depends upon the boundary conditions and eigenstate of the plate; for example, whether the ends are clamped or free to pivot.

It is easy to see that if internal loss mechanisms are kept to a minimum, such devices could operate as timing devices. If the device is put in an accelerating frame, with a reference mass at one end, it is also easy to see why they operate as accelerometers. Lastly, since density appears in Eqn. 1, changing the mass of the device (via surface adsorption) also changes the frequency - hence they can function also as sensitive mass balances.

There is a technological drive to shrink the size of QCOs to the micron level. Present technology [4] makes them at the 100 micron scale, where they operate in the megahertz regime (substitution of appropriate values in Eqn. 1 confirms this). But the demand for further sensitivity is driving their size down. For example, oil companies would like gravitational detectors which are sensitive to nano-g's; present QCOs are sensitive to micro-g's. Such miniaturization is attendant with design issues: For example, increased sensitivity is achieved at the cost of increased noise. At what point does the statistical mechanical "noise" due to atomistics start to have an effect on device performance and reliability? For example, at what point does continuum elasticity theory (inherent in Eqn. 1) start to fail (and in what way) in determining device characteristics?

When a flexural plate shrinks to the micron scale (longitudinal axis) with typical aspect ratios of 10:1, the oscillation frequency is in the gigahertz regime. The number of atoms in such a system is in the tens of millions. Thus system sizes and time scales are appropriate for direct atomistic simulation on large scale parallel platforms such as those at DoD/DoE/NSF centers. Although existing interatomic potentials for SiO$_2$ are imperfect, they nevertheless contain enough of the physics and chemistry of the bonding that a direct simulation of a micron sized device should reveal generic behavior which transcends the details of the potential and which would apply to a real device.

These, then, are the reasons for the present study. In this paper, we report results for the temperature dependent behavior of the bulk density and elastic constants. The latter are required to predict device frequencies from continuum elasticity theory. We continue by reporting results for the effective Young's modulus as a function of cross-sectional area of the SiO_2 plate (from which we develop a simple model to predict the asymptotic approach of the modulus to its bulk value) as well as frequencies obtained by direct simulation of a small device containing approx. 20,000 atoms, achievable on a dedicated work station. We conclude by describing the effect of vacancies on device performance.

A future paper will give results for direct simulation of a micron scale device performed on a parallel platform, namely the Air-force IBM SP2 at Maui. It will also present details of the domain decomposition for the parallel implementation.

In section II, we describe the SiO_2 interatomic potential and the construction of the unit cell for both the bulk and device calculations. Section III gives an account of the multiple time step, constant pressure, constant temperature algorithm that we adopted for the bulk simulations. Section IV presents results for the density, Young's modulus and Poisson ratio as functions of the temperature for the bulk. The ability of the model to describe the bulk α to β quartz phase transition will also be discussed. Section V reports the Young's modulus versus cross-sectional area of the SiO_2 flexural plate obtained via simulation. The section continues by fitting these to a simple model which allows asymptotic behavior versus cross-sectional area to be predicted. Section VI gives the frequencies obtained via direct simulation for a 17 nanometer plate and compares them with those predicted via continuum mechanics (CM). Finally, Section VII presents our conclusions.

2. SYSTEM AND INTERATOMIC POTENTIAL

The stable low temperature, low pressure phase of SiO_2 is that of alpha-quartz. It has a hexagonal unit cell comprising three silicons and six oxygens. Table 1 gives the internal parameters of the basis according to Wycoff [5]. The particular orientation of the system in which we are interested is the so-called "Z" cut [6]. This way of cutting quartz ensures that the temperature dependence of the Young's modulus (or the linear expansivity) is zero near room temperature. Hence operating frequencies are insensitive to temperature with this cut. In Miller Index parlance, this means that the faces of the plate will be: $(\bar{1}, 2, \bar{1}, 0)$, (0001) and $(1, 0, \bar{1}, 0)$. Throughout the rest of this paper the y-direction will represent the longitudinal direction of the plate; and the x-direction will be that in which the plate has maximal amplitude (and in which the plate is thinnest). Thus the yx plane is the (0001), the yz plane is the $(\bar{1}, 2, \bar{1}, 0)$ and the xz is the $(1, 0, \bar{1}, 0)$. Such an orientation allows, in the absence of fluctuations, a rectangular computational cell. For the purpose of the present bulk calculations, we simulated a system comprising 2592 atoms, with periodic boundary conditions, approximately cubic in shape with computational cell lengths near 30 A°. In order to achieve a rectilinear cell, we double the original nine atom unit cell and then use 6x4x6 of the subsequent unit cells to produce the 2592 atoms. We expect finite size effects of the thus-obtained densities and elastic constants to be small and certainly within the precision necessary for meaningful comparison with frequencies obtained via direct simulation.

J.Q. Broughton, C.A. Meli, P. Vashishta and R. K. Kalia

Table 1. Basis vectors for experimental and simulated α-SiO$_2$.

Atom	a	b	c
Experimental [5]:			
Si			
	-0.4650	0.4650	0.3333
	0.4650	0.0000	0.0000
	0.0000	0.4650	0.6667
O			
	0.4150	0.2720	0.1200
	-0.1430	-0.4150	0.4533
	-0.2720	0.1430	0.7867
	0.1430	-0.2720	-0.1200
	0.2720	0.4150	0.5467
	-0.4150	-0.1430	0.2133
Relaxed 0 K:			
Si			
	-0.4729	-0.4729	0.3333
	0.4729	0.0000	0.0000
	0.0000	0.4729	0.6667
O			
	0.4198	0.2640	0.1224
	-0.1558	-0.4198	0.4557
	-0.2640	0.1558	0.7891
	0.1558	-0.2640	-0.1224
	0.2640	0.4198	0.5442
	-0.4198	-0.1558	0.2109
Ensemble Averaged 300 K:			
Si			
	-0.4757	-0.4757	0.3333
	0.4757	0.0000	0.0000
	0.0000	0.4758	0.6667
O			
	0.4214	0.2586	0.1277
	-0.1629	-0.4215	0.4610
	-0.2586	0.1629	0.7943
	0.1629	-0.2586	-0.1277
	0.2586	0.4215	0.5390
	-0.4214	-0.1629	0.2056

For the construction of the plate, we use the same unit cell as for the bulk, but with basis atoms shifted such that the plate surfaces contain oxygens as their outermost atoms; such surfaces have the lowest energy for this potential. Vapor surrounds the flexural plate in the x and z directions, whereas periodic boundary conditions are employed in the y direction. (In reality, the x and z are also periodic, there being 40 A° of vapor surrounding the plate in these dimensions).

Turning now to the interatomic potential, a wide range of such have been employed in the literature to describe the structure and phonon frequencies of crystalline silica. Some are simple pair potentials [16, 17, 22, 21], others are shell-models [19] and still others are three-body [20, 18]. Our potential is of the latter kind. We chose to use the form due to Nakano, Kalia and Vashishta [7] because it is able to describe many of the properties of molten, crystalline and amorphous quartz. Our potential is identical in form to theirs except that values of a few constants are slightly different. These differences were tuned so that specifically crystalline properties would be obtained more accurately. The total potential energy is given as the sum of two-body and three-body terms:

$$V = \sum_{m<n} v_{mn}^{(2)}(r_{mn}) + \sum_{l,(m<n)} v_{mln}^{(3)}(\mathbf{r}_{lm}, \mathbf{r}_{ln}) \tag{2}$$

where r is the distance between atoms, and m and n designate atom indices. The two-body term is written as:

$$v_{mn}^{(2)}(r) = A\left(\frac{\sigma_m + \sigma_n}{r}\right)^{\eta_{mn}} + \frac{Z_m Z_n}{r}e^{-r/\lambda}$$
$$- \frac{\alpha_m Z_n^2 + \alpha_n Z_m^2}{2r^4}e^{-r/\varsigma} \tag{3}$$

where the three terms represent short range repulsion, Coulomb interaction due to ionicity, and a charge-induced dipole interaction caused by atomic polarizabilities, respectively. Note that the Coulomb and polarization terms are damped. Z is the nominal charge on the atom, α is the polarizability and σ is a distance of "closest approach".The two-body term is truncated at $r=r_c$ for ease of computation. For $r< r_c$, $v_{mn}^{(2)}$ is replaced by $v_{mn}^{(2)}(r) - v_{mn}^{(2)}(r_c) - (r - r_c)dv_{mn}^{(2)}/dr_c$, so that the value and its first derivative are zero at the cut-off. The three-body term is given by:

$$v_{mln}^{(3)}(\mathbf{r}_{lm}, \mathbf{r}_{ln}) = B_{mln} \exp\left(\frac{\xi}{r_{lm} - r_0} + \frac{\xi}{r_{ln} - r_0}\right).$$
$$\left(\frac{\mathbf{r}_{lm} \cdot \mathbf{r}_{ln}}{r_{lm}r_{ln}} - \cos\bar{\theta}_{mln}\right)^2 \Theta(r_0 - r_{lm})\Theta(r_0 - r_{ln}) \tag{4}$$

where Θ is a heavyside function. Notice that only the bond bending terms of O-Si-O and Si-O-Si triplets are present in the Hamiltonian. Since diffusion is not important in this system, other triplets are unlikely to be important. The constants here are chosen to get the atomic positions and the bulk modulus right at T=0. Table 2 reports all the potential constants. The only ones differing from Nakano et al. [7] are A (0.30608 eV), Z_{Si} (0.7872 e), Z_O (-0.3936 e) and $\bar{\theta}_{Si-O-Si}$ (143.70 degrees). Their potential was optimized for amorphous silica; the changes here are required for the better description

Table 2. Parameters for SiO_2 interaction potential. The electron charge is denoted as e.

A(eV)	$\lambda(\text{Å})$	$\zeta(\text{Å})$	$r_c(\text{Å})$	$r_0(\text{Å})$	$\xi(\text{Å})$
0.30608	4.43	2.50	5.50	2.60	1.00
	$\sigma_m(\text{Å})$	$Z_m(e)$	$\alpha(\text{Å}^3)$		
Si	0.47	0.7872	0.00		
O	1.20	-0.3936	2.40		
	η_{mn}				
Si-Si	11				
Si-O	9				
O-O	7				
	$B_{mln}(eV)$	$\bar{\theta}_{mln}$			
O-Si-O	4.993	109.47			
Si-O-Si	19.972	143.70			

Table 3. Elastic Constants of α-SiO_2. Units of 10^5 bar.

	Bechmann [13]	Mason [14]	Present Potential
c_{11}	8.674	8.605	6.945
c_{12}	0.699	0.505	1.927
c_{13}	1.191	1.045	2.121
c_{14}	-1.791	1.825	-0.175
c_{33}	10.720	10.710	7.459
c_{44}	5.794	5.865	3.222
c_{66}	3.988	4.050	2.401

of crystalline quartz. The experimental (Wycoff) and relaxed T=0 basis vectors agree very well and are given in Table 1. Table 3 compares two sets of experimental elastic constants with the fully relaxed T=0 values due to this potential. The overall bulk modulus of 3.736×10^5 bar is fit exactly; whereas trends in the individual constants (c_{11} through c_{66}) are obtained qualitatively correctly. Note that the two experimental values for c_{14} are essentially equal but opposite in sign; this difference is still not resolved. For the purposes to which this potential will be put (namely, studying generic QCO behavior), we deem this potential satisfactory.

3. ALGORITHMIC DETAILS

Since we want to obtain the Young's modulus and density of our SiO_2 as a function of T at a constant pressure of 1 bar, it will be necessary to perform constant T, constant P simulations for the bulk. Also, since the range of the two-body and three-body terms are so very different (the three-body terms truncate after nearest neighbors), there will be a tremendous computational speed advantage in using multiple time step methods.

This section describes our (N,P,T) ensemble algorithm including the implementation of a multiple time step method. We use this for the bulk SiO_2 simulations. The method generalizes to the (N,V,E) ensemble used for the flexural plate simulations by setting the masses of the thermostat and cell degrees of freedom to infinity.

3.1. Constant Temperature and Pressure

We use the isothermal-isobaric (N,P,T) ensemble method due to Lill and Broughton [8]. Our equations of motion are generalizations of their non-linear elastic pressure bath to a purely linear elastic theory. Since we are interested in single phases and not the path from one phase to another, the linear theory suffices. These equations allow the computational cell lengths and angles to fluctuate. They also obey the virial theorem and are invariant under a modular transformation. Also, although we use just one Nose-Hoover thermostat [9, 10] for the atomic degrees of freedom, we found that we had to use a Nose-Hoover chain (due to Martyna, Klein and Tuckerman [11]) of length 10 to achieve satisfactory thermostating of the lattice degrees of freedom. Once this was done, the atomic coordinates and lattice degrees of freedom equilibrated to the same required temperature.

The conserved Hamlitonian is then:

$$
\begin{aligned}
\mathcal{H} = & \left(\frac{1}{2} \sum_{n=1}^{N} m_n \dot{q}_i^{(n)} g_{ij} \dot{q}_j^{(n)} \right) + V + \frac{1}{2} W \dot{\epsilon}_{ij} \dot{\epsilon}_{ij} \\
& + |a| P_{ij}^{req} \epsilon_{ij} + f_a k T^{req} \lambda_a + \frac{1}{2} Q_a \dot{\lambda}_a^2 \\
& + f_c k T^{req} \lambda_{c1} + \sum_{p=2}^{N_{chn}} k T^{req} \lambda_{cp} \\
& + \frac{1}{2} \sum_{p=1}^{N_{chn}} Q_{cp} \dot{\lambda}_{cp}^2
\end{aligned}
\tag{5}
$$

where $\{q\}$ are the scaled atomic coordinates; N is the number of atoms, N_{chn} is the length of the Nose-Hoover chain; m, Q, and W are atomic, lattice and thermostat masses respectively; subscripts a and c refer to atomic and cell degrees of freedom respectively; ϵ is the symmetrized strain matrix; g is the metric; $|a|$ is the volume of the computational cell; λ is a thermostat variable; T^{req} and $\mathbf{P}^{req}$ are the externally imposed required T and P; k is Boltzmann's constant; and f represents number of degrees of freedom. f_a has the value (3N-3) because linear total momentum is conserved and f_c is equal to 6 because we specifically work with pure shears (no rotations). i and j represent components of a vector, matrix or tensor. We use a summation convention for repeated indices. Although the Hamiltonian of Eqn. 5 is conserved by the equations of motion, the Hamiltonian of the sub-space in which we are really interested (namely the atomic degrees of freedom) will, of course, fluctuate. But these fluctuations are canonical; that is they represent the (N,P,T) ensemble.

More specifically, the $\{q\}$ are defined by:

$$r_i^{(n)} = a_{ij}q_j^{(n)} \tag{6}$$

where $\{r\}$ are the atomic Cartesian coordinates and a_{ij}, here, represents the appropriate components of the dynamic lattice vector matrix defined by:

$$a = \begin{pmatrix} a_{x1} & a_{x2} & a_{x3} \\ a_{y1} & a_{y2} & a_{y3} \\ a_{z1} & a_{z2} & a_{z3} \end{pmatrix} \tag{7}$$

a_{ij} is the ith Cartesian component of the jth crystalline lattice vector. The metric g_{ij} is equal to $a_{ki}a_{kj}$.

The resulting equations of motion are:

$$\ddot{q}_i^{(n)} = F_i^{(n)} - [\kappa_{ij} + \dot{\lambda}_a\delta_{ij}]v_j^{(n)} \tag{8}$$

where $\mathbf{F}^{(n)}$ is the force on atom n (due to the interatomic potential) and κ is given as:

$$\kappa_{ij} = 2a_{ik}^{-1}\dot{\epsilon}_{kp}a_{pj} \tag{9}$$

These strains are symmetric. By starting the simulation with a symmetric strain matrix (together with its associated velocities), all strains at later times are obtained from the symmetrized associated accelerations. These accelerations are symmetrized from the full strain matrix accelerations $\ddot{\epsilon}$ directly after calculation of the latter:

$$\ddot{\epsilon}_{ij} = \frac{1}{2}[\ddot{\epsilon}_{ij} + \ddot{\epsilon}_{ji}] \tag{10}$$

$$\ddot{\epsilon}_{ij} = \frac{1}{W}[P_{ij} - P_{ij}^{req}]|a| - \dot{\lambda}_{c1}\dot{\epsilon}_{ij} \tag{11}$$

In this way, improper rotations of the lattice are prevented. P_{ij} is the instantaneous virial pressure tensor within the simulation. The lattice vectors are obtained from:

$$a_{ij} = [\delta_{ik} + \epsilon_{ik}]a_{kj}^0 \tag{12}$$

Here, the suprscript zero refers to the value of the lattice vector matrix at the very beginning of the simulation.

Finally, in this sub-section, we turn to the thermostats:

$$\ddot{\lambda}_a = \frac{1}{Q_q} \left[\left(\sum_{n=1}^{N} m_n \dot{q}_i^{(n)} g_{ij} \dot{q}_j^{(n)} \right) - f_a k T^{Req} \right] \tag{13}$$

$$\ddot{\lambda}_{c1} = \frac{1}{Q_c} \left[Q_c \dot{\epsilon}_{ij} \dot{\epsilon}_{ij} - f_c k T^{Req} \right] - \dot{\lambda}_{c1} \dot{\lambda}_{c2} \tag{14}$$

$$\ddot{\lambda}_{cp} = \frac{1}{Q_c} \left[Q_c \dot{\lambda}_{c(p-1)}^2 - k T^{Req} \right] - \dot{\lambda}_{cp} \dot{\lambda}_{c(p+1)} \tag{15}$$

where the index p in Eqn. 15 runs from 2 to N_{chn}. When considering the final thermostat in the chain, the rightmost term in Eqn. 15 is omitted.

3.2. Multiple Time Step

The multiple time step method that we employ is due to Tuckerman, Berne and Martyna [12]. The reader is referred there for a fuller discussion. Briefly, coordinates are advanced through time via a Trotter expansion of the Liouville propagator. A short time approximation to such allows an inner-loop, with short time step, to be established within an outer-loop of longer time step. The inner-loop represents an approximation to the true trajectory; the outer-loop then corrects the approximate path. The algorithmic trick is to put most computational load into the infrequently performed outer-loop step. The particular short-time approximation that is employed [12] is time reversible.

The Liouville operator, $\mathcal{L}$, is defined:

$$i\mathcal{L} = \sum_{n=1}^{N_f} \left[\dot{x}^{(n)} \frac{\partial}{\partial x^{(n)}} + F^{(n)} \frac{\partial}{\partial p^{(n)}} \right] \tag{16}$$

where {x,p} are generic coordinates together with their conjugate momenta. N_f is the number of degrees of freedom in the system. Thus, when we consider atomic and cell and thermostat degrees of freedom, they are all represented in Eqn. 16. Then, the Liouville operator propagtes {x,p} through time:

$$\Gamma(t) = e^{i\mathcal{L}t} \Gamma(0), \qquad \Gamma \equiv \{x, p\} \tag{17}$$

where t is time and (0) represents the state of the system at time zero. If the Liouvillian is written as the sum of two terms (not necessarily the two types inherent in Eqn. 16):

$$i\mathcal{L} = i\mathcal{L}_1 + i\mathcal{L}_2 \tag{18}$$

then Tuckerman et al. [12] showed that a good time reversible approximation to the propagator would be:

$$e^{i\mathcal{L}\Delta t} = e^{i\mathcal{L}_1(\Delta t/2)} \left[e^{i\mathcal{L}_2(\Delta t/h)} \right]^h e^{i\mathcal{L}_1(\Delta t/2)} \tag{19}$$

We see here the nature of the multiple time step algorithm. The innermost propagator is associated with a shorter timestep than the outer ones. If $\mathcal{L}_2$ varies rapidly in time, it requires a shorter time step to integrate it. Short time approximations to the individual components of $\mathcal{L}$ are:

$$\left[e^{\Delta t \dot{x}(0)\frac{\partial}{\partial x}}\right] x(0) = x(0) + \Delta t \dot{x}(0) \tag{20}$$

$$\left[e^{\Delta t F(0)\frac{\partial}{\partial p}}\right] p(0) = p(0) + \Delta t F(0) \tag{21}$$

In advancing the system through time according to Eqns. 17 and 19, we apply the propagators one at a time, starting with the rightmost one and progressing to the left, updating the system sequentially at each stage. If the index h in Eqn. 19 equals unity, then Tuckerman *et al.* [12] showed that the resulting equations are identical, for an (N,V,E) ensemble, to the velocity Verlet algorithm. If h is greater than unity, positions and momenta are updated in the inner-loop while only momenta are updated in the outer-loop.

The algorithmic advantage is obtained by decomposing the forces into short and long range components. It is the near neighbor forces which, due to short range hard-body collisions, determine the shortest timescale in the problem. If we define a switching function, $S(r_{nm})$, which varies monotonically between 0 and 1 such that:

$$\mathbf{F}^{(n)} = -\sum_m \frac{\partial v^{(2)}(r_{nm})}{\partial \mathbf{r}_n} S(r_{nm})$$
$$-\sum_m \frac{\partial v^{(2)}(r_{nm})}{\partial \mathbf{r}_n}[1 - S(r_{nm})]$$
$$-\frac{\partial V^{(3)}}{\partial \mathbf{r}_n} \tag{22}$$

and define the short and long range forces thus:

$$\mathbf{F}_s^{(n)} = -\sum_m \frac{\partial v^{(2)}(r_{nm})}{\partial \mathbf{r}_n} S(r_{nm})$$
$$-\frac{\partial V^{(3)}}{\partial \mathbf{r}_n}$$
$$\mathbf{F}_l^{(n)} = -\sum_m \frac{\partial v^{(2)}(r_{nm})}{\partial \mathbf{r}_n}[1 - S(r_{nm})] \tag{23}$$

Here, $v^{(2)}$ is the two-body interatomic potential and $V^{(3)}$ is the *total* three-body potential of the system. Notice that S only operates upon the two-body part of the potential. Then we can write $\mathcal{L}_1$ and $\mathcal{L}_2$ as:

$$i\mathcal{L}_1 = \sum_{n=1}^{N_f} \left[F_l^{(n)} \frac{\partial}{\partial p^{(n)}}\right] \tag{24}$$

$$i\mathcal{L}_2 = \sum_{n=1}^{N_f} \left[\dot{x}^{(n)} \frac{\partial}{\partial x^{(n)}} + F_s^{(n)} \frac{\partial}{\partial p^{(n)}}\right] \tag{25}$$

The range of switching function (i.e. where it varies most rapidly between 0 and 1) is set between nearest and next nearest neighbors. Thus near neighbor forces are calculated properly within the innermost loop and corrections to the trajectory, due to long range interactions with further neighbors, are made in the outer-loop. Properties should be evaluated after the outer-loop.

In standard MD calculations, neighbor lists are created infrequently (depending upon the diffusion constant) in order to improve computational efficiency. In the present multiple timestep algorithm, the long range forces are calculated at this stage. The neighbor lists will be short for use in the innermost loop because the switching function assures that only nearest neighbors interact. We use the same form of the switching function advocated by Tuckerman *et al.* [12]:

$$
S(r) = \begin{cases} 1, & r < (r_{sw} - \delta) \\ 1 + R^2(2R - 3), & (r_{sw} - \delta) \leq r \leq r_{sw} \\ 0, & r_{sw} < r \end{cases} \tag{26}
$$

where $R \equiv [r-(r_{sw} - \delta)]/\delta$ and where r_{sw} is the cut-off distance of S. δ is the healing length over which S changes from 1 to 0. Fig. 1 shows the radial distribution functions $(g(r))$ for SiO_2 calculated with this potential at 300K. We use three different switching functions, one for each type of pair interaction; i.e. Si-Si, O-O, Si-O. The values of r_{sw} are 3.4, 3.0 and 2.6 A$^\circ$ respectively and δ in each case is 0.6 A$^\circ$. Notice that this value of 2.6 A$^\circ$ for Si-O pair ensures that all three-body terms can be obtained from the Si-O neighbor list. Three-body forces are calculated in the inner-most loop.

Tuckerman *et al.* [12] show how to implement these multiple time-step ideas in the presence of a Nose Thermostat, i.e. an (N,V,T) ensemble. Here we show how to apply them to an (N,P,T) ensemble with Nose-Hoover chain. Care is required with the updating of the cell degrees of freedom. Eqn. 11 shows that the fluctuations in the strain variables depend upon the difference between the internal virial pressure and the externally imposed $\mathbf{P}^{Req}$. One scenario, in the same spirit as having long and short ranged pair forces, is to update strains and their momenta in the inner-loop due to the short-range virial (the virial contains derivatives similar to the force). The strain momenta would then be further updated in the outermost loop due to the long-range virial. Conceptually, there is nothing wrong with this idea, but operationally it does not work. The feedback between $\mathbf{P}$ and $\mathbf{P}^{Req}$ is sufficiently sensitive that breaking the strain update into two parts does not give an average value for $< \mathbf{P} - \mathbf{P}^{Req} >$ of zero to sufficient accuracy.

Instead, we recognize that the pressure is a global property and its fluctuations vary on a slow timescale. Thus all the pressure control is placed in the outer-loop; that is, we use a full ranged virial with no switching function for this control. The value of virial is obtained during the calculation of neighbor lists. The resulting approximation for the entire propagator is found to conserve energy very well; achieves excellent pressure control; and maintains atomic and cell temperatures extremely accurately:

$$
e^{i\mathcal{L}\Delta t} = [e^{(\Delta t/2)F_{lq}\frac{\partial}{\partial p_q}}][e^{(\Delta t/2)[F_\epsilon - \lambda_c p_\epsilon]\frac{\partial}{\partial p_\epsilon}}]
$$
$$
\{[e^{(\Delta t/2h)[F_\lambda - p_\lambda\lambda]\frac{\partial}{\partial p_\lambda}}][e^{(\Delta t/2h)[F_{sq} - \lambda_a p_q - \kappa p_q]\frac{\partial}{\partial p_q}}]
$$

$$\left.\left.\begin{array}{c}[e^{(\Delta t/2h)\lambda\frac{\partial}{\partial\lambda}}][e^{(\Delta t/2h)[\dot{q}\frac{\partial}{\partial q}+\dot{\varepsilon}\frac{\partial}{\partial\varepsilon}]}][e^{(\Delta t/2h)\lambda\frac{\partial}{\partial\lambda}}]\\[2mm] [e^{(\Delta t/2h)[F_{sq}-\lambda_a p_q-\kappa p_q]\frac{\partial}{\partial p_q}}][e^{(\Delta t/2h)[F_\lambda-p_\lambda\lambda]\frac{\partial}{\partial p_\lambda}}]\end{array}\right\}^h\right.$$

$$[e^{(\Delta t/2)[F_\varepsilon-\lambda_c p_\varepsilon]\frac{\partial}{\partial p_\varepsilon}}][e^{(\Delta t/2)F_{lq}\frac{\partial}{\partial p_q}}]$$

$$(27)$$

where significant license with shorthand notation has been employed for brevity. Comparison of Eqn. 27 with those in the equations of motion subsection above should, however, facilitate clarity. F_q, F_ε and F_λ represent the forces associated with the atomic, strain and thermostat degrees of freedom, respectively. F_q pertains to derivatives of the potential energy with respect to atomic displacement; F_ε refers to the $[P-P^{Req}]$ term of Eqn. 11; and F_λ refers to the feedback terms (i.e. those in the [ ]) in Eqns. 13, 14 and 15 depending upon the type of Nose-Hoover thermostat. Note that λ_a and the end-of-chain λ_c have no $p_\lambda\dot{\lambda}$ correction to F_λ in Eqn. 27. p_q, p_ε and p_λ are momenta. The subscripts l and s, as before, refer to the long and short range forces. The next subsection should clarify Eqn. 27 further.

3.3. Final Algorithm

After combining the equations of motion of the two prior sub-sections, the following loop structure is obtained. For each time step (Δt), we first update $\{\dot{q}, \dot{\varepsilon}\}$ sequentially, then we perform an inner-loop in which we update $\{\dot{\lambda}_a, \dot{\lambda}_c, \dot{q}, \lambda_a, \lambda_c, q, \varepsilon, \dot{q}, \dot{\lambda}_c, \dot{\lambda}_a\}$ in sequence; and finally in the outer loop once more, we update $\{\dot{\varepsilon}, \dot{q}\}$ again, in sequence. The entire algorithm follows:

a) Outer loop:

$$\dot{\mathbf{q}}^{(n)} = \dot{\mathbf{q}}^{(n)} + \frac{\Delta t}{2m_n}\mathbf{F}_{lq}^{(n)} \tag{28}$$

Now update $\mathbf{F}_\varepsilon$. Then:

$$\dot{\varepsilon} = [\dot{\varepsilon} + \frac{\Delta t}{4W}\mathbf{F}_\varepsilon]e^{-(\Delta t/2)\lambda_\varepsilon} + \frac{\Delta t}{4W}\mathbf{F}_\varepsilon \tag{29}$$

Now update $\mathbf{F}_\lambda$ for λ_a and for the beginning-of-chain λ_c. Also update κ.

b) Inner loop:
First the velocities:

$$\dot{\lambda}_a = \dot{\lambda}_a + \frac{\Delta t}{2hQ_a}\mathbf{F}_\lambda \tag{30}$$

$$\begin{aligned}
\dot{\lambda}_{cp} &= \dot{\lambda}_{cp}e^{-(\frac{\Delta t}{4h})\dot{\lambda}_{c(p+1)}}, \qquad p=1...(N_{chn}-1)\\[2mm]
F_\lambda^{(cp)} &= [Q_c\dot{\lambda}_{c(p-1)}^2 - kT^{Req}], \qquad p=2...N_{chn}\\[2mm]
\dot{\lambda}_c &= \dot{\lambda}_c + \frac{\Delta t}{2hQ_c}F_\lambda\\[2mm]
\dot{\lambda}_{cp} &= \dot{\lambda}_{cp}e^{-(\frac{\Delta t}{4h})\dot{\lambda}_{c(p+1)}}, \qquad p=1...(N_{chn}-1)
\end{aligned} \tag{31}$$

$$\dot{\mathbf{q}}^{(n)} = \dot{\mathbf{q}}^{(n)} + \frac{\Delta t}{4hm_n}\mathbf{F}_{sq}^{(n)}$$

$$\dot{q}_i^{(n)} = \dot{q}_i^{(n)} - \frac{\Delta t}{4h}\sum_{j\neq i}\kappa_{ij}\dot{q}_j^{(n)}$$

$$\dot{q}_i^{(n)} = \dot{q}_i^{(n)}e^{-(\Delta t/2h)(\lambda+\kappa_{ii})} \tag{32}$$

$$\dot{q}_i^{(n)} = \dot{q}_i^{(n)} - \frac{\Delta t}{4h}\sum_{j\neq i}\kappa_{ij}\dot{q}_j^{(n)}$$

$$\dot{\mathbf{q}}^{(n)} = \dot{\mathbf{q}}^{(n)} + \frac{\Delta t}{4hm_n}\mathbf{F}_{sq}^{(n)}$$

Then the coordinates:

$$\lambda_a = \lambda_a + \frac{\Delta t}{h}\dot{\lambda}_a \tag{33}$$

$$\lambda_c = \lambda_c + \frac{\Delta t}{h}\dot{\lambda}_c \tag{34}$$

$$\mathbf{q}^{(n)} = \mathbf{q}^{(n)} + \frac{\Delta t}{h}\dot{\mathbf{q}}^{(n)} \tag{35}$$

$$\varepsilon = \varepsilon + \frac{\Delta t}{h}\dot{\varepsilon} \tag{36}$$

Calculate short-ranged atomic forces, update κ and return to the velocities again: Update $\{q\}$ using Eqn. 32; and then update $\{\lambda_c\}$ and λ_a using Eqns. 31 and 30.

c) Outer loop:
Generate new neighbor lists, calculate $\mathbf{F}_{lq}$ and evaluate $\mathbf{F}_\varepsilon$. Then $\dot{\varepsilon}$ and $\dot{\mathbf{q}}$ are evaluated in sequence using Eqns. 29 and 28.

This completes the outer loop. The next iteration begins by jumping back to (a) above. The inner loop value of h is typically 10; that is, the inner loop is executed ten times. The equations are evaluated in the sequence given and the meaning of each is the same as that in FORTRAN; i.e. the new value of a variable (that on the left) is set equal to the expression on the right. By the time that step (c) is completed, all coordinates and velocities have been updated by Δt. The total energy/atom of Eqn. 5 is conserved to the fifth decimal place when reported in electron volts. We found it unnecessary to distinguish between Q_a and Q_c; i.e they were set equal.

The value of $(\Delta t/h)$ that we used was 5×10^{-16} sec. And the values of Q and W were 6.0×10^{-26} eV sec^2.

4. BULK PROPERTIES

4.1. Elastic Properties

The quantities of interest are density, Young's modulus and Poisson ratios. The latter are defined:

$$Y = L_y \left(\frac{\partial P_{yy}}{\partial L_y}\right)_T \tag{37}$$

$$u_x = -\sigma_x u_y$$
$$u_z = -\sigma_z u_y \qquad (38)$$

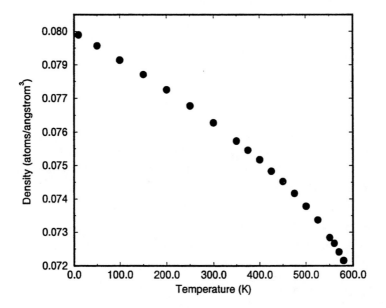

Figure 1. Density versus temperature.

where, as we have said, the direction of compression / elongation is along the y-axis and the x and z directions are allowed to "breathe" in response to the 1 bar external hyrostatic pressure. The Poisson ratio measures the degree to which compression/elongation along one axis affects the length along another. Since SiO_2 is not isotropic the Poisson ratios, σ_x and σ_z, in the x and z directions will not be equal. u is the relative displacement ($\Delta L/L$ etc.). The derivatives inherent in Eqns. 37 and 38 are evaluated numerically. At each temperature, three simulations were performed corresponding to $P_{yy} = 1$ and ± 3000 bar. In each case, off diagonal components of the pressure tensor were set to zero while P_{xx} and P_{zz} were set to unity. A least squares linear fit is then performed on these results to obtain Y, σ_x and σ_z. Each simulation required a 3,000 Δt equilibration followed by a 6,000 Δt statistics phase. Elasticity results are not given above 580K because a phase transition occurs to the β phase (see below).

Fig. 1 shows the dependence of density upon temperature. The error bars here are easily within the size of the symbols. Throughout the entire temperature range the angles between the cell vectors remain at 90 degrees.

Fig. 2 illustrates what happens to the computational cell lengths as a function of temperature and pressure. (For the sake of clarity-of-viewing, these distances are not represented as unit cell lengths; to do so, the x,y and z lengths of Fig. 2 should be divided by 6,4 and 6 respectively). There are three curves for each direction corresponding to applied P_{yy} pressures of -3000,1 and 3000 bar. The positive P_{yy} pressure gives rise to a smaller y-length and larger x- and z-lengths. The slope of each line is positive, in contradistinction with what is observed in experiment. The reason why

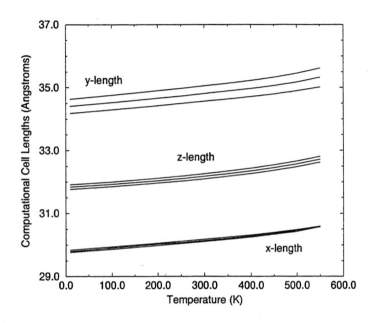

Figure 2. Computational cell lengths versus temperature. See text.

QCOs are oriented with the "Z-cut" is that near room temperature, the thermal linear expansivity goes through an extremum which translates into a Young's modulus which is invariant to temperature [6]. In other words extremely good thermal stability is achievable under ambient conditions for these QCOs. We note that our system shows no such behavior. We have two explanations for this difference; the most obvious is that our interatomic potential, as we have already noted, is imperfect. But a more likely explanation is that these simulations are purely classical, whereas the phonon population at these temperatures is almost certainly determined by the Planck distribution. (A quick look at the high frequency cut-off of any experimental quartz phonon spectrum indicates that the Debye temperature will be in excess of 1000K [15]). Although the statistical mechanics of long wavelength acoustic modes is likely to be treated properly in these MD simulations, the high frequency modes are almost certainly incorrectly treated and it may be their temperature dependent population which leads to the experimentally observed invariance of the Young's modulus near 300K.

The cell length information of Fig. 2, with the use of Eqns. 37 and 38, produces the Young's modulus and x and z-Poisson ratios of Figs. 3 and 4. Notice that there is quite a strong variation of modulus with temperature, something that we require to know for accurate analysis of our direct device frequency evaluations.

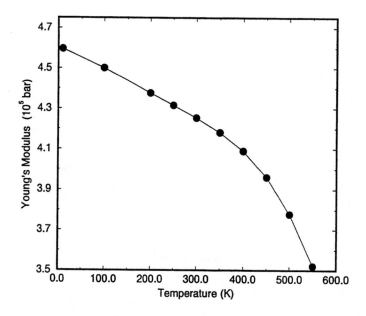

Figure 3. Young's Modulus versus temperature. Lines are to guide the eye.

4.2. β-Phase Vectors

Above 580K and at a pressure of 1 bar, changes in the slope of energy and density versus temperature indicate a phase transition to have occurred. This transition occurs without hysteresis; upon cooling, the same path for intensive quantities is retraced. Fig. 5 shows the heat capacity versus temperature for the system. (This property is obtained simply by numerical differentiation of the enthalpy versus temperature data). There is a clear discontinuity between 580 and 590K. This (within the precision achievable by MD), combined with the enthaply data, indicates a second order phase transition. Fig. 6 illustrates that second neighbor distributions change for each of the Si-O, O-O and Si-Si pair correlation functions in going from 300 to 700K. That these observations are at least compatible with the known α to β transition in quartz is clearly shown in Tables 1 and 4. Table 1 compares basis vectors for the system at 0K and 300K. The latter are ensemble averages. Using Wycoff's [5] nomenclature, the parameter which controls the Si positions, u, changes from 0.473 to 0.476 with temperature. His x,y,z parameters, which determine O positions, change from 0.420, 0.264, 0.122 to

0.421, 0.258, 0.128 respectively with temperature. Table 4, in contrast, compares the experimental basis positions of the high temperature β-phase with those obtained via the present simulation at 700K. The β-phase, like that of the α, is also hexagonal. Note the excellent agreement between the two. Well defined fractions, like 1/2, 1/3, 1/6 and 5/6 are obtained with high accuracy from the 700K ensemble. And the only internal parameter (which again Wycoff calls "u") has an experimental value of 0.197 which is to be contrasted with a value of 0.213 in the present simulation. Wycoff notes that "when a single crystal of low-quartz is cautiously heated above 575°C, it gradually and without violence changes into a single crystal of the high-temperature form". Apart from a difference of 270K between the experimental and simulation transition temperatures, the results reported here are in accord with this depiction of the phase-change.

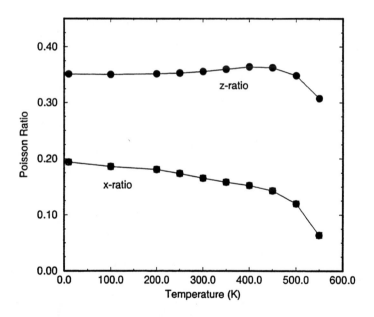

Figure 4. Poisson ratios versus temperature. Error bars $\pm$ 0.02 at the highest temperature. Lines are to guide the eye.

Several models have been advanced in the literature to explain the observed long wavelength phonon mode softening thought to be responsible for the α to β transition. The ramifications of such models and how they fit the neutron scattering data are to be found in references 23 and 24. Whether the precise pathway for the transition is accurately captured by our Hamiltonian and system size is beyond the scope of this paper.

J.Q. Broughton, C.A. Meli, P. Vashishta and R. K. Kalia

Table 4. Basis vectors for experimental and simulated β-SiO$_2$.

Atom	a	b	c
Experimental [5]:			
Si			
	0.5000	0.5000	0.3333
	0.5000	0.0000	0.0000
	0.0000	0.5000	0.6666
O			
	0.3940	0.1970	0.1666
	-0.1970	-0.3940	0.5000
	-0.1970	0.1970	0.8333
	0.1970	-0.1970	0.8333
	0.1970	0.3940	0.5000
	-0.3940	-0.1970	0.1666
Ensemble Averaged 700 K:			
Si			
	0.4999	0.4997	0.3333
	0.5003	0.0000	0.0000
	0.0000	0.5001	0.6667
O			
	0.4262	0.2128	0.1667
	-0.2134	-0.4260	0.5002
	-0.2126	0.2134	0.8338
	0.2135	-0.2123	0.8329
	0.2125	0.4256	0.4997
	-0.4262	-0.2135	0.1665

5. PLATE MODULUS AND ANALYSIS

Systems in the sub-micron regime have large surface to volume ratios. A significant surface contribution to the effective Young's modulus of the device can be expected. It is the actual Young's modulus of the specific device which is required to predict, via continuum mechanics, the frequency of oscillation of the flexural plate. To quantify this, we evaluated the Young's modulus of flexural plates of varying cross sectional area.

We used a constant x:z aspect ratio of approximately 1:2 and evaluated the Young's modulus at T=10 and 300K. Specifically, a constant number of seven y unit cells were employed and we varied the number of x and z cells subject to the constraint that the number of z cells was twice that of the x. The cells are defined in section II above.

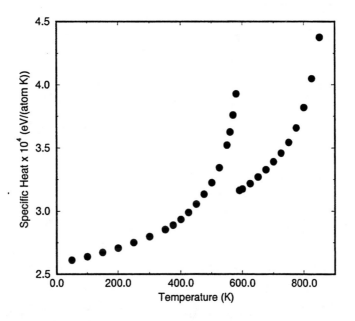

Figure 5. Specific heat versus temperature.

For each system size and temperature, the plate was equilibrated over 50,000 time steps for y cell lengths near $P_{yy}=0$. A further 50,000 steps were then employed to gather pressure statistics. A linear fit was then performed on the pressure versus length data to extract (see Eqn. 37) the Young's modulus. Note that in evaluating the pressure, the volume used for the virial expression is that of the plate itself which is obtained (a) from the known length of the plate and (b) the cross-sectional area, A, defined by examining the density profiles in the x and z directions. Linear density profiles show oscillatory behavior in crystals; the outer maximum in each direction is used to define the width of the plate. For the purposes of the present analysis, any ambiguity in the definition of the volume of the plate is of little consequence in the analysis which follows. System sizes ranged up to 100,800 atoms (20x7x40 unit cells).

Fig. 7 presents our raw data. Included also are results for a system with 1% of vacancies. Such vacancies are placed at random throughout the system and comprise both Si and O (in the ratio 1:2 to maintain charge neutrality). Such systems are of interest because commercial grade quartz, such as that used in many oscillator applications contain defects (hydrogen, aluminum etc) at similar levels (see [4]). While we do not currently have useful Si, O, Al, H interacting interatomic potentials, we *can* ascertain the gross effects of defects at these concentrations by the simple expedient of using random vacancies.

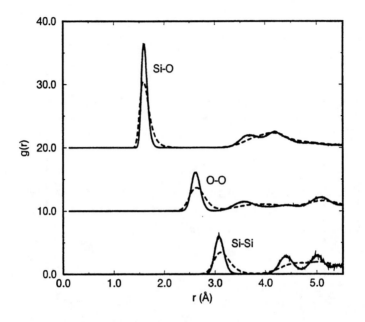

Figure 6. Radial Distribution Functions. Full line 300K. Dashed line 700K.

In order to obtain the asymptotic behavior of modulus with increasing system size, it is necessary to fit this data to an appropriate analytic expression. In developing such, we imagine that the core of the plate has the bulk value (at given temperature) of the Young's modulus. We then assume that the surface skin has a different value for its modulus. We tried the following *ansatz* for the modulus of the bar:

$$Y = Y_B \left(\frac{A_B}{A_{Tot}} \right) + Y_S \left(\frac{A_S}{A_{Tot}} \right) \tag{39}$$

where subscripts B and S refer to bulk and surface respectively. The cross-sectional area associated with the surface skin, assumes a surface skin depth. Of course, the sum of A_B and A_S must equal A_{Tot}. This equation makes no distinction between the yx and yz surfaces; rather, it treats them as an average. For a given set of data (eg. the 10K data of Fig. 7), there is no unique best fit for Eqn. 39 if both the skin depth and Y_S are allowed to vary as fitting parameters. Rather, one must be fixed. When this is done, the equation does not fit the data adequately.

The reason, we believe, is that corners (i.e. edges) of the plate are not properly treated. We therefore modified Eqn. 39 thus:

$$Y = Y_B \left(\frac{A_B}{A_{Tot}} \right) + Y_S \left(\frac{A_S}{A_{Tot}} \right) + Y_C \left(\frac{A_C}{A_{Tot}} \right) \qquad (40)$$

where subscript C refers to the corners. The sum of A_B, A_S and A_C now equals A_{Tot}. Again, for a given data set, there is no unique best fit if the skin depth, Y_S and Y_C are allowed to vary, but there is a best fit if the skin depth is fixed and the surface and corner moduli are allowed to vary. We set the skin depth to a value of 10 A°. This model now does fit the data satisfactorily, the result being shown as continuous lines in Fig. 7. The horizontal dashed lines represent the bulk asymptotes (from section IV) to which the moduli must tend.

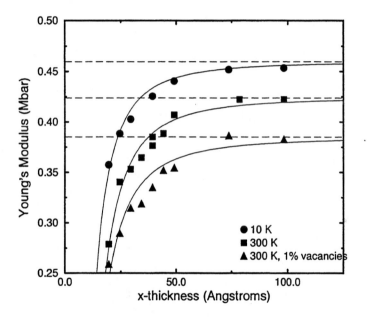

Figure 7. Young's modulus versus x-width for three different plates; 10K, 300K, and 300K with 1% vacancies. x:z aspect ratio constant at 1:2. Continuous lines are from Eqn. 40. Horizontal dashed lines are bulk limit.

Table 5 gives the values of the fit to Eqn. 40. Note that the arbitrary, but reasonable, choice of 10 A° for the skin depth produces surface Young's moduli slightly in excess of the bulk values. This should not be taken too literally; this is a fit to an *ansatz*, albeit one, we feel, which has physical basis. Larger values of the skin depth act to decrease Y_S.

The conclusion from this analysis is that for a plate with a typical aspect ratio of 1:14:2, the effective modulus of the bar will achieve 90% of the bulk value at 5.3×10^4, 95% at 1.8×10^5 and 99% at 1.5×10^6 atoms. These conclusions are independent of temperature or of the vacancy concentration. These self-same atom numbers correspond to device lengths of 0.09, 0.13, 0.26 microns respectively. Thus the asymptotic approach to bulk elastic behavior is occurring in precisely the size regime to which next generation QCOs aspire. This general observation is unlikely to change with future improvements in the form of the SiO_2 interatomic potential.

Table 5. Young's modulus fit for flexural plate according to Eqn. 40. Moduli in units of Mbar.

	10K Plate	300K Plate	300K Plate 1% Vacancies
Y_B	0.460	0.424	0.385
Y_S	0.463	0.429	0.386
Y_C	0.223	0.101	0.082

Whether the actual oscillator frequencies converge on macroscopic behavior in the same size regime is the subject of a subsequent paper. Such calculations require the use of a parallel machine, but below we study an 18,000 atom (17.2 nm) plate whose simulation can be performed on a desktop workstation.

6. DIRECT SIMULATIONS OF PLATES

The system comprised 5x20x10 unit cells in the x,y and z dimensions. The surfaces, as in the plate elastic constant calculations described above, were all oxygen terminated. We considered both perfect and 1% vacancy SiO_2 plates. Periodic boundary conditions are applied in all three directions, but with a vapor region surrounding the plate in the x and z dimensions. The center of gravity (CG) of the y *ends* of the plate are "constrained" in the x and z directions by adding two harmonic spring terms to the Hamiltonian of the system. Since the unperturbed plate's center of gravity is placed at (0,0,0) in our coordinate system, the two additional terms are $K.(CG_x)^2$ and $K.(CG_z)^2$ where K is a judiciously chosen spring constant. The center of gravity used in these expressions pertain to a y width of two unit cells at the ends of the plate (i.e one at low values of y, the other at high). The effect of these soft constraints is to mimic the effect of clamps at either end of the plate. It would be possible to mimic further the effect of clamps by adding a random and dissipative heatbath to the ends, but this we chose not to do here.

After equilibration of each system at 10 and 300K, such that the mean value of P_{yy} is near zero bar, their oscillatory motion was initiated by deforming (i.e. plucking) them in the x-direction according to the continuum mechanical solution for a plate with clamped ends (which is distinct from a plate with freely pivoting ends). Such x-displacement is of the form:

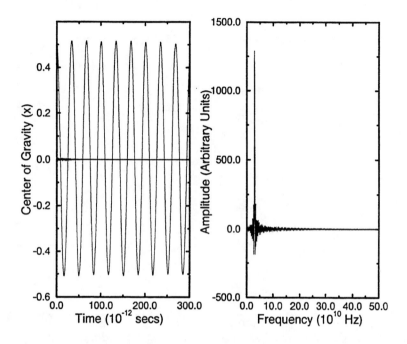

Figure 8. Oscillation behavior of 10K perfect plate. (a) Center of gravity versus time. Horizontal line is center of gravity of ends of plate. (b) Cosine transform.

$$x(y) = Q[(\sin \kappa L_y - \sinh \kappa L_y)(\cos \kappa y - \cosh \kappa y) \\ -(\cos \kappa L_y - \cosh \kappa L_y)(\sin \kappa y - \sinh \kappa y)] \tag{41}$$

and is applied to all atoms of the equilibrated undeflected state. Q is an amplitude factor. κ is obtained from the roots of:

$$\cos \kappa L_y \cosh \kappa L_y = 1 \tag{42}$$

the first three of which are 4.7300, 7.8532 and 10.9956. The characteristic frequencies are given (in radians/sec) by:

$$\omega = \frac{\kappa^2 d}{(\sqrt{12})L_y^2}\sqrt{\frac{Y}{\rho}} \tag{43}$$

3/3

The appropriate Young's modulus to include here, of course, should be that which pertains to the plate, of given temperature and cross-sectional area, under consideration. Typical displacements at the mid-point of the plate are 1 A°, the plate being 172 A° long. This corresponds to less than 1° of deflection.

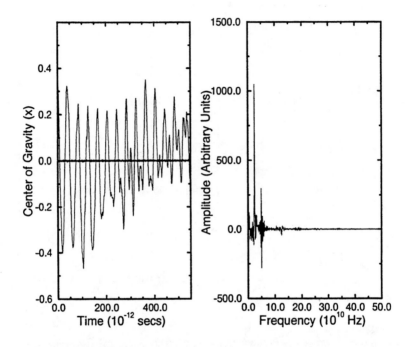

Figure 9. Oscillation behavior of 300K perfect plate. (a) Center of gravity versus time. Horizontal line is center of gravity of ends of plate. (b) Cosine transform.

Figs. 8 and 9 give the x-center of gravity versus time and the associated cosine transform for the perfect plates at 10 and 300K. (The center of gravity is for the entire plate). Notice firstly that the 10K perfect system behaves very harmonically. The cosine transform gives a clean single peak at 3.05×10^{10} Hz. In contrast, the 300K perfect plate's CG behavior is very clearly anharmonic and the Fourier transform indicates a principle peak at 2.44×10^{10} Hz with a smaller but still significant one at 4.95×10^{10} Hz. There is also a feature at 12.45×10^{10} Hz. By dividing the plate into 20 y-sections and monitoring the CG of each of these versus time, it is possible to determine to what modes these frequencies relate. Fig. 10 shows the modes for the three frequencies. Note that the 2.44 and 4.95×10^{10} Hz states have a single maximum in the center of the plate, whereas the 12.45×10^{10} has two nodes. The 4.95×10^{10} state is therefore a second harmonic and the 12.45×10^{10} state represents the third excited state of the plate. No second excited state is present because it is disallowed by symmetry. Second harmonics are only allowed when significant anharmonicity is present [3]. The difference in behavior for the two temperatures is truly dramatic. We strongly suspect that, at the higher temperature, the motion of the surface atoms is highly anharmonic which, such is the large surface to volume ratio for these system sizes, couples significantly to the oscillatory dynamics of the whole plate.

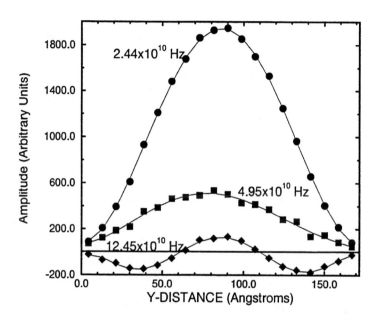

Figure 10. Flexural Modes of 300K Perfect Plate. Data points fit with a sixth order polynomial to guide the eye.

Turning now to the systems with vacancies. Fig. 11 shows, side by side, the dynamics of the 1% vacancy systems at both 10K and 300K. The 10K system appears very harmonic, but notice what has happened to the baseline. The amplitude fluctuations should be distributed equally around zero; that they are not shows that after the initial displacement, a memory of the event is stored in the system. In other words, significant plasticity occurs and the plate is permanently curved. In contrast, at 300K, not only is the plate permanently curved, but also the oscillatory behavior is extremely anharmonic. The Fourier transforms (not shown) of these two systems produce a clean single peak spectrum and a noisy multiple peak spectrum respectively.

The frequency spectra obtainable via direct simulation can be compared with those predicted from continuum mechanics. The comparison is shown in Table 6. The Young's modulus quoted is specific to the plate that was simulated. For this system size, the discrepancy between the continuum prediction and the frequency obtained via direct simulation would have been much larger if the *bulk* value of the modulus had been employed. As it is, we still see significant differences. At 10K, there is a 10% difference, while at 300K the difference is approximately 30%. The third state which is just determinable at 300K (see Fig. 9) at 12.45×10^{10} Hz is to be compared with a prediction of 17.22×10^{10} Hz which constitutes a 38% difference.

Table 6. Comparison of MD frequencies and those predicted by Continuum Mechanics. Moduli in Mbar. Frequencies in 10^{10} Hz.

System	Y	ω_{MD}	ω_{CM}
Perfect Plate 10K	0.388	3.05	3.36
Perfect Plate 300K	0.335	2.44	3.19
Vacancy Plate 10K	0.343	2.82	3.17

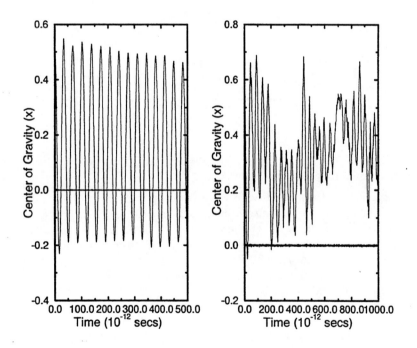

Figure 11. Oscillation behavior of 1% vacancy plate. (a) Center of gravity versus time at 10K. (b) Center of gravity versus time at 300K. Horizontal lines are centers of gravity of ends of plate.

7. CONCLUSIONS

This paper has described an interatomic model suitable for the simulation of bulk crystalline quartz and for the description of quartz crystal oscillators. Even though this potential may not duplicate all the details of experimental quartz, there is sufficient physics contained therein that the qualitative details of how QCOs behave as a function of system size is likely to be correct. The paper has also described a multiple time step algorithm suitable for an (N,P,T) ensemble appropriate for bulk systems described by three-body potentials.

The principal objective here has been to obtain elastic properties germane to the analysis of the "Z-cut" oscillator frequencies directly evaluated in the latter part of the paper. These properties were obtained for bulk systems as well as for plate-geometries as a function of plate cross-sectional area. We predict that asymptotic approach to within 1elastic value occurs at device sizes just below 1 μ. Also, we find that for oscillators that are approximately 20 A^o long, estimates of oscillator frequency based upon continuum mechanics, even when elastic constants appropriate for the specific device size under consideration are used, are in error by 10-30 % depending upon the temperature.

A by-product of this work has been to show that Poisson-ratios in the directions normal to the "Z-cut" are not isotropic and that the α to β transition is faithfully reproduced with our interatomic potential albeit at a temperature 270K below experiment.

We have shown that evaluation of device frequency is possible via direct atomistic simulation and that whereas device behavior is very harmonic at T=10K, the surface plays an important role in introducing anharmonicity at 300K for 20 A^o devices. Behavior of 0.1 μ devices is presently being studied by methods similar to those employed here on parallel machines. Such studies will be the subject of a future paper. We further showed that effects on device characteristics can be observed via atomistic simulation when defects at the 1 % level are present.

ACKNOWLEDGMENTS

JQB wishes to thank ONR for partial support of this work and CAM wishes to acknowledge the National Research Council for a postdoctoral fellowship.

REFERENCES

[1] J.W.Grate, S.J.Martin and R.M.White Anal. Chem. **65**, 940A (1993).

[2] J.W.Grate, S.J.Martin and R.M.White Anal. Chem. **65**, 987A (1993).

[3] L.D.Landau and E.M.Lifshitz, *Theory of Elasticity*, Pergamon Press, Oxford, (1984).

[4] See for example, *Proceedings of the IEEE International Frequency Control Symposium, (1995)*, San Francisco. IEEE#95CH-375-2.

[5] R.W.G. Wyckoff, *Crystal Structures*, John Wiley & Sons, New York, Vol. 1, (1963).

[6] V.E. Bottom, *Introduction to Quartz Crystal Unit Design*, Van Nostrand Reinhold, New York, (1982).

[7] A.Nakano, R.K.Kalia and P.Vashishta, J.Non-Cryst.Sol. **171**, 157 (1994).

[8] J.V.Lill and J.Q.Broughton, Phys.Rev. **B49**, 11619 (1994).

[9] S.Nose Mol.Phys. **57**, 187 (1986).

[10] W.G.Hoover, Phys.Rev. **B34**, 2499 (1986).

[11] G.J.Martyna, M.L.Klein and M.Tuckerman, J.Chem.Phys. **97**, 2635 (1992).

[12] M.Tuckerman, B.J.Berne and G.J.Martyna, J. Chem.Phys. **97**, 1990 (1992).

[13] R.Bechmann, Phys.Rev., **110**, 1060 (1958).

[14] W.P.Mason *Piezoelectric Crystals and their Applications to Ultrasonics*, Van Nostrand, New York, (1950), 84.

[15] D.Strauch and B.Dorner, J.Phys.:Condens.Matt. **5**, 6149 (1993).

[16] M.M.Elcombe, Proc.Phys.Soc. **91**, 947 (1967).

[17] T.H.K.Barron, C.C.Huang and A.Pasternak, J.Phys.C. **9**, 3925 (1976).

[18] M.E.Striefler and G.R.Barsch, Phys.Rev. **B12**, 4553 (1975).

[19] H.Schober, D.Strauch, K.Nutzel and B.Dorner, J.Phys. Condens. Matter **5**, 6155 (1993).

[20] T.Pilati, F.Demartin and C.M.Gramaccioli, Acta Cryst. **B50**, 544 (1994).

[21] S.Tsuneyuki, M.Tsukaada, H.Aoki and Y.Matsui, Phys.Rev.Lett. **61**, 869 (1988).

[22] R.G.D.Valle and H.C.Andersen, J.Chem.Phys. **94**, 5056 (1991).

[23] H.Boysen, B.Dorner, F.Frey and H.Grimm, J. Phys. C **13**, 6127 (1980).

[24] J.Bethke, G.Dolino, G.Eckold, B.Berge, M.Vallade, C.M.E.Zeyen, T.Hahn, H.Arnold and F.Moussa, Europhys. Lett. **3**, 207 (1987).

ELECTRONIC STRUCTURE OF CONDUCTING CARBON NANOTUBES

J. W. Mintmire, R. A. Jishi,[1] and C. T. White
Theoretical Chemistry Section, U. S. Naval Research Laboratory
Washington, DC 20375, U. S. A.

ABSTRACT

Recent developments using synthetic methods typical of fullerene production have been used to generate graphitic nanotubes with diameters on the order of fullerene diameters: "carbon nanotubes". Recent work by Thess, *et al.*suggests that [10,10] carbon nanotubes with diameters of 1.38 nm have been observed with high conductivities. This structure belongs to the same serpentine ([n,n]) family of nanotubes as the [5,5] carbon nanotube theoretically shown to have electronic states appropriate for high conductivity in our earlier work. We have calculated the electronic structure for this high-symmetry [10,10] carbon nanotube using a first-principles, self-consistent, all-electron Gaussian-orbital based local-density functional approach to calculate the band structure. We find that the electronic structure results are similar in band width and behavior near the Fermi level to our earlier results for the [5,5] nanotube.

1. INTRODUCTION

Current research efforts on carbon nanotubes have expanded rapidly over the last few years as recent experimental results have confirmed early theoretical predictions of their anomalous electronic and structural properties [1]. The major efforts over the last few years to synthesize, characterize, and understand the basic properties of carbon nanotubes result from the conjunction of two important areas of chemical research that have evolved over the last several decades: fullerenes and carbon fibers. Fullerene science has developed rapidly over the past decade [2-4], dating from the initial observations by Kroto, Curl, Smalley, and coworkers that laser vaporization of graphite

---

[1]*Permanent address*: Department of Physics, California State University, Los Angeles, CA 90032 USA.

produced uniquely abundant, stable C_{60} clusters in supersonic carbon cluster beams [5]. This experimental observation led to their suggestion of a high-symmetry, truncated icosahedral structure forming a closed cage of spherically-arranged carbon atoms, and denoted as buckminsterfullerene in honor of Buckminster Fuller and his geodesic domes, with this work subsequently leading to the 1996 Nobel prize in chemistry. Buckminster-fullerene, C_{60}, is the most prominent member of a family of structurally related, hollow, carbon cages composed solely of five- and six-membered rings formed from three-fold coordinated carbon atoms. The second major area of graphite fibers has been an active area of research for several decades, dating back to the 1960's and earlier [6]. In 1991 Iijima [7] reported a novel synthesis based on the techniques used for fullerene synthesis of substantial quantities of multiple-shell graphitic nanotubes with diameters of nanometer dimensions. These nanotube diameters were more than an order of magnitude smaller than those typically obtained using routine synthetic methods for graphite fibers [8, 9]. This work has been widely confirmed in the literature, with other subsequent work demonstrating the synthesis of bulk quantities of these materials [10] and the further synthesis of abundant amounts of single-shell graphitic nanotubes with diameters on the order of one nanometer [11, 12]. Early theoretical work by Mintmire, *et al.* [13], published concurrently with the experimental work by Iijima, speculated that the serpentine high-symmetry conformations of carbon nanotubes would be highly conductive, basing these suggestions on first-principles band structure calculations of a 0.7 nm diameter carbon nanotube. Other subsequent theoretical work suggested that carbon nanotubes could have other conformations with semiconducting properties [14, 15] and anomalous structural properties [16]. Recent experimental work on single-wall nanotubes has strongly supported these early suggestions [17-19], with an increased interest in these materials' potential applications as constituents in novel nanoscale materials and devices [17-21]. The recent work by Thess, *et al.* [18] is of particular interest because it suggests that [10,10] carbon nanotubes with diameters of 1.38 nm have been synthesized and have a measured resisitivity of less than 10^{-4} Ω-cm at 300K. This structure belongs to the serpentine ([n,n]) family of nanotubes that we originally predicted to be metallic [13, 14]. In the original work, we reported detailed first-principles results for the [5,5] nanotube. Stimulated by the results by Thess, *et al.* [18] we have undertaken a detailed study of the [10,10] nanotube. Herein we report preliminary first-principles results for the [10,10] nanotube. These results will be the basis for further calculations on properties of interest such as optical response and photoelectron spectra for these novel materials. We describe our results for the [10,10] nanotube using a first-principles, self-consistent, all-electron Gaussian-orbital based local-density functional approachto calculate the band structure, and discuss some of the implications of the rotational and helical symmetry on the band structure. We have recently extended a first-principles local-density functional (LDF) approach, developed originally to treat the electronic properties of chain polymers with translational periodicity [22], to calculate the total energies and electronic structures of chain polymers with helical symmetry [23]. This method calculates the total energy and the electronic structure using local Gaussian-type orbitals within a one-dimensional band structure approach. Such an approach is particularly appropriate for the nanotubes, because while the nanotubes can be constructed with one-dimensional translational symmetry, the translational repeat distance leads to extremely large unit cell sizes.

2. APPROACH

Each nanotube can be visualized as a conformal mapping of a two-dimensional honeycomb lattice (depicted in Fig. 1) onto the surface of a cylinder subject to periodic boundary conditions both around the cylinder and along its axis [7,13-16,24-25]. The proper boundary condition around the cylinder can only be satisfied if the cylinder circumference maps to one of the Bravais lattice vectors of the graphite sheet [24]. Thus each real lattice vector of the two-dimensional hexagonal lattice (the Bravais lattice for the honeycomb) defines a different way of rolling up the sheet into a nanotube. Each such baseline lattice vector, $\mathbf{B}$, can be defined in terms of the two primitive lattice vectors $\mathbf{R}_1$ and $\mathbf{R}_2$ and a pair of integer indices $[n_1,n_2]$, such that $\mathbf{B} = n_1\mathbf{R}_1 + n_2\mathbf{R}_2$. The point group symmetry of the honeycomb lattice will make many of these equivalent, however, so truly unique nanotubes are only generated using a one-twelfth irreducible wedge of the Bravais lattice. Within this wedge only a finite number of nanotubes can be constructed with a circumference below any given value.

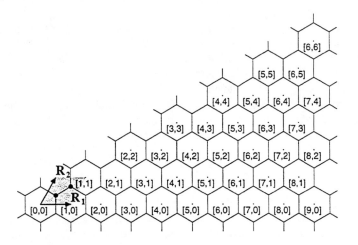

Figure 1. Two-dimensional graphite lattice structure. Primitive lattice vectors $\mathbf{R}_1$ and $\mathbf{R}_2$ are depicted in inset.

The construction of the nanotube from a conformal mapping of the graphite sheet shows that each nanotube can have up to three inequivalent (by point group symme- try) helical operations derived from the primitive lattice vectors of the graphite sheet. Thus while *all* nanotubes have a helical structure, nanotubes constructed by mapping directions corresponding to lattice translation indices of the form $[n,0]$ and $[n,n]$ will possess a reflection plane. These high-symmetry nanotubes will therefore be achiral [16, 24]. For convenience we have labeled these special structures based on the shapes made by the most direct continuous path of bonds around the circumference of the nanotube [16]. Specifically, the $[n, 0]$-type structures were labeled as sawtooth and the $[n, n]$-type structures as serpentine. For other conformations, the nanotubes are chiral and have three inequivalent helical operations.

Because the primitive reciprocal lattice vectors of the hexagonal lattice (the Bravais lattice of the honeycomb lattice) are scalar multiples of real lattice vectors, each nanotube thus generated can be shown to be translationally periodic down the nanotube axis [14, 16, 24]. However, even for relatively small diameter nanotubes, the minimum number of atoms in a translational unit cell can be quite large. For example, if $n_1 = 10$ and $n_2 = 9$ then the radius of the nanotube is less than 0.7 nm, but the translational unit cell contains 1084 carbon atoms [24]. The rapid growth in the number of atoms that can occur in the minimum translational unit cell makes recourse to the helical and any higher point group symmetry of these nanotubes practically mandatory in any comprehensive study of their properties as a function of radius and helicity. These symmetries can be used to reduce to two the number of atoms necessary to generate any nanotube [24]. For example, the matrices that have to be diagonalized in a calculation of the nanotube's electronic structure can be reduced to a size no larger than that encountered in a corresponding electronic structure calculation of two-dimensional graphite.

We have examined the electronic structure of fully three-dimensional nanotubes using both all-valence empirical tight-binding calculations and a first-principles local-density functional (LDF) method originally developed for chain polymers [22] and adapted for systems with helical symmetry [23]. Borrowing concepts from solid-state band-structure theory, we can define a helical chain polymer as a nuclear lattice constructed from a finite basis of atoms and a screw operation $S(h, \varphi)$. For mathematical convenience we define the screw operation in terms of a translation h units down the z axis in conjunction with a right-handed rotation φ about the z axis. That is,

$$S(h, \varphi)\, \mathbf{r} \equiv \begin{pmatrix} x \cos \varphi - y \sin \varphi \\ x \sin \varphi + y \cos \varphi \\ z + h \end{pmatrix}, \qquad (1)$$

with the screw operation henceforth denoted as S, with arguments h and φ implicitly understood. Because the symmetry group generated by the screw operation S is isomorphic with the one-dimensional translation group, Bloch's theorem can be generalized so that the one-electron wavefunctions will transform under S according to

$$S^m \psi_i(\mathbf{r}; \kappa) = e^{i\kappa m}\, \psi_i(\mathbf{r}; \kappa). \qquad (2)$$

The quantity κ is a dimensionless quantity which is conventionally restricted to a range of $-\pi < \kappa \leq \pi$, a central Brillouin zone. For the case $\varphi = 0$ (i.e., S a pure translation), κ corresponds to a normalized quasimomentum for a system with one-dimensional translational periodicity; i.e., $\kappa \equiv kh$, where k is the traditional wavevector from Bloch's theorem in solid-state band-structure theory. In the previous analysis of helical symmetry, with $\mathbf{H}$ the lattice vector in the graphene sheet defining the helical symmetry generator, κ corresponds similarly to the product $\kappa = \mathbf{k} \cdot \mathbf{H}$.

The one-electron wavefunctions ψ_i are constructed from a linear combination of Bloch functions φ_j, which are in turn constructed from a linear combination of nuclear-centered functions $\chi_j(\mathbf{r})$ (for the LDF calculations these are Gaussian-type orbitals constructed from products of Gaussians and the real solid spherical harmonics),

$$\psi_i(\mathbf{r}; \kappa) = \sum_j c_{ji}(\kappa)\, \varphi_j(\mathbf{r}; \kappa) \tag{3}$$

$$\varphi_j(\mathbf{r}; \kappa) = \sum_m e^{-i\kappa m}\, \mathcal{S}^m\, \chi_j(\mathbf{r}). \tag{4}$$

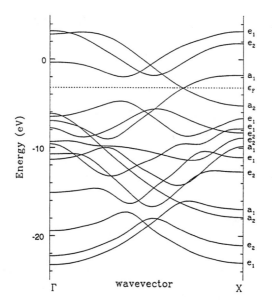

Figure 2. LDF valence band structure of [5,5] serpentine nanotube. The Fermi level at ε_F is denoted with a dotted line. Γ and X correspond to the dimensionless wavenumber coordinate κ ranging from 0 to π.

3. RESULTS

We initially calculated the electronic structure of an infinitely long nanotube using the first-principles LDF method [13, 14] for the high symmetry [5,5] serpentine nanotube. The structure considered was generated by a planar ring of 10 carbon atoms with D_{5h} symmetry arranged in 5 pairs; the distance between interior members of adjacent pairs was fixed at twice the nearest-neighbor separation typical of fullerenes and other graphitic systems of 0.142 nm. The nanotube structure was generated with a screw operation having a twist of π radians and a translational shift of 0.123 nm chosen to yield nearest-neighbor separations between rings equal to the in-ring values. The one-electron states are Bloch functions generated by repeated application of the screw operation, and belong to irreducible representations of the screw symmetry group with the dimensionless analog of the wave vector, κ. In the calculations we used 20 evenly-spaced points in the one-dimensional Brillouin zone ($-\pi < \kappa \leq \pi$) and a carbon 7s3p Gaussian basis set.

We depict our calculated valence band structure in Fig. 2. All of the operations of the C_{5v} point group commute with the screw-symmetry helical operation used to generate the band structure. We thus label all bands according to the 4 irreducible representations of the C_{5v} group: the rotationally-invariant a_1 and a_2 representations, and the doubly-degenerate e_1 and e_2 representations. For this lattice structure we find the nanotube is a metal, with the a_1 bands and a_2 bands crossing at a position in the Brillouin zone roughly 0.69 of the width of the half-Brillouin zone from the origin. The Fermi level, ε_F, coincides with this crossing. These states are analogous to those we would obtain for a single sheet of graphite if we impose Born-von Karman boundary conditions over a width equivalent to the circumference around the nanotube for a single axis direction in the graphitic system. Continuing this analogy, we find that the a_1 and a_2 states in the vicinity of ε_F are predominantly p-orbital states aligned with the local normal of the nanotube surface, similar to the π-like states in graphite. We obtain four states at the Fermi level (a_1 and a_2 states at k_F and $-k_f$), which are related to the four inequivalent (by translational symmetry) states at ε_F found in two-dimensional graphite.

We have currently carried out a band structure calculation for the [10,10] structure possibly observed to have a high single-nanotube conductivity [18]. Similar to the [5,5] structure, the structure in our current work was generated by a planar ring of 20 carbon atoms with D_{10h} symmetry arranged in 10 pairs. The spacing between carbon atoms corresponds to a carbon-carbon bond distance of 0.144 nm in a graphene sheet, with the graphene sheet then conformally mapped to the surface of a cylinder. As we noted in our study on the rotational and helical symmetries of carbon nanotubes, the resulting nanotube structure can then be generated with a translational shift of 0.125 nm and a helical twist of $\pi/10 \pm 2\pi n/10$. All ten of these different possible primitive helical twists will generate the correct lattice structure, but will produce different depictions of the band structure. We note that in the [5,5] structure, the choice of a helical twist angle of π radians results in a helical operator that commutes with the reflection plane of the C_{5v} point group, and we can label the one-electron wavefunctions according to the irreducible representations of the C_{5v} point group. In the [10,10] structure none of the possible helical operations commute with the reflection operation in the C_{10v} point group, thus the e_m irreducible representations will not be doubly degenerate but will split into two related but nondegenerate bands. Likewise the b representation will not split into the b_1 and b_2 representations. A similarity transformation on the helical operation using the reflection operator results in a helical operation with a helical twist angle of the opposite sign. Operating on e_m or b states, use of these two operators gives results different by a phase shift of $\pi m/5 \pm 2\pi nm/5$ ($n = 5$ for the b states), but identical results are obtained on the rotationally symmetric a states. This implies that the a_1 and a_2 irreducible representations of the C_{10v} point group are still good labels for the one-electron wavefunctions, and we expect the Fermi level to be defined by the crossing of a_1 and a_2 states.

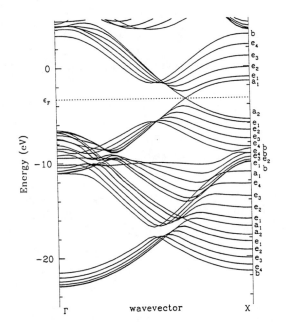

Figure 3. LDF valence pseudoband structure of [10,10] serpentine nanotube. The Fermi level at ε_F is denoted with a dotted line. Γ and X correspond to the dimensionless wavenumber coordinate κ ranging from 0 to π.

Because of the multiple nonunique ways of depicting the band structure of the [10,10] carbon nanotube, we depict in Fig. 3 what we denote as the "pseudo" band structure. This band structure was obtained by shifting all the e_m bands in κ by a quantity corresponding to a phase factor of $\pi m/10 \pm 2\pi mn/10$, and the b bands by a phase factor of $\pi/2$. This choice of phase shifts the e_m bands to doubly degenerate bands, and leads to a depiction of the band as equivalent as possible to the bands that would be present if the system had translational periodicity with a repeat distance of the helical unit cell. That is, if we considered a continuum of values of the local phase factor for the irreducible representations of the C_{10} point group, then this labeling corresponds to defining κ in terms of the phase shift by stepping along the helical axis direction by the helical step size but with zero helical twist angle. Observing that the [10,10] nanotube has translational symmetry over a length twice the helical unit cell length, we note that a depiction for the band structure that would result from use of the translational symmetry can be obtained directly by folding over the band structure in Fig. 3.

Recent work by Thess, *et al.* [18] suggests that [10,10] carbon nanotubes with diameters of 1.38 nm have been synthesized and have a measured resisitivity of less than 10^{-4} Ω-cm at 300K. This structure belongs to the same serpentine ([n,n]) family of nanotubes as the [5,5] carbon nanotube theoretically shown to have electronic states appropriate for high conductivity in our earlier work [13, 14]. We have calculated the electronic structure for this high-symmetry [10,10] carbon nanotube using a first-principles, self-consistent, all-electron Gaussian-orbital based local-density functional approach to calculate the band structure. We find that the electronic structure results are similar in band width and behavior near the Fermi level to our earlier results for the [5,5] nanotube. These results will become the basis for detailed future calculations of other properties of these nanotubes.

ACKNOWLEDGMENTS

This work was supported by the U.S. Office of Naval Research, both directly and through the Naval Research Laboratory. RAJ acknowledges support under the ASEE-NRL Summer Faculty program. A grant of computational time was provided under the DoD High Performance Computing Initiative Program at the Vicksburg CEWES center.

References

[1] J. W. Mintmire and C. T. White, in *Carbon Nanotubes: Preparation and Properties*, Ed., Thomas W. Ebbesen (CRC, Boca Raton, 1997), pp. 191–224, and references therein.

[2] See, e.g., *Buckminsterfullerenes*, (Editors, W. E. Billups and M. A Ciufolini) VCH, New York, (1993).

[3] M. S. Dresselhaus, G. Dresselhaus, and P. C. Eklund, J. Mater. Res. 8, 2054 (1993).

[4] J. W. Weaver and D. M. Poirier, in *Solid State Physics*, v. 48, Eds., H. Ehrenreich and F. Spaepen (Academic Press, New York, 1994),

[5] H. W. Kroto, J. R. Heath, S. C. O'Brien, R. F. Curl, and R. E. Smalley, Nature 318, 162 (1985).

[6] M. S. Dresselhaus, G. Dresselhaus, K. Sugihara, I. L. Spain, and H. A. Goldberg, *Graphite Fibers and Filaments*, (Springer-Verlag, Berlin, 1988).

[7] S. Iijima, Nature **354**, 56 (1991).

[8] G. G. Tibbetts, J. Crystal Growth **66**, 632 (1983).

[9] J. S. Speck, M. Endo, and M. S. Dresselhaus, J. Crystal Growth **94**, 834 (1989).

[10] T. W. Ebbesen and P. M. Ajayan, Nature (London) **358**, 220 (1992).

[11] S. Iijima and T. Ichihashi, Nature (London) **363**, 603 (1993).

[12] D. S. Bethune, C. H. Klang, M. S. DeVries, G. Gorman, R. Savoy, J. Vazquez, and R. Beyers, Nature (London) **363**, 605 (1993).

[13] J. W. Mintmire, B. I. Dunlap, and C. T. White, Phys. Rev. Lett. **68**, 631 (1992).

[14] J. W. Mintmire, D. H. Robertson, B. I. Dunlap, R. C. Mowrey, D. W. Brenner, and C. T. White, in *Electrical, Optical, and Magnetic Properties of Organic Solid State Materials*, Eds., L. Y. Chiang, A. F. Garito, and D. J. Sandman, MRS Symposia Proceedings No. 247 (Materials Research Society, Pittsburgh, 1992), 339.

[15] N. Hamada, S. Sawada, and A. Oshiyama, Phys. Rev. Lett. **68**, 1579 (1992).

[16] D. H. Robertson, D. W. Brenner, and J. W. Mintmire, Phys. Rev. B **45**, 12592 (1992).

[17] M. M. J. Treacy, T. W. Ebbesen, and J. M. Gibson, Nature **381**, 678 (1996).

[18] A. Thess, R. Lee, P. Nikolaev, H. Dai, P. Petit, J. Robert, C. Xu, Y. H. Lee, S. G. Kim, A. G. Rinzler, D. T. Colbert, G. E. Scuseria, D. Tománek, J. E. Fischer, and R. E. Smalley, Science **273**, 483 (1996).

[19] S. J. Tans, M. H. Devoret, H. Dai, A. Thess, R. E. Smalley, L. J. Geerligs, and C. Dekker, submitted for publication.

[20] T. W. Ebbesen, Physics Today **49**, 26 (1996).

[21] W. A. deHeer, Châtelain, and D. Ugarte, Science **270**, 1179 (1995).

[22] J. W. Mintmire and C. T. White, Phys. Rev. Lett. **50**, 101 (1983); Phys. Rev. B **28**, 3283 (1983).

[23] J. W. Mintmire, in *Density Functional Methods in Chemistry*, (Editors, J. Labanowski and J. Andzelm) Springer-Verlag, New York, (1991), pp. 125–138.

[24] C. T. White, D. H. Robertson, and J. W. Mintmire, Phys. Rev. B **47**, 5485 (1993).

[25] R. Saito, M. Fujita, G. Dresselhaus, M. S. Dresselhaus, Phys. Rev. B **46**, 1804 (1992); Mater. Res. Soc. Sym. Proc. **247**, 333 (1992); Appl. Phys. Lett. **60**, 2204 (1992).

OPTICAL PROPERTIES OF SINGLE-WALLED CARBON NANOTUBES AND SILICON CLATHRATES

Ernst Richter, Madhu Menon, and K.R. Subbaswamy
Department of Physics and Astronomy
University of Kentucky
Lexington, KY 40506, U.S.A.

ABSTRACT

Computations relating to the geometrical and physical properties of two novel materials currently of experimental interest are described: single-walled nanotubes of carbon, and silicon clathrates. Relaxed geometries are computed in a constant-pressure molecular dynamics scheme in which forces are obtained from a non-orthogonal tight-binding electronic energy calculation. Vibrational modes and Raman spectra are calculated and compared with available experiments.

1. INTRODUCTION

Experimental methods of synthesis away from thermodynamic equilibrium, such as laser ablation, arc discharge, etc., have revolutionized materials research in recent years. Even time-honored staples such as carbon have been synthesized into highly metastable, new morphologies with very interesting geometries and physical properties. The isolation of novel materials in single phase is a very lengthy and arduous process. Initial attempts always yield multiphase materials, and through laborious trial and error, experimentalists either isolate particular phases, or prepare enriched samples through various chemical and physical means. Thus, for theory to be particularly useful in materials research, it must go beyond determination of geometries of pure phases, and must be willing and able to tackle the measurable attributes (e.g., various spectroscopic signatures) of multiphase samples. This may, in many cases, call for abandoning the purity of *ab initio* approaches and stooping to empirical or semi-empirical methods. In this paper we describe the discovery of a very interesting size dependent resonant Raman scattering effect in single-walled nanotubes (SWNT)

of carbon through a joint experimental and theoretical effort [1] . We also make predictions for Raman scattering from pure silicon clathrates which are on the verge of
being synthesized.

2. NON-ORTHOGONAL TIGHT-BINDING SCHEME

We have introduced [2] a transferable non-orthogonal tight-binding scheme containing only four *adjustable* parameters, and a simple exponential distance dependence
for the parameters with no artificial cut-off in the interactions. The parameters were
fitted to experimental bond lengths and vibrational frequencies of the dimer and the
bulk crystal in such a way as to minimize errors at both ends.

The details of the technique can be found in earlier literature [2, 3, 4] . Here we
give a brief summary. In the non-orthogonal tight-binding scheme the characteristic
equation for obtaining the eigenvalues is given by

$$(\mathbf{H} - E_n \mathbf{S})\mathbf{C}^n = 0, \tag{1}$$

where $\mathbf{H}$ and $\mathbf{S}$ are the Hamiltonian and overlap matrices respectively [2, 5] .

The Hellmann-Feynman expression for obtaining the electronic part of the force is
given by [2] ,

$$\frac{\partial E_n}{\partial x} = \frac{\mathbf{C}^{n\dagger}(\frac{\partial \mathbf{H}}{\partial x} - E_n \frac{\partial \mathbf{S}}{\partial x})\mathbf{C}^n}{\mathbf{C}^{n\dagger}\mathbf{S}\mathbf{C}^n}. \tag{2}$$

In the Slater-Koster scheme the Hamiltonian matrix elements are obtained from
the parameters $V_{\lambda\lambda'\mu}$ in terms of the bond direction cosines l, m,n [2, 6, 7] . In the MS
scheme the parameters $V_{\lambda\lambda'\mu}(r)$ are taken to decrease exponentially with r [2] :

$$V_{\lambda\lambda'\mu}(r) = V_{\lambda\lambda'\mu}(d_0)e^{-\alpha(r-d_0)}, \tag{3}$$

where d_0 is the sum of the covalent radii of the pair of interacting atoms and α is an
adjustable parameter. The scaling of the repulsive term is also taken to be exponential:

$$\phi(r) = \phi_0 e^{-\beta(r-d_0)}, \tag{4}$$

where $\beta = 4\alpha$ [2] .

In the non-orthogonal scheme, the overlap matrix is calculated in the spirit of
extended Hückel theory [8] by assuming a proportionality between $\mathbf{H}$ and $\mathbf{S}$ [5] :

$$S_{ij} = \frac{2}{K}\frac{H_{ij}}{H_{ii} + H_{jj}}. \tag{5}$$

The diagonal elements of H_{ij}, as in the orthogonal theory, are taken to be the valence
s and p energies. The off-diagonal inter-atomic matrix elements are given in terms of
the Hamiltonian matrix elements in orthogonal theory, V_{ij}, by

$$H_{ij} = V_{ij}[1 + \frac{1}{K} - S_2^2], \tag{6}$$

where,

$$S_2 = \frac{(S_{ss\sigma} - 2\sqrt{3}S_{sp\sigma} - 3S_{pp\sigma})}{4} \qquad (7)$$

is the nonorthogonality between sp^3 hybrids [5] . The quantities $S_{\lambda\lambda'\mu}$ in turn are determined from

$$S_{\lambda\lambda'\mu} = \frac{2V_{\lambda\lambda'\mu}}{K(\varepsilon_\lambda + \varepsilon_{\lambda'})}. \qquad (8)$$

We take a simple exponential distance dependence in the nonorthogonality coefficient, K,

$$K(r) = K_0 e^{\sigma(r-d_0)^2}. \qquad (9)$$

3. SINGLE-WALLED NANOTUBES OF CARBON

With the recent progress in methods for synthesizing samples containing high concentrations of single-walled nanotubes (SWNT) of carbon [9] it has become possible to probe the properties of these quasi-one dimensional crystals in some detail [1] . The major component of these samples appears to be the so-called "armchair" achiral tubes designated as (10,10) tubes in the standard notation (see Fig. 1) [10, 11] .

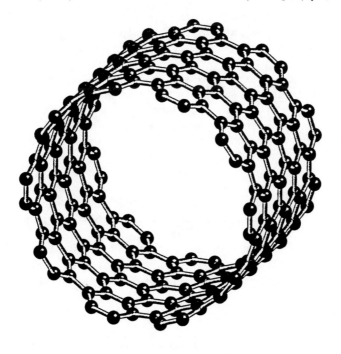

Figure 1. The (10,10) nanotube.

In particular, the measurement of the Raman spectra of SWNT [1] as a function of exciting laser wavelength has revealed some peculiar results: not only is there evidence for a resonant process (i.e., variation in the relative intensities of the spectral features), but the frequencies of the observed vibrational modes themselves appear to shift. The results were interpreted as evidence for the presence of a distribution of SWNT of different diameters, whose vibrational modes were selectively excited due to an electronic inter-band resonance which itself is diameter dependent. This analysis was based solely on computed electronic density of states with no attention to the relevant Raman matrix elements. In a recent paper [12] we computed the matrix elements for the Raman process [13] for SWNT within a simple tight-binding model which is known to give an accurate description of the π-band structure of graphite.

For our calculation we use the simple, zone-folding tight-binding model introduced in this context by Dresselhaus, et al. [14, 10] :

$$E(k_t, k) = \pm\gamma_0 \left\{ 1 + 4\cos\left(\frac{\sqrt{3}k_t a_0}{2}\right)\cos\left(\frac{ka_0}{2}\right) + 4\cos^2\left(\frac{ka_0}{2}\right) \right\}^{1/2}, \quad (10)$$

where γ_0 is the nearest-neighbor C-C overlap integral, and $a_0 = 2.68 \times \sqrt{3}$ a. u. is the lattice constant of the 2D graphene sheet. k is the wave vector of the one dimensional lattice. We take $\gamma_0 = -2.5$ eV which is suggested by comparison with first-principles calculations [15] .construction of the tube involves the restriction of the tangential wave vectors k_t corresponding to the two dimensional lattice of the graphene sheet to a small number of discrete, quantized values, dependent on the diameter. Thus, for instance, for the (N,N) armchair tubes, the tangential wave vector values (k_t) are restricted to $2\pi q/(N\sqrt{3}a_0)$, with $q = 1, 2, ..., n$. From this restriction, one obtains the one-dimensional electronic bands. The band structure and electronic density of states obtained in this manner for the (10,10) case are shown in Figs. 2a and 2b. Using band folding it can be shown [10] that all (N,N) tubes are semi-metals.

For calculating Raman spectra, we also need the phonon modes of the system. For this purpose, we have used empirical (nearest- and next-nearest neighbor) force-constant parameters derived by Jishi, et al. [16] by fitting to the phonons of a graphene sheet. The parameters have been transferred without change form the plane to the cylindrical geometry. The number of symmetry allowed Raman active zone center modes, which is independent of the tube diameter, is 16 for even values and 15 for odd values of N. Frequencies of Raman allowed modes for tubes with various diameters are listed in [1] .

Raman scattering (inelastic scattering of photons due to the emission or absorption of phonons) The scattering cross-section is proportional to the squared modulus of the suitably contracted Raman tensor [13] whose contributions are of the form

$$R_\sigma^{ij} = \sum_{\alpha,\beta} \frac{p_{0\beta}^i p_{\beta\alpha}^j \Xi_{\sigma,\alpha 0}}{(\omega_I - \omega - \omega_\beta)(-\omega - \omega_\alpha)}, \quad (11)$$

where $p_{\alpha\beta}^i$ refer to the electron momentum matrix elements, *matrixelements*,respectively, to the incident photon frequency, the scattered photon frequency, and phonon frequency. There are five other contributions. The electron-phonon matrix elements are related to the deformation potential, and we obtain them by introducing a displacement dependence to the nearest-neighbor C-C overlap integral:

$$\gamma_0(|\vec{R} + \vec{u}|) = \gamma_0(|\vec{R}|) + \Delta \frac{\vec{R}}{|\vec{R}|} \cdot \vec{u}, \tag{12}$$

with $\vec{R}$ being the equilibrium bond vector. In this simple model the constant Δ enters the calculation only as an overall scaling factor for the Raman intensities and, therefore, can be chosen arbitrarily. Another parameter needed is the excited electron lifetime; we take this to be 0.10 eV [17] .

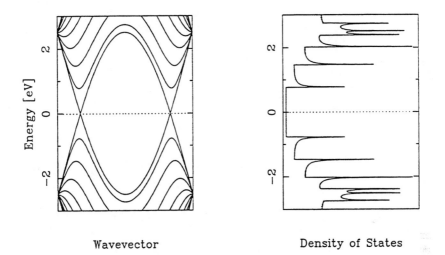

Figure 2. (a) Electronic bands near the Fermi level, and (b) the density of states, for a (10,10) tube.

In Fig. (3) the variation of the intensity of the Raman active modes in a (10,10) tube are shown as a function of the incident frequency. The resonance effect is clearly displayed. Finally, In Fig. 4(a) and (b) the spectra corresponding to a mixture of (8,8) (9,9), and (10,10) tubes in equal proportion are shown at two different laser frequencies.

We have shown by explicit calculation within a simple, but realistic model, that the unique, one-dimensional nature of the single-walled carbon nanotubes give rise to diameter-dependent electronic resonance effects that are clearly discernible in Raman scattering. Even in an inhomogeneous sample, incident laser frequency may be tuned to select out the vibrational modes of specific tube-types. Measurement of the resonance shape can be used to extract information on the electron-phonon coupling in these systems.

In Fig. (4a) the (8,8) tubes are close to a resonance, and their modes are much more prominent in the spectrum. In (4b) the (10,10) tubes are close to resonance, and

E. Richter, M. Menon and K.R. Subbaswamy

their modes are dominant in the spectrum. Thus, when the incident laser wavelength is varied, not only do the Raman intensities change, but the peak positions also change. This is a clear demonstration of the diameter-selective resonance effect that was invoked as a possible explanation of the Raman scattering from the purified SWNT samples observed by Rao, et al. [1] .

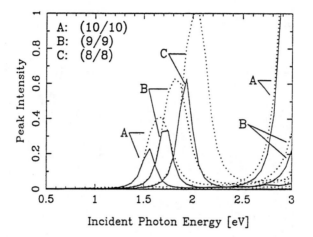

Figure 3. Variation of the Raman peak intensity as a function of incident laser energy for several vibrational modes in a (10,10) tube.

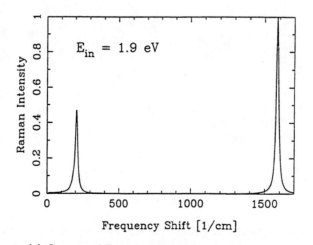

Figure 4. (a) Computed Raman spectrum at 1.9 eV for a mixture of tubes.

4. SILICON CLATHRATES

The possibility of altering band-gaps of semiconductors is of great interest because it can lead to control over the wavelength of light emitted, making it possible to generate photoluminence in the visible region with important technological implications. A number of mechanisms towards this end have been proposed.

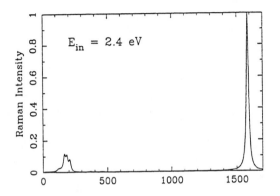

Figure 4. (b) Same as (a), for 2.4 eV excitation.

They include making Si in "porus" form [18, 19] , which causes the band gap to be increased. Although confinement effect has been suggested as a possible cause for this increase [20] , the full mechanism of the band gap increase is not yet fully understood. Other mechanisms have also been suggested for the case of porus Si [21] . Formation of alloys with Si as one of the component provides an alternative route towards altering the band gap. The widely studied $Si_{1-x}Ge_x$, however, has smaller band gap when compared to that of diamond Si.

The most promising method of achieving an increase in the band gap appears to be changing the geometry of bulk diamond Si. It was found that a whole new class of materials called clathrates can be generated by altering the bond angles from their ideal tetrahedral value in the diamond structure [22, 23, 24, 25] . The clathrates form local minima of the total energy and some of them are only slightly less stable than the diamond phase. The silicon clathrate compounds are not attacked even by strong acids except for HF. Importantly, the band gaps of these structures are substantially larger than that of Si in the diamond structure and well into the visible region. It is worth noting that the high pressure phases of bulk Si such as β-Sn, on the other hand, are metallic with no gap. The clathrate structures are relatively open, allowing for endohedral doping with impurity atoms.

In this section we give a summary of the structural and vibrational properties of the lowest energy clathrates [26] ; specifically a clathrate structure containing 34 atoms in an face-centered cubic (fcc) unit cell and another structure containing 46 atoms in a simple-cubic (sc) unit cell, using our tight-binding molecular dynamics scheme. We denote these clathrate structures by Clath(34) and Clath(46) respectively. Fig. 5 shows drawings of these structures.

The equilibrium bond lengths for the dimer and bulk diamond structure are obtained to be 4.23 a. u. and 4.46 a. u., respectively, in our method [4] . These compare very favorably with the experimental values of 4.23 a. u. and 4.44 a. u. respectively [27] . When applied to the bulk phases of Si, our method has reproduced binding energy differences between the high-pressure metallic phases and the diamond phase in very good agreement with the schemes based on the local density approximation (LDA) [4] .

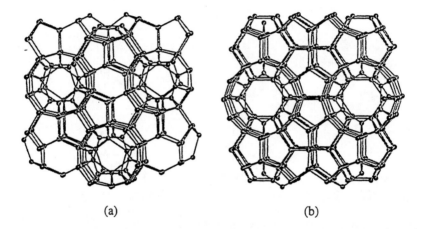

(a)   (b)

Figure 5. (a) The Clath(34) structure and, (b) Clath(46) structure.

For the treatment of clathrate structures, a constant pressure ensemble method was incorporated into our scheme to allow for realistic lattice relaxation. This is necessitated by the fact that these structures contain a large number of atoms in the unit cell, making the structure dependent on the choice of the lattice parameters. The constant pressure MD method was first introduced by Andersen [28] and subsequently extended by Parrinello and Rahman [29] .

For Clath34 we consider a face centered cubic (fcc) lattice with a 34 atom basis. The minimum energy configuration is obtained by a symmetry unconstrained molecular dynamics relaxation with a simultaneous symmetry unrestricted relaxation for the lattice parameters. We use a uniform grid of 8 k-points in the zone. Convergence was checked by increasing k-points. The fully relaxed lattice maintained its fcc symmetry with a cohesive energy difference of 0.03 eV when compared with diamond and a lattice constant of $a = 27.76$ a. u.. This is in very good agreement with the LDA based calculations that predict the lattice constant a to between 27.33 a. u. and 28.09 a. u., depending on the choice of the basis (e.g., plane wave or local orbital) [22] . Four distinct bond lengths were obtained for this structure ranging from 4.428 a. u. to 4.508 with an average bond length of 4.472 a. u.. This compares favorably with the experimentally determined bond length of 4.479 a. u. for this clathrate [22] . The volume per atom is found to be increased by 14% when compared to diamond. In Table I we summarize our structural results and compare with LDA results.

Raman Intensity

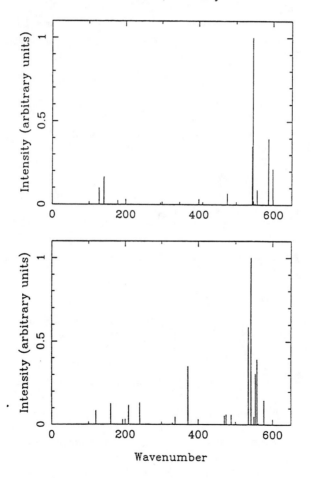

Figure 6. Computed Raman spectra of (a) Clath(34) (top panel) and (b) Clath(46) (bottom panel).

Table 1: Structural properties of Si clathrates. V_0 is the volume of the diamond structure at it's minimum energy

	Lattice	Average bond length	V/V_0	a (a. u.)
Diamond	fcc	4.46 (4.44)[1]	1.000	10.30 (10.26)[2]
Clath(34)	fcc	4.472 (4.48)[1]	1.14 (1.17)[2]	27.76 (27.33-28.09)[2]
Clath(46)	sc	4.474	1.14 (1.17)[1]	19.28 (19.56)[2]

[1]Ref. [27]
[2]Ref. [22]

We obtain an increase in the band gap of 0.61 eV for Clath34 over that for Si in bulk diamond structure. This value compares very favorably with the value obtained by the LDA based scheme which finds an increase of 0.7 eV [22] . This widening of the gap has important technological implications when one considers the fact that Clath34 differs from diamond structure in only the bond angles.

For Clath46 we consider a simple cubic (sc) lattice with a 46 atom basis. The fully relaxed lattice maintained it sc symmetry with a cohesive energy difference of 0.035 eV when compared with diamond and a lattice constant of $a = 19.28$ a. u.. This is in very good agreement with the LDA based calculations that predict the lattice constant of $a = 19.56$ a. u.. Four distinct bond lengths were obtained for this structure ranging from 4.436 a. u. to 4.527 a. u. with an average bond length of 4.474 a. u.. The volume per atom is found to be increased by 14% when compared to diamond.

We obtain an increase in the band gap of 0.65 eV for Clath46 over that for Si in bulk diamond structure. This value compares very favorably with the value obtained by the LDA based scheme which finds an increase of 0.7 eV [22] .

The force constants for the evaluation of vibrational modes are obtained by explicitly calculating analytic second derivatives of the electronic structure Hamiltonian matrix elements [3] .

The phonon spectra were obtained by using precisely the same tight-binding parameters that were used in optimizing the geometries of the structures by molecular dynamics relaxation. In our calculations, we find 8 k−points in the full zone to be sufficient for obtaining converged results for phonons in both Clath34 and Clath46. The phonon dispersion curves, while displaying the expected zone-folding effects, also contain many interesting features. The most striking feature is the presence of a gap in the optical region. This feature may be understood on general grounds by a consideration of the distribution of bond lengths in the unit cell. The presence of more than one bond length in the unit cell is effectively analogous to having more than one force constant, as is the case, for example, for hetero-atomic systems. The optical bands are almost flat with only a small dispersion. Another interesting feature is the increase in frequency for the transverse optical (Γ_{TO}) mode in Clath(34) when compared to the diamond phase. This is perhaps not surprising when one considers the fact that even though the average bond length is larger than in diamond phase, one of the bonds is shorter (4.428 a. u.). The largest frequency (Γ_{TO}) obtained within the present scheme has a value of 599 cm^{-1}.

A bond polarizability model [30, 31] was used, in conjunction with the phonon eigenvectors, to calculate the Raman spectra of these clathrates. The five fitting parameters for calculating the spectra were taken from Ref. [30] for Si . In Fig. 6(a) and (b) we show the computed Raman spectra of Clath(34) and Clath(46), respectively, in the $(\perp, \perp)$ scattering geometry. Such calculations should help in the analysis of multiphase samples as the attempts to produce pure Si clathrates proceed.

ACKNOWLEDGMENTS

This research was supported in part by USDOE contract DE-FC22-93PC93053, by NSF grant OSR 94-52895, and by the University of Kentucky Center for Computational Sciences.

References

[1] A.M. Rao, E. Richter, S. Bandow, B. Chase, P.C. Eklund, K.A. Williams, S. Fang, K.R. Subbaswamy, M. Menon, A. Thess, R.E. Smalley, G. Dresselhaus, and M.S. Dresselhaus, Science **275**, 187 (1997).

[2] M. Menon and K.R. Subbaswamy, Phys. Rev. B **50**, 11577 (1994).

[3] M. Menon, E. Richter and K. R. Subbaswamy, J. Chem. Phys. **104**, 5875 (1996).

[4] M. Menon and K.R. Subbaswamy, Phys. Rev. B **55**, 9231 (1997).

[5] M. van Schilfgaarde and W. A. Harrison, Phys. Rev. B **33**, 2653 (1986).

[6] J. C. Slater and G. F. Koster Phys. Rev. **94**, 1498 (1954).

[7] W. A. Harrison, *Electronic Structure and the Properties of Solids* (Freeman, San Francisco, 1980), and references therein.

[8] R. Hoffmann, J. Chem. Phys. **39**, 1397 (1963).

[9] A. Thess, *et al.* Science **273**, 483 (1996).

[10] M.S. Dresselhaus, G.F. Dresselhaus, and P.C. Eklund, *Science of Fullerenes and Carbon Nanotubes* (Academic Press, New York, 1996)

[11] The numbers (n,m) refer to the integer multiples of the unit vectors of the two-dimensional graphite sheet needed to construct the spanning vector for the graphite strip from which the tube is derived by folding.

[12] Ernst Richter and K.R. Subbaswamy (unpublished).

[13] W.Hayes and R. Loudon, *The Scattering of Light by Crystals* (John Wiley, New York, 1976); M.V. Klein in *Light Scattering in Solids III*, ed. M. Cardona and G. Güntherodt (Springer, Berlin, 1982).

[14] M.S. Dresselhaus, G.F. Dresselhaus, and R. Saito, Solid State Commun. **84**, 201 (1992).

[15] J.W. Mintmire and C.T. White, Carbon **33**, 893 (1995).

[16] R.A. Jishi, L. Venkataraman, M.S. Dresselhaus, and G. Dresselhaus, Phys. Rev. B **51**, 11176 (1995).

[17] P.C. Eklund, G.D. Mahan, J.G. Spolar, J.M. Zhang, E.T. Arakawa, and D.M. Hoffman, Phys. Rev. B **37**, 691 (1988).

[18] L. T. Canham, Appl. Phys. Lett. **57**, 1046 (1990).

[19] A. G. Cullis and L. T. Canham, Nature (London) **353**, 335 (1991).

[20] F. Buda, J. Kohanott, and M. Parrinello, Phys. Rev. Lett. **69**, 1272 (1992).

[21] M. S. Brandt, M. D. Fuchs, M Stutzmann, J. Weber and M. Cardona, Solid State Commun. **81**, 307 (1992).

[22] G. B. Adams, M. OKeeffe, A. A. Demkov, O. F. Sankey and Y.-M. Huang, Phys. Rev. B **49**, 8048 (1994).

[23] A. A. Demkov, O. F. Sankey, K. E. Schmidt, G. B. Adams, and M. OKeeffe, Phys. Rev. B **50**, 17001 (1994).

[24] A. A. Demkov, W. Windl, and O. F. Sankey, Phys. Rev. B **53**, 1288 (1996).

[25] E. Galvani, G. Onida, S. Serra, and G. Benedek, Phys. Rev. Lett. **77**, 3573 (1996).

[26] M. Menon, Ernst Richter, and K.R. Subbaswamy (unpublished).

[27] G. Nilsson and G. Nelin Phys. Rev. B **6**, 3777 (1972).

[28] H. C. Andersen, J. Chem. Phys. **72**, 2384 (1980).

[29] M. Parinello and A. Rahman, Phys. Rev. Lett. **45**, 1196 (1980).

[30] S. Go, H. Bilz and M. Cardona, Phys. Rev. Lett., **34**, 580 (1975).

[31] W. Weber, Phys. Rev. B **15**, 4789 (1977).

CONDUCTANCE IN ULTIMATE NANOSTRUCTURES

Inder P. Batra
IBM Almaden Research Center
San Jose, California 95120-6099, U.S.A

ABSTRACT

Wires of atomic-size widths represent the ultimate limit of point contacts and define one-dimensional transport. The advent of scanning tunneling microscopy (STM) and atomic force microscopy (AFM) has enabled researchers to fabricate and analyze transport through such nanowires. Many authors have shown that for an ideal one dimensional channel under ballistic transport conditions the conductance is transversally quantized in steps of $2e^2/h$. The transverse dimension is the constriction width, and can be varied. We present an elementary analysis of quantum transport through an ultimate nanostructure. By invoking the uncertainty principle in conjunction with the definition of transport in the quantum limit we readily obtain conductance quantization. The finite resistance of a perfect conductor thus has a quantum mechanical origin in the uncertainty principle. Having analyzed the quantization using simple arguments, we observe that the quantized Hall resistance shows similar step structure, and hence may be similarly motivated. We close by speculating on the possibility of observing fractional quantization in quantum transport, in analogy with the fractional quantum Hall effect.

1. INTRODUCTION

The conductance through constrictions and the subtle role played by the contacts and the measuring probes are clearly laid out in a 1989 paper by Landauer[1]. He has given a balanced review and readers are encouraged to consult his paper for other references. Landauer[1] has spelled out the conditions under which various formulae relating conductance to transmission are valid. For a one-dimensional sample (in the absence of any dissipative scattering) connecting two reservoirs at different electrochemical potentials through ideal conductors, the conductance G is given by,

$$G = (2e^2/h)(T/R) \tag{1}$$

where T and R are the transmission and reflection probabilities of the sample. The quantization unit involves the two fundamental constants, Planck's constant, h and the electronic charge, e. The factor of 2 arises because the spin-up and spin-down modes are degenerate.

The experimental manifestation of equation (1) came through the discovery[2, 3] of a sequence of steps in the conductance of a small constriction, in a two-dimensional (2D) electron gas in semiconductor heterojunction, as its width, W, was varied by varying the gate voltage. The steps, near-integer multiples of $G_0 = 2e^2/h$, are explained[4, 5, 6] by the opening of yet another conducting channel. Each channel, characterized by a different group velocity of electrons at E_f, carries the same current[4]. Hence the conductance is nG_0, where n is the number of conducting channels.

To observe steps in conductance, the constriction width has to be in the range of λ_f, the wavelength of the electron at the Fermi energy. Whereas for semiconductors this number is large, for metals, λ_f tends to be in the angstrom regime. Hence the contact between two metal electrodes must be formed by wires of atomic dimensions. Wires of such atomic scale dimensions represent the ultimate in nanostructure technology. Earlier using the scanning tunneling microscope (STM)[7] and later using the atomic force microscope (AFM)[8] wires consisting of just a single atom were fabricated. The measured conductance has been analyzed by several authors[9, 10, 11]. Lang[9] obtained quantitative values for the resistance of wires consisting of only a few atoms from fully self-consistent-field calculations. His predictions have been verified[9] by Yazdani et al using novel STM techniques. The effect of geometry and bonding arrangements on conductance has also been considered[12].

Landauer[1] has pointed out some of the issues connected with equation (1). For example, if the sample has perfect transmission, we expect a finite value for the quantum conductance, $G \sim 2e^2/h$. However, equation (1) yields an infinite conductance (zero resistance) when $T = 1$ because then the reflection probability $R = 0$. Interestingly enough, the result $G = 2e^2T/h$ does indeed follow from equation (1) but only in the other extreme limit of $T \ll 1$ because R can then be replaced by unity. Therein lies the contradiction. But it is now well understood[13, 14] that if the electrochemical potential is measured in the reservoirs, the formula is $G = 2e^2T/h$ is indeed applicable over the entire range of T.

2. THEORETICAL ANALYSIS

We present an elementary analysis of quantum conductance through an ideal 1D ballistic conductor (that is, a conductor with no scattering). We employ the uncertainty principle in conjunction with the definition of transport in the quantum limit to derive the conductance formula. Following the standard definition, conductance is a measure of current through a sample connecting two reservoirs at different potentials,

$$G = \frac{\Delta I}{\Delta V}. \tag{2}$$

The current is given by the rate of charge flow,

$$\Delta I = \frac{\Delta Q}{\Delta t}. \tag{3}$$

Now the charge is quantized in units of elementary charge, e. Hence in the extreme quantum limit, setting $\Delta Q = e$, we get

$$\Delta I = \frac{e}{\Delta t} \tag{4}$$

Combining equations (2) and (3), and using the fact that the potential difference, ΔV is equal to electrochemical potential difference ΔE divided by the electronic charge, e we get,

$$G = \frac{e^2}{\Delta E \Delta t} \tag{5}$$

Next invoking Heisenberg's uncertainty principle, $\Delta E \Delta t = h$, one gets for ballistic conductance,

$$G = \frac{e^2}{h}. \tag{6}$$

The experimental step size is $G = 2e^2/h$ because of the spin degeneracy and has a value of $8 \times 10^{-5} \Omega^{-1}$. Finally, if the transmission probability is less than unity due to scattering etc., we multiply the above result by T, where $0 < T \leq 1$ to get the celebrated formula,

$$G = \frac{2e^2 T}{h}. \tag{7}$$

This equation also supports the notion[4, 14] that "conduction is transmission". A note of caution is perhaps in order here. The analysis presented above is not a strict derivation of the relationship. In a sense, we have simply performed a dimensional analysis. Since the uncertainty principle is expressed in the form of an inequality, $\Delta E \Delta t \geq h$, our answers may only be correct to within factors of 2, π, etc.

It is implicit in the above analysis that the constriction is so narrow that there is only one transverse state occupied. If there are n transverse states below E_f then the conductance is n times G_0. To demonstrate it explicitly we consider an ideal 2D conductor of a small width W and follow the analysis by Büttiker[15]. The Hamiltonian is

$$H = \frac{p_x^2}{2m} + \frac{p_y^2}{2m} + V(y), \tag{8}$$

where x is the direction of current flow and y is the transverse coordinate where electrons are confined in a strip of width W. Assuming an infinite square well potential, $V(y)$, the transverse eigenvalues become quantized. The lowest eigenvalue can be estimated from the uncertainty relation, since $\Delta y \approx W$. The momentum must then be of the order of h/W. Hence the lowest eigenvalue is $E_1 = \frac{h^2}{2mW^2}$. Higher eigenvalues will increase as n^2 since we could apply the uncertainty relation to progressively smaller regions, W/n. An exact calculation, however, leads to transverse eigenvalues, $E_n = \frac{n^2h^2}{8mW^2}$. The total energy of the state,

$$E(n, k_n) = E_n + \frac{\hbar^2 k_n^2}{2m} \tag{9}$$

is the sum of the transverse energy and the energy associated with the longitudinal motion.

The maximum number of transverse occupied states, N, is obtained by setting $E(N, 0) = E_f$. Such a procedure gives

$$N = W\sqrt{\frac{8mE_f}{h^2}}, \tag{10}$$

which can be rewritten in the form, $N = Wk_f/\pi$. As expected, the number of transverse occupied states increases with the width of the constriction. For each transverse occupied state, $n \leq N$, one can find a corresponding value of k_n and hence a contribution to the total current. The absolute maximum momentum is associated with the lowest transverse eigenstate $n = 1$, and is given by

$$k_1 = \sqrt{\frac{2m}{\hbar^2}(E_f - E_1)}. \tag{11}$$

The longitudinal momentum thus spans the range $0 <| k | \leq k_1$. It turns out that the current carried by each channel is independent of k_n. This is due to[4] the density of states at E_f canceling the k_n factor in the expression for current. The faster the carrier the lesser its weight! Thus each transverse occupied state contributes $2e^2/h$ to conductance. Hence the conductance of an n channel wire is $2ne^2/h$.

To obtain the conductance of an n channel nanostructure from our simple analysis based on the uncertainty relation, we set $\Delta Q = ne$ in equation (3) rather than e. Everything proceeds as before. The final result for conductance then has a multiplicative factor of n leading to the celebrated Landauer formula for conductance,

$$G = \frac{2e^2}{h}nT. \tag{12}$$

We note, in passing, that using equation (10) for the number of modes in a wide conductor of macroscopic length, Landauer formula reverts to standard Ohm's law. Some prefer to express the results in terms of quantum resistance,

$$R = \frac{h}{2e^2}\frac{1}{n}, n = 1, 2, 3.... \tag{13}$$

The numerical value for the quantum of resistance, $h/2e^2$ is 12949 Ω which for ease of recall may be approximated by 12345 Ω. Finite resistance for a perfect conductor thus seems to be a strictly quantum mechanical effect which has its origin in the uncertainty principle. Thus we have provided an alternate explanation for the origin of non-zero resistance in a ballistic conductor. Previously, it has been understood in terms of contact effects.

It is interesting to note that the expression for the quantized Hall resistance in high magnetic field is

$$R_H = \frac{h}{ne^2}, n = 1, 2, 3....$$ (14)

The quantity h/e^2 has been called von Klitzing's constant, R_K in honor of Klaus von Klitzing[16]. It has a value which is twice as large as the quantum of resistance. The quantized Hall effect may be viewed as an extension of the quantum transport where the confining potential, V(y) is provided by an externally applied magnetic field B along the z-direction. The electrons in the 2D layer (xy-plane) move in cyclotron orbits and execute a harmonic-oscillator type of motion along the y-direction. The transverse quantized states then are the Landau levels with eigenvalues, $E_n = \hbar\omega_c(n + \frac{1}{2})$, where $\omega_c = eB/mc$. Thus the role of a quantum channel is replaced by a Landau level. Since the magnetic field also lifts the spin degeneracy, a factor of two is missing in the expression for Hall resistance.

3. SUMMARY

We have motivated by using the uncertainty principle that the finite resistance in a perfect conductor is a quantum mechanical effect. The jumps in quantum conductance which are integer multiples of $2e^2/h$ may be viewed as a consequence of the quantization of charge into elementary charges e. Hall quantization, in addition, involves flux quantization in units of h/e. In analogy with the fractional quantum Hall effect, which has been interpreted on the basis of elementary excitations of quasiparticles with a charge $e^* = e/3, e/5, e/7$, etc., we speculate that fractional jumps may also be observable in standard quantum transport in a 2D electron gas. This speculation is based on rather simplistic arguments. First of all, the fractional quantum Hall effect has its origin in a novel many-body ground state as formulated by Laughlin[17]. The other being the crucial role of the magnetic field in Hall quantization. As it turns out, the magnetic field suppresses[15] backscattering by localizing states carrying current in different directions to the opposites edges of the sample. This gives the sample the 'appearance' of a ballistic conductor even in the presence of impurities. One may never be able to produce samples of sufficiently high purity to observe fractional quantization in quantum transport. But we are often surprised by the ingenuity of our experimental colleagues!

ACKNOWLEDGMENTS

I am most grateful to Professor S. Ciraci of Bilkent University in Ankara, Turkey for introducing me to this subject. It is my pleasure to acknowledge useful discussions with Dr. R. Landauer, Dr. N. Lang, and Dr. A. K. Rajagopal. Finally my sincere thanks to Professors B. Rao, P. Jena, and S. N. Behera for inviting me to participate in this exciting workshop on novel materials.

DEDICATION

As a graduate student at the University of Delhi (1962-1964), I had the good fortune to attend a course taught by Professor D. S. Kothari on Quantum Mechanics. I vividly remember the emphasis he placed on simple dimensional arguments to explain complex phenomena. In understanding conductance through wires of atomic dimensions, I have tried to follow Professor Kothari's foot steps. Since the workshop where this presentation was made was held in India, it is my pleasure to dedicate this paper to Professor Kothari in the form of a "shradhanjali" from a student to his teacher. For the record, Professor Kothari was born on July 6, 1906 and died on February 4, 1993.

References

[1] R. Landauer, J. Phys.: Condens. Matter 1, 8099 (1989).

[2] B. J. van Wees, H. van Houten, C. W. J. Beenacker, J. G. Williams, L. P. Kouwen-howen, D. van der Marel, and C. T. Foxon, Phys. Rev. Lett. 60, 848 (1988).

[3] D. A. Wharam, T. J. Thorton, R. Newburg, M. Pepper, H. Ritchie and G.A.C Jones, J. Phys. C 21, L209 (1988).

[4] H. van Houten and C. W. J. Beenaker, Physics Today 49, 22 (1996).

[5] M. Büttiker, Y. Imry, R. Landauer and S. Pinhas, Phys. Rev. B 31, 6207 (1985).

[6] E. Tekman and S. Ciraci, Phys. Rev. B 43, 7145 (1991).

[7] J. K. Gimzewski and R. Möller, Phys. Rev. B 36, 1284 (1987); N. Agrait, J. G. Rodrigo, and S. Vieira, Phys. Rev. B 47, 12345 (1993); J. M. Krans, C. J. Muller, I. K. Yanson, Th. C. M. Govaert, R. Hesper, and J. M. van Ruitenbeek Phys. Rev. B 48, 14721 (1993).

[8] E. S. Snow, D. Park, and P. M. Campbell, Appl. Phys. Lett. 69, 269 (1996).

[9] N. D. Lang, Phys. Rev. B 36, 8173 (1987); Phys. Rev. B 52, 5335 (1995); A. Yazdani, D. M. Eigler and N. D. Lang, Science 272, 1921 (1995).

[10] S. Ciraci and E. Tekman, Phys. Rev. B 40, 11969 (1989).

[11] L. Escapa and N. Garcia Appl. Phys. Lett. 56, 901 (1990). P. Garcia-Mochales, P. A. Serena, N. Garcia, and J. L. Costa Krämer, Phys. Rev. B 53, 10268 (1996); A. L. Yetani, A. M. Rodero, and F. Flores, Phys. Rev. B (in press).

[12] H. Meherz, S. Ciraci, A. Buldum, and I. P. Batra, Phys. Rev. B **55**, R1981 (1997).

[13] Y. Imry in *Directions in Condensed Matter Physics*, edited by G. Grinstein and G. Mazenko (World Scientific, Singapore, 1986), p 101. Also see, P. W. Anderson, D. J. Thouless, E. Abrahams, and D. S. Fisher, Phys. Rev. B **22**, 3519 (1980).

[14] R. Landauer, *Proceedings of the NATO Advanced Research Workshop on Nanowires*, eds. P. A. Serena and U. Landman, (Kluwer Academic Publishers, Dordrecht) to be published.

[15] M. Büttiker, Phys. Rev. B **38**, 9375 (1988).

[16] K. von Klitzing, Rev. Mod. Phys. **58**, 519 (1986); K. von Klitzing, G. Dorda, and M. Pepper, Phys. Rev. Lett. **45**, 495 (1980).

[17] R. B. Laughlin, Phys. Rev. Lett. **50**, 1395 (1983).

ELECTRONIC AND MAGNETIC PROPERTIES OF TRANSITION METAL CLUSTERS ON METAL SURFACES

W.Hergert[a], V.S. Stepanyuk[a], P. Rennert[a]
K. Wildberger[b], R. Zeller[b], and P.H. Dederichs[b]
[a] Fachbereich Physik, Martin-Luther-Universität Halle-Wittenberg
D-06099 Halle, Germany
[b] Institut für Festkörperforschung, Forschungszentrum Jülich
D-52425 Jülich, Germany

1. INTRODUCTION

Magnetic low-dimensional structures are discussed today in connection with applications in storage devices, sensors, and other microelectronic components. If magnetic nanostructures could be produced on semiconductor surfaces, a direct link to the classical electronics would be possible. A new field named "Magnetoelectronics" is seen to grow up in the future and magnetic nanostructures should play an important role.

In low-dimensional metallic structures new effects are discovered in the last decade like: (i) exchange coupling of magnetic layers over a nonmagnetic spacer [1] , (ii) giant magnetoresistance [2], and (iii) magnetism in low-dimensional metallic systems built from metals, which are non-magnetic in bulk form. [3, 4]

We will concentrate our considerations on metallic nanostructures on metallic surfaces. The artificial structures which we investigate by means of *ab initio* electronic structure calculations can be built experimentally in different ways. Methods to assemble small metallic clusters on surfaces are:

(i) **Assembling of individual atoms by means of the scanning tunneling microscope (STM) :** Crommie *et al.* [5, 6] have built a "quantum corral" made of 48 Fe adatoms on the Cu(111) surface using an adatom "sliding" process. Microscopic electron confinement structures can be built in this way. A series of different methods to manipulate individual atoms with the STM are available now.[7]

(ii) **Diffusion-controlled aggregation :** In this method it is used that nucleation and growth are competing processes at the surfaces. The final structure depends on the surface diffusion and the deposition flux. The average adatom diffusion length is temperature dependent. Therefore the whole process can be controlled by external parameters. Linear, two-dimensional or fractal structures of nanometer dimension with a high number density ($10^{12} - 10^{14}$ cm^{-2}) can be grown in this way. [8] Ag dimers, compact islands of 6 to 26 atoms are grown on Pt(111). If the anisotropy of fcc(110) surfaces is used, chains of atoms of various length can be grown (Cu on Pd(110), cf. [9])

(iii) **Decoration of steps :** Wire structures a few nanometers wide can be fabricated by decoration of step edges at vicinal surfaces. Fe and Co on Cu(111) can be used to form magnetic wire structures. [10, 11]

Especially the last two methods are able to create metallic nanostructures of the same shape and size with a very high number density. Therefore such structures can be investigated with conventional surface sensitive analytic techniques. It will be possible in the near future to investigate artificially created supported metallic nanostructures in detail, as it is already done for free TM clusters.[12]

Our calculations give first insight in the magnetic properties of small TM clusters on metallic fcc(001) surfaces. The first experimental confirmation of the magnetism of single $4d$ adatoms on noble metal surfaces , which was theoretically predicted [13], was reported recently.[14]

2. METHOD OF CALCULATION

The calculations are carried out with the Korringa-Kohn-Rostocker (KKR) Green's function method. The method is based on density functional theory (LDA) in the local spin density approximation (LSDA). The calculation starts with the selfconsistent determination of the Green's function (GF) of that material, which will be used as substrate material. The bulk GF is transformed into a Wannier-Bloch basis taking into accout whether an (100), (110) or (111) surface will be considered later. Now the atomic potentials of several layers are removed to create two practically uncoupled half-crystals. According to the different distances of atomic planes at different surfaces, the number of "vacuum" layers used to decouple the half-crystals depends on the crystallographic surface. For fcc(100) surfaces 7 layers are removed. The GF of the ideal surface $G_{LL'}^{s,jj'}\left(\mathbf{k}_{\parallel}, E\right)$ ($\mathbf{k}_{\parallel}$ - wave vector parallel to the atomic planes, E - energy) will be determined by a self-consistent solution of the Dyson equation.

$$G_{LL'}^{s,jj'}\left(\mathbf{k}_{\parallel}, E\right) = G_{LL'}^{b,jj'}\left(\mathbf{k}_{\parallel}, E\right) + \sum_{j''L''} G_{LL''}^{b,jj''}\left(\mathbf{k}_{\parallel}, E\right) \Delta t_{L''}^{j''}(E) G_{L''L'}^{s,j''j'}\left(\mathbf{k}_{\parallel}, E\right) \quad (1)$$

$G_{LL'}^{b,jj'}$ is the GF of the bulk crystal in Wannier-Bloch basis, j, j' are layer indices and L, L' indicate the angular momentum ($L = (l, m)$). $\Delta t(E)$ represents the changes in the t-matrices of the corresponding layers, i.e. the change of the scattering properties of atoms in layers near the surface with respect to the bulk.

Clusters on the surface destroy the two-dimensional translational symmetry. Therefore the GF of the ideal surface has to be transformed in a site representation. The

Dyson equation is used again to calculate the Green's function $G_{LL'}^{c,nn'}(E)$ of the cluster on the surface:

$$G_{LL'}^{c,nn'}(E) = G_{LL'}^{s,nn'}(E) + \sum_{n''L''} G_{LL''}^{s,nn''}(E)\Delta t_{L''}^{n''}(E)G_{L''L'}^{c,j''j'}(E) \tag{2}$$

The sum runs over all sites n'' and angular momenta L'' for which the perturbation $\Delta t_L^n(E)$ between the t-matrices of the ideal surface problem and the surface with cluster are important. In our calculations perturbations on the nearest neighbor sites of the cluster are taken into account. Fig. 1 demonstrates which system is finally represented by the selfconsistently calculated GF $G_{LL'}^{c,nn'}(E)$. The method of successive application of the Dyson equation solves the surface problem exactly. Only the region near the supported cluster is shown in the figure.

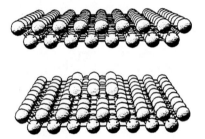

Figure 1. Cluster of 9 atoms (square island) on the fcc(001) surface. The configuration of atoms near the supported cluster which is used in the KKR-calculation is shown.

The full charge density is taken into account using a multipole expansion up to $l_{max} =6$. Coulomb and exchange correlation energies are evaluated using $l_{max}= 12$. For the exchange correlation potential the local functional of Vosko *et al.* [15] is used. Potentials are assumed to be spherical symmetric inside the Wigner-Seitz sphere.

3. RESULTS AND DISCUSSION

To study the electronic and magnetic properties of supported clusters on metallic surfaces in detail a series of different systems including different substrates (Cu, Ag, Pd, Pt) and clusters built from different metals ($3d$-, $4d$-, $5d$-transition metals (TM)) are investigated. The calculations give information about

- the magnetism of supported TM-clusters (magnetism in supported clusters built from nonmagnetic TM), [13, 16, 18]

- the influence of the substrate on the magnetic properties of supported clusters (comparison of different substrates and with free clusters), [19, 20]

- metamagnetism in metallic nanostructures, [21, 22, 23]

- influence of the real structure and of impurities on the magnetic properties of small clusters.

3.1. Magnetism of supported clusters

It was demonstrated by Wildberger *et al.* [16] that $4d$ nanostructures (dimers, linear chains, plane islands) on Ag(001) have a sizable magnetic moment, even if the structures are built from non-magnetic metals. Similar calculations for $3d$- and $5d$-TM came to the same conclusion. As an example the magnetic moments in the ferromagnetic state of $3d$ adatoms, dimers, trimers, and small islands (I4) on Cu(001) are given in Fig.2.

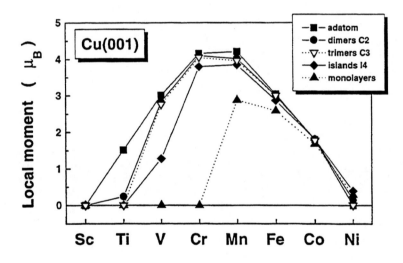

Figure 2. Magnetic moments of the ferromagnetic state for $3d$ adatoms, dimers, trimers, and tetramers on Cu(001). The magnetic moments of the complete monolayer [24] are given for comparison.

For $4d$-TM on Ag(001) a pronounced shift of the maximum moment to larger valencies is obtained, if islands are compared with chain structures. (cf. [16]) This tendency is also visible here, but not so pronounced.

As a general trend we find for $3d$- and $4d$-TM on Cu and Ag surfaces, that nanostructures of elements at the end of the series (Fe,Co,Ru,Rh) are less sensitive to cluster size and shape than structure built of elements from the beginning of the series. Fe and Co structures on Cu(001) show practically no change of the moment per atom from the single adatom to the full monolayer. This is a consequence of the fact, that in all these cases the majority band is practically filled. Otherwise for Ti the moment is strongly supressed if we go from the dimer to the trimer.

3.2. Influence of the substrate

The influence of the substrate on the magnetic properties of supported clusteres can be studied in two ways. Firstly, we study the properties of a given nanostructure on different substrates. This gives information how changes in lattice constants and hybridization change the properties of the cluster. Secondly, we compare the supported cluster with a free cluster of the same structure and the free relaxed cluster.(cf. [17])

The lattice constant of Cu is about 10 % smaller than the lattice constant of Ag, which increases the sp-d hybridization considerably. As a result, the magnetic moments of $3d$-, $4d$- and $5d$-adatoms are always smaller than those on the Ag surface.[18] Compared to the Ag and Cu surfaces, the interaction between the adatoms and the Pd and Pt substrates supresses the magnetic moments at the beginning of the d-series and enhances the moments at the end of the d-series. Magnetic adatoms (Fe,Co,Ru,Os) induce sizable magnetic moments (0.15-0.2 μ_B) on neighboring Pd and Pt atoms at the surfaces.

Further insight in the cluster-substrate interaction can be obtained by the comparison of supported and free clusters. Free clusters can be calculated within the KKR-method by placing the cluster in the middle of the vacuum layer (cf. Fig.1). In this calculation the structure of the free cluster is fixed to that of the supported cluster. The V dimer has a magnetic moment/atom of 2.85 μ_B on the Cu(001) surface. If the free dimer is calculated for the same interatomic distance the moment is increased to 3.45 μ_B/atom. This is attributed to the loss of the hybridization with the substrate. If we increase the interatomic distance to the nearest neighbor distance of Ag the moment is increased to 3.58 μ_B/atom. If this dimer is supported on Ag the hybridization with the substrate decreases the moment to 3.38 μ_B/atom. This is higher than on Cu from reasons discussed above. A SCF-LCAO-MO calculation gives a spin septet configuration for free V dimers with interatomic distances correpondig to Cu and Ag nearest neighbor distances. The free dimer has a bond length of 1.71 which is much smaller than the Ag or Cu nearest neighbor distance. The free relaxed V dimer has a magnetic moment of 1 μ_B/atom.

3.3. Metamagnetism in metallic nanostructures

So far we have discussed only ferromagnetic and antiferromagnetic arrangements of the magnetic moments. The existence of different magnetic states like high spin ferromangetic (HSF), low spin ferromagnetic (LSF) and antiferromgnetic (AF) is well known for bulk systems. In a series of papers Moruzzi et al. investigated the volume dependence of the magnetic and electronic properties of transition metals.[25] A theoretical investigation of Zhou *et al.* [26] shows up to five different magnetic states for γ-Fe at a given lattice constant.

Our calculations have demonstrated that $3d$ nanostructures on Cu(001) can show a large variety of magnetic states like it is known for bulk systems and free clusters. The energy differences between different states can be so small that magnetic fluctuations occur.

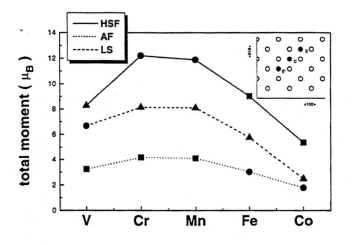

Figure 3. Total moment of $3d$-TM trimers on Cu(001). The total energy is indicated by the symbols □ - ground state, △ - state with highest energy, o - state with intermediate energy.

Fig. 3 shows the total magnetic moment for the different magnetic states of $3d$-trimers on the Cu(001) surface. A fourth state, the so called antisymmetric (AS) one (moments $M_E = -M_{E'}$, $M_c = 0$) is not shown here, because the total moment is zero and the total energy of this state is less than the energy of the paramagnetic state, but considerably higher than the energy of the states shown in the figure. The ground state changes from AF for V, Cr, Mn to HSF for Fe, Co. The energetic balance between the magnetic states is particularly delicate for V. The AF state has an energy only 8 meV/atom lower than the LSF state, which itself has a slightly lower energy than the HSF state. Another interesting case is Mn trimers, where the energy difference between the AF ground state and the HSF state is only 2 meV/atom (corresponding to a temperature of 25 K) Thus already small variations of external parameters like temperature or applied field could lead to transitions between these states, which in the case of Mn would change the total moment of the cluster by 7.8 μ_B.

Table 1: Local moments of the atoms at the center (1) at the edge (2) and at the corner (3) for ferromagnetic (FM) and antiferromagnetic (AFM) states of the Mn_9 cluster on Ag(001) in units of μ_B. Energy differences per atom are given to the antiferromagnetic groundstate.

Magnetic Phase		Magnetic Moment/Atom			ΔE (meV)
	Atom Type	1	2	3	
AFM ↑↓↑		4.19	-4.32	4.36	0
AFM ↑↑↓		4.15	4.28	-4.38	40
AFM ↑↓↓		4.37	-4.19	-4.32	64
FM ↑↑↑		4.15	4.24	4.34	73

Effects of metamagnetism are also obtained in other supported clusters. Mn_9 compact islands on Ag(001) show magnetic bistability, i.e antiferromagnetic and ferromagnetic states are nearly degenerate. [23]

The ground state of the Mn_9 cluster is an antiferromagnetic configuration, where the spins of the nearest neighbors are opposit to each other. The energy differences of all the possible states are given in Tab. 1. The paramagnetic solution is about 1eV/atom above the antiferromagnetic phase and is clearly energetically unfavorable. Magnetic bistability is observed experimentally recently in free ligated Mn_{12} metal-ion clusters. [27]

3.4. Defects in supported clusters

All kinds of possible technical applications need a suffcient stability of the nanostructures in use. Therfore it is important to investigate the influence of defects like impurities and vacancies of the electronic and magnetic properties on metallic nanostructures.

Semi-empirical calculations [28] and later *ab initio* calculations [29] have shown that intermixture at the Rh-Ag interface will strongly decrease the magnetic moments of Rh atoms in a Rh layer on Ag(001). Therefore it is interesting to study the effect of impurities and vacancies on the magnetic properties of supported Rh clusters on Ag(001).

We started to study the influence of vacancies on the magnetism of small Fe and Ni clusters. The properties of Fe or Ni atoms in a plane island (Fe_9, Ni_9) on Cu(001) are practically not influenced by vavcancies.

For Rh on Ag(001) a strong influence on the magnetic properties by such real structure effects was found. The coordination of Rh atoms in the cluster can be changed in such a way, that impurities and vacancies lead to an increase of the moment/atom in the cluster. This probem is subject of more detailed investigations now.

4. CONCLUSIONS

We have demonstrated that a variety of interesting magnetic properties is connected to metallic nanostructures on metal surfaces. Non-magnetic metals can form magnetic clusters on metal surfaces.

Detailed predictions about the magnetic properties of small supported metallic nanostructures can be made by means of the *ab inito* calculations. The results give hints for experimental investigations of such magnetic nanostructures.

For a detailed comparison with experimetal results larger clusters have to be studied. The *ab initio* results may serve as starting point for semi-empirical calculations for larger systems.

AKNOWLEDGMENTS

Financial support of the Human Capital and Mobility Program "Ab inito (from electronic structure) calculation of complex processes in materials" of the European Union is greatfully aknowledged. The computations are performed on Cray computers of the Forschungszentrum Jülich and the German supercomputer center (HLRZ).

References

[1] S.S.P. Parkin, N. More, and K.P. Roche, Phys. Rev. Lett. 64, 2304(1990).

[2] M.N. Baibich, J.M. Broto, A. Fert, F. Nguyen Van Dam, F. Petroff, P. Etienne, G. Creutzet, A. Friedrich, and J. Chazelas, Phys. Rev. Lett. 61, 2472 (1988); G. Binasch, P. Grünberg, F. Saurenbach, and W. Zinn, Phys. Rev. B39, 4828 (1988).

[3] R.-Q. Wu and A.J. Freeman, Phys. Rev. B 45, 7222 (1992).

[4] S. Blügel, Europhys. Lett. 18, 257 (1992); S. Blügel, Phys. Rev. Lett. 68, 851 (1992).

[5] M.F. Crommie, C.P. Lutz, and D.M. Eigler, Science 262, 218 (1993).

[6] D.M. Eigler and E.K Schweitzer, Nature 344, 524 (1990), J.A. Stroscio and D.M. Eigler, Science 254, 1319 (1991).

[7] G. Meyer, B. Neu, and K.H. Rieder, phys. stat. sol. (b) 192, 313 (1995);

[8] H. Röder, E. Hahn, H. Brune, J.-P. Bucher, and K. Kern, Nature 366, 141 (1993).

[9] H. Brune, H. Röder, C. Boragno, and Klaus Kern, Phys. Rev. Lett. 73, 1955 (1994).

[10] T. Jung, R. Schlittler, J.K Gimzewski, and F.J. Himpsel, Applied Physics A 61, 467 (1995).

[11] F.J. Himpsel, Y.W. Mo, T. Jung, J.E. Ortega, G.J. Mankey, and R.F. Willis, Superlatt. and Microstruc. 15, 237 (1994).

[12] J.P. Bucher, D.C. Douglas, and L.A. Bloomfield, Phys. Rev. Lett. 66, 3052 (1991); J.P. Bucher and L.A. Bloomfield, Int. J. mod. Phys. B7, 1079 (1993); I.M.L. Billas, A. Chatelain, and W.A. de Heer, Science 265, 1682 (1994).

[13] P. Lang, V.S. Stepanyuk, K. Wildberger, R. Zeller, and P.H. Dederichs, Solid State Commun. 92, 755 (1994).

[14] H. Beckmann, R. Schäfer, Wenqi Li, and G. Bergmann, Europhys. Lett. 33 563 (1996); R. Schäfer, and G. Bergmann, Solid State Commun. 98, 45 (1996).

[15] S.H. Vosko, L. Wilk, M. Nusair, J. Can. Phys. 58, 1200 (1980).

[16] K. Wildberger, V.S. Stepanyuk, P. Lang, R. Zeller, and P.H. Dederichs, Phys. Rev. Lett. 75, 509 (1995).

[17] S.E. Weber, B.K. Rao, P. Jena, V.S. Stepanyuk, W. Hergert, K. Wildberger, R. Zeller, and P.H. Dederichs, submitted to J. Phys.: Cond. Matter.

[18] V.S. Stepanyuk, W. Hergert, K. Wildberger, R. Zeller, and P.H. Dederichs, Phys. Rev. B 53, 2121 (1996).

[19] V.S. Stepanyuk, W. Hergert, P. Rennert, K. Wildberger, R. Zeller, and P.H. Dederichs, Phys. Rev. B 54, 14121 (1996).

[20] S.K. Nayak, S.E. Weber, P. Jena, K. Wildberger, R. Zeller, P.H. Dederichs, V.S. Stepanyuk, and W. Hergert, submitted to Phys. Rev. B

[21] V.S. Stepanyuk, W. Hergert, P. Rennert, K. Wildberger, R. Zeller, and P.H. Dederichs, J. Magn. Magn. Mater. 165, 272 (1997).

[22] V.S. Stepanyuk, W. Hergert, P. Rennert, K. Wildberger, R. Zeller, and P.H. Dederichs, Solid State Commun. 101,559 (1997).

[23] V.S. Stepanyuk, W. Hergert, K. Wildberger, S.K. Nayak, and P. Jena, submitted to Phys. Rev. B.

[24] S. Blügel, Solid State Commun. 84, 621 (1992).

[25] V.L. Moruzzi, P.M. Marcus, K. Schwarz, and P. Mohn, Phys. Rev. B 34, 1784 (1986); V.L. Moruzzi, Phys. Rev. Lett. 57, 2211 (1986); V.L. Moruzzi, P.M. Marcus and P.C. Pattnaik, Phys. Rev. B 37, 8003 (1988); V.L. Moruzzi and P.M. Marcus B 42, 8361 (1990).

[26] Yu-mei Zhou, Nuen-ging Zhang, Lie-ping Zhong, and Ding sheng Wang, J. Magn. Magn. Mater. 145, L237 (1990).

[27] R.Sessoli, D. Gatteschi, A. Caneschi, and M.A. Novak, Nature **365**, 141 (1993); J.R. Firiedman, M.P. Sarachik, J. Tejada, and R. Ziolo, Phys. Rev. Lett. **76**, 3830 (1996).

[28] W. Hergert, P. Rennert, C. Demangeat, H. Dreyssé, Surf. Rev. Lett. **2**, 203 (1995).

[29] I. Turek, J. Kudrnovsky, M. Sob, V. Drchal, and P. Weinberger, Phys. Rev. Lett. **74**, 2551 (1995).

FIRST-PRINCIPLES AND MOLECULAR-DYNAMICS SIMULATIONS OF MARTENSITIC TRANSFORMATIONS IN TRANSITION-METAL ALLOYS: TRACING OUT THE ELECTRONIC ORIGIN[*]

P. Entel[†],[1] H. C. Herper,[1] K. Kadau,[1] R. Meyer,[1], E. Hoffmann,[1] G. Nepecks,[1] M. Acet,[2] and W. Pepperhoff[2]

[1] *Theoretische Tieftemperaturphysik*
[2] *Experimentelle Tieftemperaturphysik*
Gerhard-Mercator-Universität Duisburg, 47048 Duisburg, Germany

ABSTRACT

Results are presented of first-principles total-energy calculations and molecular-dynamics simulations of structural transformations in magnetic transition metal alloys. While first-principles calculations allow to identify those structures having the lower total energy, molecular-dynamics simulations can be used to trace out the dependence of the transformation on temperature and concentration. We use the semi-empiric embedded-atom potential in the molecular-dynamics simulations. The simulations allow to describe supercooling and superheating associated with the change of structure as a function of temperature and concentration.

1. INTRODUCTION

Research on martensitic phase transformations [1] in ferrous alloys covers a period of more than 100 years and has led to substantial understanding of the kinetics of the transformation (for a recent review see [2]). Martensitic transformations occur in

---

[*]Gewidmet Herrn Prof. Dr. E. F. Wassermann anlässlich seines 60. Geburtstags.
[†]Email: entel@thp.uni-duisburg.de

alloys of different metallic elements. Usually the alloys are substitutional mixtures, where atoms of the host element are partially replaced by different kinds of metal atoms. This can lead to reduced or extra electronic charge. In the Hume-Rothery alloys this is responsible for the appearance of a definite crystal structure if the ratio of the valence electron number to atom number gets a certain value. But band filling, screening and atomic disorder effects connected with the extra charges do not only lead to martensitic transitions: They are responsible for a whole class of phenomena like precipitation, spinodal decomposition (and ordering) and other phase separation phenomena. The task of theory is to give a unified description of the electronic origin. In spite of many systematic theoretical investigations (for example, see [3]) this problem has not been solved completely. In this article we try to address a few questions associated with martensitic transformations in ferrous alloys with the help of first-principles calculations.

The martensitic phase transformation and structural trends of Hume-Rothery alloys are rather well understood. They are driven by the van Hove singularities in the densitiy of states arising from band gaps at specific Brillouin zone boundaries. Is the new crystal structure for a critical electron number so that the new Brillouin zone can accomodate all electrons, a large energy gain can be expected. However, detailed calculations show that in addition the renormalization of one-particle properties is essential due to the logarithmic singularity in the slope of the Lindhard function at $q = 2 k_F$. Since this singularity is driven by the ratio of valence electron number to atom number, the Hume-Rothery alloys are called *electron phases*. In Hume-Rothery alloys the phase transition is of weakly first order from a high-temperature less-close packed structure to a low-temperature close-packed one. Hysteresis associated with the transformation and the influence of atomic disorder is less well understood. Systematic first-principles investigations have shown that in many Hume-Rothery alloys compositional short-range ordering occurs which is again related to details of the Fermi surface touching the Brioullin zone. This determines the behavior of the atomic pair correlation function in an essential way [4, 5]. Atomic ordering effects seem also to be important for the shape-memory alloys which usually show stoichiometric order, and for metals showing rubber-like behavior [6]. Rubber-like behavior is caused by atomic short-range ordering of atoms with time via vacancies leading to superelastic behavior. There is some qualitative understanding but no microscopic theory of kinetic details of the martensitic transformation like nucleation of martensite at the surface or in the bulk at nucleation centers, concentration of martensitic embryos in the melt, role of diffusion processes via vacancies etc.

Structural transformations in ferrous alloys, where magnetic long-range order and spin fluctuations interfere, are less well understood. The experimental observation is that in alloys with a ferromagnetic ground state, the less-close packed crystal structure can be stabilized at low temperatures. Hysteresis, supercooling, and superheating

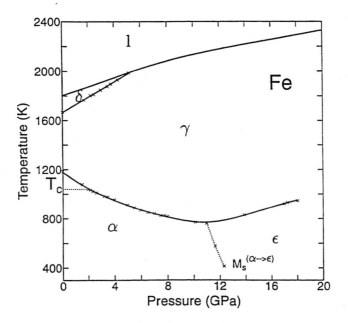

Figure 1. Pressure-Temperature phase diagram of iron, showing the range of stability of the bcc structure (α and δ), of the fcc structure (γ), and of the hcp structure (ε).

effects are usually more pronounced in the magnetic alloys. There are only a few stoichiometrically ordered phases. The element iron itself is interesting, since it can exist as bcc (α, δ), fcc (γ) or hcp (ε) phase. Fig. 1 shows the experimental pressure-temperature phase diagram. Iron-based alloys including Fe-C, Fe-Ni, Fe-Mn etc. usually exhibit the same richness of phases but also show systematic changes due to changes in the valence-electron number. *Ab initio* calculations of the total energy and of magnetic moments of the bcc, fcc and hcp phase of iron as a function of volume explain the ferromagnetic bcc $\rightarrow$ paramagnetic hcp transition around 10 GPa. Under pressure the d band widens and the density of states at ε_F decreases until the Stoner criterion for ferromagnetism is no longer fulfilled [7]. There also is a very strong volume dependence of the fcc and hcp moments near the normal atomic volume. The stability of the bcc structure at low temperatures is due to long-range ferromagnetic order. The physical picture at finite temperatures is less clear. It has been argued very early that the occurrence of the γ-phase at higher temperatures is driven by large ferromagnetic spin fluctuations as first assumed by Weiss [8]. Interpretation of the behavior of the specific

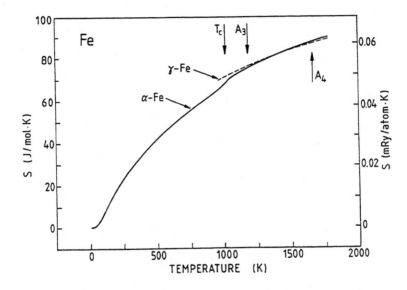

Figure 2. Variation of the entropy of elemental iron extrapolated from specific heat and enthalpy data [9]. The jump at A_3 is $T\Delta S = 0.606$ mRy/atom.

heat of the α- and γ-phase leads to the same conclusion [9]. At still higher temperatures the return to the bcc δ-phase is assumed to arise from entropy contributions of the softer δ-phase phonons or from loss of magnetic short-range order with increasing temperature [10]. The assumption of a magnetically driven α → γ transition with increasing temperature has recently been questioned [11]. Investigation of the phonon dispersion of α-Fe reveals strong softening of the entire $T_1[\xi\xi 0]$ and $T_1[\xi\xi 2\xi]$ branch when approaching the α → γ transition. The eigenvectors of these phonons are in the direction of displacements needed for the γ-phase. This has been interpreted by the authors as a dynamical precursor for the martensitic transformation. The data also suggest that the high-temperature phase is mainly stabilized by the increase in vibrational entropy, the entropy difference being $\Delta S^{\alpha\gamma}_{\text{vib}} = 0.142$ k_B/atom $= 0.887 \cdot 10^{-3}$ mRy/atom·K or $T\Delta S^{\alpha\gamma}_{\text{vib}} = 1.05$ mRy/atom at $T = A_3 = 1184$ K which is larger than the value extrapolated from [9] giving $T\Delta S^{\alpha\gamma}_{\text{vib/mag}} = 0.606$ mRy/atom. Fig. 2 shows results for the entropy obtained from the specific heat and enthalpy data of Bendick and Pepperhoff [9]. Thus the controversy of what stabilizes the high-temperature γ-phase of elemental iron needs further inspection. The transition is not completely driven by the entropy of ferromagnetic spin fluctuations; the vibronic part of the entropy might be of equal importance which would confirm the conjecture of Petry [11].

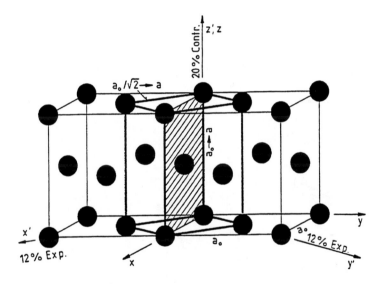

Figure 3. Bain's model for the formation of bcc martensite by appropriate contraction and expansion of the parent austenite cell [17].

Further problems arise when we try to describe the martensitic transformation on a microscopic level. While first-principles total-energy calculations give reliable results for the energy difference between different crystal structures at zero temperature for elemental iron and Fe-Ni alloys [12, 13, 14] or alkali elements [15, 16]), and also allow to simulate the structural transformation from the fcc → bcc structure along the Bain path [17] in Fig. 3, the continuation to finite temperatures in form of first-principles molecular-dynamics simulations is still restricted to so small numbers of atoms that structural changes cannot be simulated. Unsolved also is the question of how to incorporate magnetism at finite temperatures. Therefore, in the present work we have combined first-principles total-energy calculations and semi-empiric molecular

dynamics simulations based on the embedded atom method introduced by Daw and
Baskes [18]. The latter method allows to handle a sufficiently large number of atoms
and to simulate premartensitic behavior and the martensitic transformation [19, 20,
21, 22]. Magnetism can be dealt with indirectly by using potential functions which
reproduce the correct elastic behavior of the magnetic alloys at low temperatures. The
simulations then allow to study the growth of martensite, for example, in thin Fe-
Ni films with notch-like defects which facilitate martensitic nucleation at the surface
[23]. In this paper we will discuss a few characteristic results for ferromagnetic and
antiferromagnetic Fe and Fe-Ni alloys and nonmagnetic Ni-Al alloys.

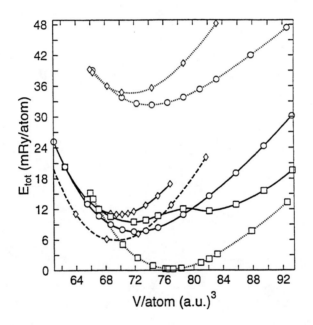

Figure 4. Total energies of elemental iron as a function of the atomic volume
obtained from FLAPW/GGAII calculations [25]. Dotted, solid, and dashed curves cor-
respond to the α-, γ- and ε-phase, respectively, whereby squares, circles, and diamonds
denote the ferromagnetic, antiferromagnetic, and nonmagnetic solution, respectively.

2. RESULTS OF FIRST-PRINCIPLES TOTAL-ENERGY CALCULATIONS

We first discuss results for elemental α- and γ-iron. Recent first-principles calculations of the total energy of the ferromagnetic and antiferromagnetic phase as function

of the volume using the WIEN95 code [24] yielded slightly different results [25] than previous calculations [26]. Fig. 4 shows that (i) α-Fe has the lowest ground-state energy,

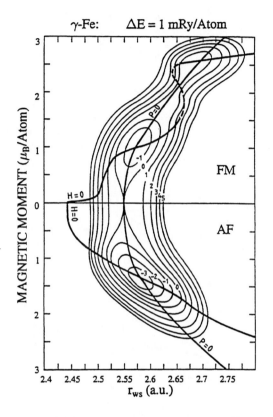

Figure 5. Ferro- and antiferromagnetic binding surface of γ-Fe as obtained by extrapolation of full-potential data [25]. Contour lines are at 1 mRy intervals. The $H = 0$ curve displays a LM $\rightarrow$ HM instability on the ferromagnetic part of the surface at larger volumes than the antiferromagnetic ground-state volume. The energy differences relative to the nonmagnetic reference state are (NM: nonmagnetic; FM: ferromagnetic; AF: antiferromagnetic; LM: low moment; HM: high moment): $E_{NM} - E_{FM}(LM) = 1.63$ mRy/atom; $E_{NM} - E_{FM}(HM) = -0.65$ mRy/atom; $E_{NM} - E_{AF}(LM) = 3.54$ mRy/atom.

(ii) γ-Fe would be antiferromagnetic, (iii) γ-Fe possesses a ferromagnetic low-moment and high-moment state as well as many close in energy lying noncollinear states (ASW calculations) which have been omitted from Fig. 4. All findings agree with experimental data. Of interest is the behavior of γ-Fe and the question whethernwe can draw from the zero-temperature calculations some reliable information about the structural transformation which occurs at higher temperatures. This can be done by discussing the corresponding binding surfaces which can be obtained for the case of γ-Fe and γ-Fe$_3$Ni by extrapolation of total energy data. Fig. 5 shows the surface of elemental iron. Energy contour lines are at 1 mRy intervals. The ferromagnetic surface displays

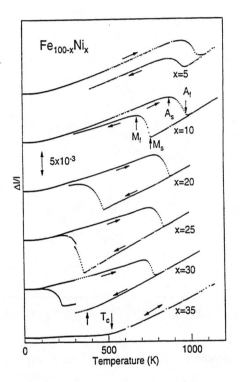

Figure 6. Experimental results for the relative volume change of various Fe-Ni alloys. $M_{S,F}$ and $A_{S,F}$ are martensite and austenite start and final temperatures, respectively; T_c is the Curie temperature.

two shallow local minima corresponding to a low moment at low atomic volume and a high moment at a larger atomic volume, and a nonmagnetic saddle point. The energies of both minima are higher than the antiferromagnetic minimum which agrees with the experimental observation that γ-Fe has an antiferromagnetic ground state at appropriate lattice spacing [27, 28]. The qualitative difference between the γ- and the α-phase is that the high-moment state of the γ-phase becomes lowest in energy when passing over to the bcc structure along the Bain path. In spite of the fact that we have not yet done first-principles caluclations at finite temperatures or simplified model calculations based on Gaussian treatment of fluctuations used earlier [29, 30, 31, 32], it is straightforward to speculate about the finite-temperature scenario. Although the antiferromagnetic solution has lowest energy, it will gradually disappear with increasing temperature, leaving at higher temperatures dominating ferromagnetic spin fluctuations which lead to an enhanced atomic volume and anti-Invar behavior. These ferromagentic spin fluctations are easy to excite and do not cost much energy. The spin fluctuations will also help to stabilize the γ-phase with respect to the α-phase at elevated temperatures in the absence of long-range ferromagentic order. This scenario is in agreement with Hasegawa's model calculation of the phase diagram of iron [10]. There is another interesting information concerning the simulation of the $\alpha \rightarrow \gamma$ transition at zero temperature in elemental iron. The ab $initio$ calculations show that the transition is associated with a substantial charge transfer from t_{2g} orbitals to e_g orbitals. The consequences of such a charge transfer would result in a lower atomic volume right after the transition in the γ-phase. This is indeed observed in experiments. Fig. 6 shows the relative volume change of various Fe-Ni alloys. Other specific features of Fe-Ni Invar alloys also depend on this type of charge transfer [33].

When discussing alloy systems, it is important to know whether they have more or less valence electrons than elemental iron. Adding extra d electrons like in the Fe-Ni Invar alloys destabilizes the bcc structure leading to Invar behavior in the γ-phase. This instability has two aspects, one is associated with the extra charge [33], the other with lattice vibrational arguments (in bcc Ni the shear modes have negative squared phonon energies [21]). Fig. 7 shows the magnetic binding surface of Fe$_3$Ni which has been extrapolated from full-potential calculations of the total energy [34]. There still is an antiferromagnetic solution, but the ferromagnetic solution is more stable. Due to the specific form of the surface, we expect strong admixture of ferromagnetic and antiferromagnetic spin fluctuations at temperatures lower than the Curie temperature. Atomic disorder will further contribute to competing spin fluctuations. The Invar instability itself is related to these fluctuations which couple strongly to phonons and local $e_g \rightleftharpoons t_{2g}$ charge fluctuations [33]. Because of the specific form of the surface, we also expect that ferromagnetic spin fluctuations will dominate at temperatures higher than the Curie temperature T_c in much the same way as in γ-iron. This would explain Invar-like behavior for $T < T_c$ and anti-Invar behavior for $T > T_c$ in the Fe-Ni alloys. When the number of d electrons is reduced like in the Fe-Mn alloys, the ground state of the γ-phase becomes antiferromagnetic leading to anti-Invar behavior above the Néel temperature as in the case of γ-iron. Both alloy systems, Fe-Ni as well as Fe-Mn, are disordered and have lower martensitic transition temperatures than elemental iron. The transition temperature decreases further with increasing concentration of Ni or Mn atoms.

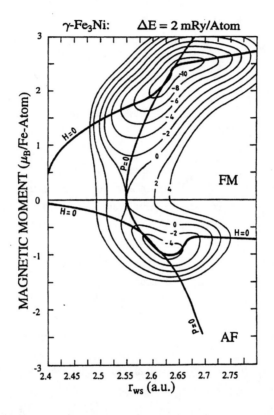

Figure 7. Ferro- and antiferromagnetic binding surface of γ-$Fe_{75}Ni_{25}$ obtained by extrapolation of full-potential data [34]. Contour lines are at 2 mRy intervals.

The strong concentration dependence of the martensitic transition temperature is a longstanding problem. As an example, we show in Fig. 8(a) the strong variation of the martensitic and austenitic transition temperatures with concentration for the Fe-Ni system. The large arrows mark supercooling and superheating behavior. It recently has been argued that the large concentration dependence could be associated with strong mass-disorder scattering of phonons [35] which, however, in the case of Fe-Ni might be doubtful since Fe and Ni have practically the same atomic mass. Disorder scattering occurs here in the spin channel because the energies of the 3d-minority-spin states of Fe and Ni are different. This leads to strong disorder effects in the minority-spin density of states curve. In order to gain more insight into the dependence on concentration, we have performed first-principles total-energy calculations by using the KKR-CPA method for the disordered alloys [14]. In addition the FLAPW/GGAII method has been used to calculate energy differences for a few stoichiometric cases. The resulting concentration dependence of energy differences shown in Fig. 8(b) agrees qualitatively with the behavior of the experimental transition temperatures in Fig. 8(a). This shows that the strong variation of M_0 (which is the thermal equilibrium martensitic transition temperature, where the free energies of the α- and γ phase cross) with concentration is an intrinsic effect and associated with the concentration dependence of the energy barrier between the α- and γ-phase. This energy difference is gained by martensitic nucleation in parent austenite, whereby the gain in energy has to overcome the energy cost associated with increasing stress during the growth of the nucleus. This partially solves the longstanding problem of what determines the concentration dependence of the martensitic transition temperature in ferrous alloys.

The remaining question to be answered is, what then is responsible for the strong concentration dependence of the small energy difference between the α and γ phase? This brings us back to the observation that it is ferromagnetic order which stabilizes the bcc structure at low temperatures. First-principles calculations for α-Fe and Fe-Ni actually prove that it is the magnetic contribution to the energy which stabilizes the less-close packed bcc structure at zero temperature [14]. At high temperatures this is no longer evident as discussed in the introduction. We thus face the interesting situation that we can have different stabilizing mechanisms acting at different temperatures in the Fe-Ni alloy system: At low T it is the magnetic pressure which stabilizes the bcc structure; at high T it is the vibrational entropy and strong ferromagnetic spin fluctuations which stabilize the fcc structure. The origin of the strong concentration dependence of the energy difference between the fcc and bcc structure is at low temperatures connected with a subtle interplay of different magnetic ordering tendencies and related occupation of antibonding and nonbonding d-orbitals. The zero-temperature magnetic binding surfaces shown in Fig. 5 and 7 for γ-Fe and γ-Fe$_3$Ni give an impression of the different ordering tendencies. Adding Ni to ferromagnetic bcc Fe with large exchange splitting between minority-spin and majority-spin states, destabilizes the bcc structure because extra d-electrons from Ni mainly go to the nonbonding states. This leads to reduced partial electronic pressure, to a decrease of the volume, and finally to ferromagnetic order in the fcc structure at still higher concentrations of Ni. At finite temperatures the contibutions of phonons to the entropy cannot be neglected and might be of the order of the magnetic contribution.

The forestanding discussion shows that the change of electronic properties with Ni concentration in the Fe matrix plays an important role. However, is this sufficient to

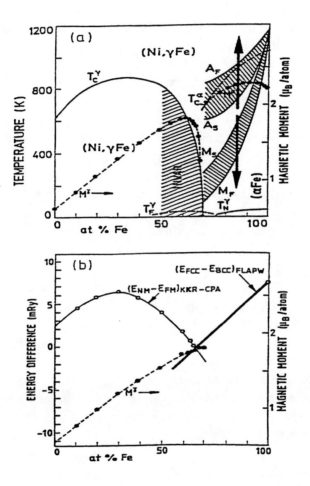

Figure 8. (a) Experimental and (b) theoretical phase diagram of Fe_xNi_{100-x}. Besides Curie temperature $T_c^{\gamma,\alpha}$ and magnetic moment $M^{\gamma,\alpha}$ of γ-Fe-Ni and α-Fe-Ni, the martensitic and austenitic start and final temperatures, $M_{S,F}$ and $A_{S,F}$, are shown in (a). Panel (b) shows zero-temperature energy differences obtained from total-energy calculations.

explain the martensitic transformation? At finite temperatures the coupling of electrons to phonons and resulting phonon softening due to Fermi-surface nesting will help to induce a structural transformation in the Fe-Ni alloys [13, 14, 25, 34, 36, 37, 38]. Of special interest is in this case the interplay of magnetic order. Evaluation of electron-phonon matrix elements show that it is mainly the coupling of minority-spin electrons to phonons which is responsible for phonon softening effects going hand in hand with the decrease of the minority-spin density of states at the Fermi energy while passing from the fcc to bcc structure along the Bain path. This leads in the bcc structure of Fe-Ni alloys to remarkable softening of the $T_1[110]$ phonon with increasing temperature as observed in the molecular-dynamics simulations [21]. The softening is most pronounced at the zone boundary and at the temperature, where the simulations show a sudden transition from bcc to fcc. However, the phonon does not completely soften at the transition. This agrees with the behavior of the corresponding phonon in α-Fe as observed by neutron scattering [11]. Similar discussion of Fermi-surface nesting and softening of transverse phonons as dynamical precursor of the austenitic or martensitic transformation in Ni-Ti and Ni-Al can be found in [39, 40, 41].

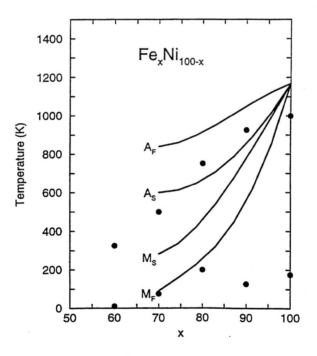

Figure 9. Phase diagram of Fe-Ni as obtained by molecular dynamics simulations (upper dots for the bcc $\rightarrow$ fcc, lower dots for the fcc $\rightarrow$ bcc transition). Solid lines are experimental data for $M_{S,F}$ and $A_{S,F}S$ temperatures.

Another problem is connected with hysteresis effects which are difficult to deal with in the *ab initio* calculations. It is known that the transformation starts with the formation of lenticular or needle-like nuclei of martensite around defects in the parent phase. These nuclei usually grow in size into *easy directions* until they meet other martensitic grains. The magnitude of supercooling, superheating and hysteresis effects is usually attributed to the degree of difficulty of the system to accomodate the continuous nucleation production and the difference of shape between the two phases. It is out of scope of the present work to discuss in detail the different contributions to the free energy in the two phases due to growing martensitic or austenitic palets. However, molecular-dynamics simulations of the transition in Fe-Ni in the presence of impurities (vacancies) discussed below, show that the temperature range of supercooling and superheating is not arbitrary or a complex function of the concentration of various lattice defects, but is characteristic for the specific alloy under consideration [23, 38]. Also the hysteresis depends strongly on the energy difference between the fcc and bcc structure. Defects will further broaden the transition; but they are not originally responsible for the structural transformation as a phase transition of first order.

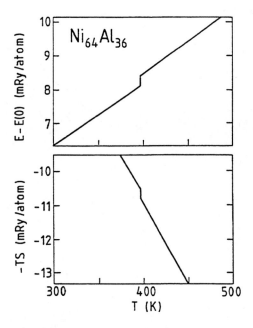

Figure 10. Molecular-dynamics simulations of the internal energy and entropy as a function of temperature of $Ni_{64}Al_{36}$ (adapted from [44]).

3. RESULTS OF MOLECULAR-DYNAMICS SIMULATIONS

With increasing computer power molecular-dynamics simulations have become a standard method of statistical mechanics. For example, they have been used successfully in investigations of dynamical properties, spinodal decomposition and pattern formation of mixtures of liquids. The low density of fluids allows to cover a time scale which is sufficient for the evaluation of the time dependence of the velocity-velocity correlation function. Integration over time then yields reliable values for diffusion constants. Even growth exponents can be predicted which agree with experiment. Similar simulations in solids give much less information since simulation times are restricted to reciprocals of typical phonon frequencies. This means that segregation effects like spinodal decomposition have not yet been simulated successfully. Therefore, it is remarkable that transformations of solids bearing traces of spinodal decomposition like the martensitic transformation, can be investigated with the help of molecular-dynamics simulations. In contrast to liquids, where simple Lennard-Jones potentials yield accuarate results, it is important that one uses more refined many-body interatomic potentials for the solids. Here the embedded-atom method [18] has been used with success to construct the potentials. Simulation of the martensitic transformation at finite temperatures is possible by using a thermostat [42] and a fluctuating volume box [43]. So different crystallographic structures can be simulated by using the same N-body interatomic potential in the different phases. In the following we will discuss a few aspects connected with molecular-dynamics simulations of the austenitic and martensitic transformation in Fe-Ni and Ni-Al alloys. Fig. 9 shows results of simulations of the phase diagram of Fe-Ni. There is some disagreement with the experimental phase diagram on the Fe-rich side because magnetism is not taken into account in an optimal way. But supercooling and superheating effects are clearly visible.

It is of interest to discuss briefly molecular dynamics simulations of martensitic transitions in other alloy systems. Analysis of the structural transition in Ni-Al [44, 45] has shown that the transition is first order, entropy driven, and follows the *Zener picture* [46] according to which the transformation is due to the competition of internal energy and entropy between the austenitic and martensitic phase. Entropy driven means that it is the higher vibrational entropy of the austenitic phase which is responsible for the transition when raising the temperature with a considerable increase of the mean square displacement of the atoms in the high-temperature phase. Fig. 10 shows results of simulations for the composition $Ni_{64}Al_{36}$ [44]. The entropy change at the transition is of the order of $T\Delta S = 0.26$ mRy/atom. But the analysis also shows that the entropy differences between different atomic configurations are very small, and that, as a result, the hypothetical equilibrium martensitic transformation temperature of the Ni-Al alloys, M_0, follows the difference of the (internal) energy of the various atomic configurations [44]. This allowed to describe the influence of disorder on the martensitic transformation temperature in a coherent way. The *ab initio* results presented in the last section for Fe-Ni alloys seem to confirm this type picture for the case of magnetic alloys. If we assume that for small changes in concentration the entropy differences between the fcc and bcc phase will not drastically change, then $M_0(x)$ should scale with $|E_{fcc} - E_{bcc}|(M_0, x) \sim M_0(x)$. Fig. 8(b) shows that the corresponding energy difference in Fe-Ni evaluated at zero temperature, indeed scales with $M_0(x)$ in the Fe-Ni alloys, although we do not know the experimental $M_0(x)$ exactly.

We now discuss results for the thermodynamic potential of Fe-Ni. The molecular dynamics simulations have been performed for the alloy $Fe_{80}Ni_{20}$ instead of simulating elemental iron. The reason is that on the iron-rich side results of simulations become less reliable since we do not know how to incorporate magnetism appropiately. Results for the free energy are shown in Fig. 11. They demonstrate that the structural change is of first order. Results for the free energy of Ni-Al behave much the same way [44]. The crossing of the free energies defines the equilibrium transition temperature M_0 which is of course different from $M_{S,F}$ and $A_{S,F}$, the martensitic and austenitic start and final temperatures, respectively. We have already discussed experimental results for the entropy of elemental iron in Fig. 2 [9] which yield an entropy jump at A_3 of the order of 0.606 mRy/atom which must be compared with the vibrational value of 1.05 mRy/atom [11]. How does this value compare with results of molecular-dynamics simulations of Fe-Ni alloys? Fig. 12 shows results for the temperature variation of the potential energy and the entropy of $Fe_{80}Ni_{20}$ (the kinetic energy has been subtracted since it is the same in both phases). As in the case of Ni-Al, the variation of TS is much larger than the variation of the energy which proves that the transition is driven by the entropy. The magnitude of the entropy change at M_0 is of the order of $T\Delta S = 1$ mRy/atom. Although the transition temperature M_0 is much smaller than A_3 of elemental Fe, the value for the entropy change agrees remarkably well with the experimental value given in Ref. [11] and less well with the value obtained from the specific heat data which is smaller. The entropy change in the simulations is mostly of vibrational origin. A magnetic part cannot be split off, since magnetism is involved only via elastic properties. However, we may conclude that both, vibronic and magnetic contributions to the free energy will be important and determine the martensitic transition at A_3 in elemental Fe and in the Fe-Ni alloys, i.e. the vibrational excess entropy of the fcc structure due to large mean square amplitudes of the atomic vibrations is at least as important as the magnetic excess entropy due to strong spin fluctuations with large mean square amplitudes. For a discussion which emphasizes more the importance of magnetic fluctuations see [47].

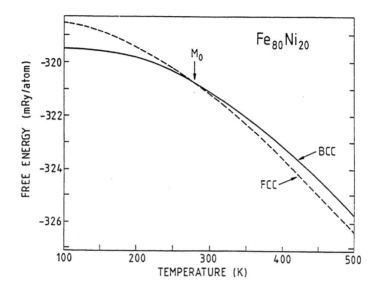

Figure 11. Molecular-dynamics simulations of the free energy as a function of temperature of α- and γ-$Fe_{80}Ni_{20}$. The crossing of the free enegies defines M_0.

We finally would like to show, how the structural transformation is actually achieved in the molecular dynamics simulations. Along with the structural transformation from bcc to fcc in the Fe-Ni alloys upon heating, twin boundaries and stacking faults are created. Fig. 13 shows results for the $Fe_{65}Ni_{35}$ alloy. The following orientational relationships can be observed: $(011)_{bcc} \approx (111)_{fcc}$, $[001]_{bcc} - [001]_{fcc} \approx 9.7°$ being close to experimental values. A detailed analysis of the influence of different kinds of defects on the transformation and growth processes of embryos and final nucleation is under current investigation. Although defects are limited for the moment to vacancies, free surfaces and surfaces with big notch-like defects, there are promising results for growth processes of martensite under the influence of these defects and under the influence of additional external strain.

Further detailed investigations have been done with respect to superelastic behavior of thin Fe-Ni films and shape-memory effects in Fe-Ni bulk systems [38]. Here we cite the main results. Thin Fe-Ni films with composition $Fe_{70}Ni_{30}$ have initially been prepared in the fcc strucure at low temperatures. In order to facilitate the nucleation of martensite, a large notch-like defect has been prepared in the film plane. After thermalizing tensile stress has been applied, leading to 6% expansion of the film and

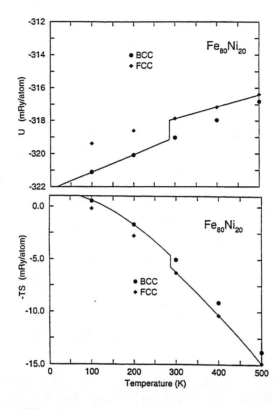

Figure 12. Molecular-dynamics simulations of the internal energy and entropy as a function of temperature of $Fe_{80}Ni_{20}$. Note that Al-Ni and Fe-Ni behave qualitatively the same way. In both systems the structural transformation is mainly driven by the entropy.

the appearance of typical martensitic texture which gradually disappears if external forces are swichted off and the system is allowed to relax, see Fig. 14. Elemental Ni, for example, behaves completely different in the simulations. Here we observe fracture when tensile stress is applied, with a crossover from brittle fracture at low temperatures to ductile fracture as plastic shear with emission of dislocations at high temperatures, in agreement with experiment. This proves that (apart from the poor treatment of magnetism) the embedded atom potentials used for the description of the Fe-Ni alloys are very accuarate. One-way shape memory effects have also been observed during the simulations [38]. If we heat $Fe_{80}Ni_{20}$ in the simulations up to 700 K, the alloy transforms to the fcc structure. We then stress the system at 800 K till an expansion of 12% is reached. By switching off the load and subsequent cooling down to 200 K, the system gains back its original bcc structure practically without formation of additional defect structures. However, in order to achieve this, a very high concentration of vacancies of the order of 2% is needed. Here further simulations are needed in order to clarify the shape memory effect in relation to defects at fixed positions required for nucleation of martensite at always the same local sites.

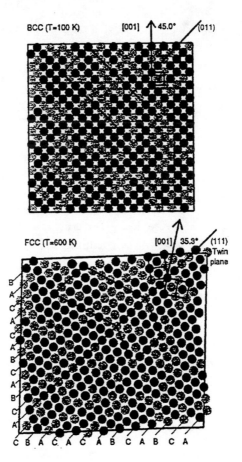

Figure 13. Transformation of the bcc structure (upper panel) to the fcc structure with twins and stacking faults (lower panel) in $Fe_{65}Ni_{35}$ (Fe: black; Ni: grey). Paper plane is (100) for bcc and $(1\bar{1}0)$ for fcc, respectively.

ACKNOWLEDGMENT

One of us (P. E.) would like to thank Prof. S. N. Behera, Director of the Institue of Physics of Bhubaneswar, for the invitation to join the symposium on *Novel Materials* and for his hospitality during the visit in India.

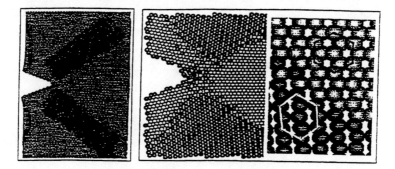

Figure 14. Results of molecular-dynamics simulations of martensitic nucleation in a thin film of $Fe_{70}Ni_{30}$ (31 570 atoms) under the influence of a large surface defect and additional strain leading to a 6% expansion. Left panel: Dark areas show nucleated martensite while lighter areas correspond to the initially prepared hcp structure (view is onto the (001) plane; the figure does not distinguish between Fe and Ni atoms). Middle panel: Enlargement close to the notch-like surface defect. Right panel: Still larger enlargement allows to distinguish between the hcp and the bcc arrangement of Fe and Ni atoms.

REFERENCES

[1] The name is due to the research done by A. Martens, VDI-Zeitschrift **22**, 205, 483 (1878).

[2] L. Delaey in *Materials Science and Technology*, Vol. 5, Eds. R. W. Cahn, P. Haasen, and E. J. Kramer (VCH, Weinheim, 1991), p. 339.

[3] D. G. Pettifor in *Materials Science and Technology*, Vol. 1, Eds. R. W. Cahn, P. Haasen, and E. J. Kramer (VCH, Weinheim, 1993), p. 61.

[4] J. B. Staunton, D. D. Johnson, and F. J. Pinski, Phys. Rev. B **50**, 1450 (1994).

First-Principles and Molecular-Dynamics Simulations of Martensitic   ... 379

[5] J. D. Althoff, D. D. Johnson, F. J. Pinski, and J. B. Staunton, Phys. Rev. B **53**, 10 610 (1996).

[6] R. W. Cahn, Nature **374**, 120 (1995).

[7] J. Madsen, O. K. Anderson, U. K. Poulsen, and O. Jepsen in *Magnetism and Magnetic Materials*, AIP Conf. Proc. **29**, Eds. J. J. Becker and G. H. Landes (AIP, New York, 1976), p. 327.

[8] R. J. Weiss, Proc. Phys. Soc. **82**, 281 (1963).

[9] W. Bendick and W. Pepperhoff, Acta Metall. **30**, 679 (1982).

[10] H. Hasegawa and D. C. Pettifor, Phys. Rev. Lett. **50**, 130 (1983).

[11] J. Neuhaus, K. Nicolaus, W. Petry, B. Hennion, and A. Krimmel, Physica B **234-236**, 897 (1997).

[12] G. L. Krasko and G. B. Olson, Phys. Rev. B **40**, 11 536 (1989).

[13] E. Hoffmann, H. Herper, P. Entel, S. G. Mishra, P. Mohn, and K. Schwarz, Phys. Rev. B **47**, 5589 (1993).

[14] M. Schröter, H. Ebert, H. Akai, P. Entel, E. Hoffmann, and G. G. Reddy, Phys. Rev. B **52**, 188 (1995).

[15] V. L. Sliwko, P. Mohn, K. Schwarz, and P. Blaha, J. Phys.: Condens. Matter, **8**, 799 (1996).

[16] P. Mohn, K. Schwarz, and P. Blaha, J. Phys.: Condens. Matter **8**, 817 (1996).

[17] E. C. Bain and N. Y. Dunkirk, Trans. AIME **70**, 25 (1924).

[18] M. S. Daw and M. I. Baskes, Phys. Rev. B **29**, 6443 (1984).

[19] M. Meyer and P. Entel, J. Physique IV, C2-123 (1995).

[20] M. Meyer and P. Entel, *submitted to ESOMAT'97*.

[21] R. Meyer and P. Entel, *submitted to Phys. Rev. B (1997)*.

[22] R. Meyer, *to be published (1997)*.

[23] K. Kadau, P. Entel, and R. Meyer, *to be published*.

[24] P. Blaha, K. Schwarz, P. Dufek, and R. Augustyn, WIEN95, TU of Vienna 1995. Improved and updated UNIX version of the original WIEN-code, published by P. Blaha, K. Schwarz, P. Sorantin and S. B. Trickey, Comput. Phys. Commun. **59**, 399 (1990).

[25] H. C. Herper, E. Hoffmann, and P. Entel, *contribution to ESOMAT'97.*

[26] V. L. Moruzzi, P. M. Marcus, K. Schwarz, and P. Mohn, Phys. Rev. B **34**, 1784 (1986).

[27] W. A. A. Macedo and W. Keune, Phys. Rev. Lett. **61**, 475 (1988).

[28] A. Onodera, Y. Tsunoda, N. Kunitomi, O. A. Pringle, R. M. Nicklow, and R. Moon, Phys. Rev. B **50**, 3532 (1994).

[29] P. Entel and M. Schröter, Physica B **161**, 160 (1989).

[30] D. Wagner, J. Phys.: Condens. Matter **1**, 4635 (1989).

[31] M. Schröter, P. Entel, and S. G. Mishra, J. Magn. Magn. Mater. **87**, 163 (1990).

[32] P. Mohn, K. Schwarz, and D. Wagner, Phys. Rev. B **43**, 3318 (1991).

[33] P. Entel, E. Hoffmann, P. Mohn, K. Schwarz, and V. L. Moruzzi, Phys. Rev. B **47**, 8706 (1993).

[34] H. C. Herper, E. Hoffmann, and P. Entel in *Properties of Complex Inorganic Solids*, Eds. A. Gonis, A. Meike, and P. E. A. Turchi (Plenum, New York, 1997), p. 213.

[35] P. A. Lindgard, J. Physique IV, C2-29 (1995).

[36] H. C. Herper, P. Entel, and W. Weber, J. Physique IV, C2-129 (1995).

[37] H. C. Herper, E. Hoffmann, P. Entel, and W. Weber, J. Physique IV, C8-293 (1995).

[38] P. Entel, E. Hoffmann, M. Clossen, K. Kadau, M. Schröter, R. Meyer, H. C. Herper, and M.-S. Yang in *The Invar Effect: A Centennial Symposium*, Eds. J. Wittenauer (TMS, Warrendale (PA), 1997), p. 87.

[39] G. L. Zhao, T. C. Leung, B. N. Harmon, M. Keil, M. Müllner, and W. Weber, Phys. Rev. B **40**, 7999 (1989).

[40] G. L. Zhao and B. N. Harmon, Phys. Rev. B **48**, 2031 (1993).

[41] G. L. Zhao and B. N. Harmon, Phys. Rev. B **45**, 2918 (1992); **47**, 8706 (1993).

[42] S. Nosé, Molecular Phys. **50**, 1055 (1983).

[43] M. Parrinello and A. Rahman, Phys. Rev. Lett. **45**, 1196 (1980).

[44] S. Rubini and P. Ballone, Phys. Rev. B **48**, 99 (1995). Erratum: Phys. Rev. B **49**, 15 428 (1994).

[45] S. Rubini and P. Ballone, Phys. Rev. B **50**, 1297 (1994).

[46] C. Zener, Phys. Rev. B **71**, 846 (1947).

[47] M. Acet, E. F. Wassermann, K. Anderson, A. Murani, and O. Schärpff, *submitted to ESOMAT'97*.

DYNAMICS OF COMPLEX MATERIALS UNDER HIGH TEMPERATURE AND PRESSURE

S. L. Chaplot
Solid State Physics Division
Bhabha Atomic Research Centre
Trombay, Mumbai 400085, India

ABSTRACT

A fundamental understanding of the behaviour of materials has to be based on the study of the microscopic structure and its dynamics. These studies can be advanced by a combination of scattering experiments and computer simulations. We briefly review our recent molecular dynamics studies on fullerene and geophysical minerals, carried out using the BARC parallel computer ANUPAM. The MD simulations are used to derive the dynamical orientational correlations in the disordered phase of fullerene and to explain the neutron- diffuse- quasielastic scattering experiment. In the case of geophysical minerals, the simulations have provided a valuable microscopic understanding of the pressure- temperature variation of the structure and dynamics including melting.

1. INTRODUCTION

Development of new and novel materials require the study of their thermodynamic properties. A fundamental understanding of these properties has to be based on the study of the microscopic structure and dynamics. Most materials of interest comprise of several atomic species or have complex structures. Study of such complex materials may be based on complementary experimental and theoretical techniques [1].

As complex materials may involve a number of atomic species interacting through a variety of interactions (e.g., ionic, covalent, van der Waals, etc.), their simulations presently have to employ empirical potentials [2-4], while the first- principles simulations [5,6] are picking up. We have developed empirical interatomic potentials in variety of complex ionic and molecular solids [7-17], tested their lattice dynamical predictions against neutron scattering and other experimental data, and used them in molecular dynamics simulations at high temperature and pressure. In this article, we review our recent work on fullerene and several mineral of interest to the physics of Earth's interior.

At Trombay, we have developed the necessary softwares for lattice dynamics (DISPR) [18] and for molecular dynamics simulations (MOLDY), which may be applicable to systems of arbitrary size and symmetry. In particular, these softwares are also applicable to systems containing molecular ions for which the long- ranged Coulomb interaction is handled via the Ewald technique. Using the BARC- parallel computer ANUPAM, we are now able to simulate fairly large systems of about 30000 particles.

2. FULLERENE

The fullerene is a nearly spherical molecule made of sixty carbon atoms. Its surface contains 30 C=C double bonds, 60 C-C single bonds, 12 pentagons and 20 hexagons. In the crystal, the molecular centres arrange in a fcc lattice (see [19] for a review). At $T > 260$ K, the molecules rotate nearly freely and no librations are observed. Below 260 K, in the simple cubic lattice, the molecules stay in two possible orientations of nearly equal energies, so their relative population varies from 60:40 at 250 K to 85:15 at $T < 90$ K. In the majority (minority) state, a double- bond centre of one molecule faces opposite a pentagon (hexagon) of a neighbouring molecule. Using detailed measurement of the neutron diffuse scattering from a single crystal and its analysis by a reverse Monte- Carlo method, we could varify the proportion of the majority and the minority orientations from 10 to 250 K [20]. We also found evidence of a scatter of the orientational angles of about 2 degrees about its average value in each of the two states.

The intermolecular interaction in fullerene is largely of the van der Waals type since the nearest carbon atoms of the neighbouring molecules are separated by about the sum of their van der Waals radii. Using previously known vdW potential, either from the organic molecules or from graphite, it is possible to nearly predict the lattice constant, bulk modulus and the phonon dispersion relation of the rigid molecular translations. However, both the orientational structure and the librational frequencies are not explained by the vdW potential. Various empirical models [19] were developed with considerable success which involved electrostatic charges. We developed a split-bond- charge model [15,16] in which a small negative charge (-0.27 e) was placed at two split positions radially inwards and outwards a double- bond centre, and a compensating positive charge (0.27 e) placed on the carbon atom. Among all the available models, this model produced the best agreement with the experimental phonon dispersion relation of the external modes and the associated specific heat at low temperatures [16].

The short range correlations in a disordered structure may be conveniently studied by the wave- vector dependence of the diffuse scattering of X-rays or neutrons from a single crystal. Above 260 K in the disordered phase, the diffuse scattering experiments [17] indicated significant short- range order of the orientations extending to several neighbours beyond 20 A°. The order was found to be much more pronounced in neutron quasielastic scattering than in X-ray scattering, which indicated its dynamical nature.

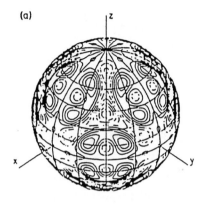

(a)

(b)

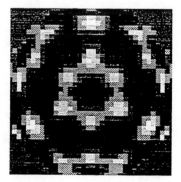

(c)

Figure 1. Atomic number density on the spherical surface of fullerene, shown as excess over the average density. (a) From the refinement of single- crystal Bragg data [21]. The full and dashed contours indicate excess and deficit densities respectively. (b and c) From our MD simulations using the split- bond- charge model and the van-der- Waals potential model respectively. The brighter and darker areas show excess and deficit densities respectively.

We have carried out MD simulation on 256 molecular periodic cells, to understand the observed pattern of the diffuse scattering and thereby reveal the nature of the short range order [15,17]. Simulation using the split- bond- charge model produces the pattern of the atomic density on the molecular surface in good agreement with that derived from the diffraction experiments (Fig. 1a,b). So also the calculated diffuse scattering from the MD simulation compares very well with the observed diffuse scattering. The simulated orientational correlations indicate preference for the double bonds on one molecule to face opposite the pentagons of the neighbouring molecules (Fig. 2a). In contrast, the MD simulation using the vdW model produces an atomic density pattern in clear disagreement with experiment (Fig. 1a,c), and predicts preference for the double bond to hexagon correlations (Fig. 2b).

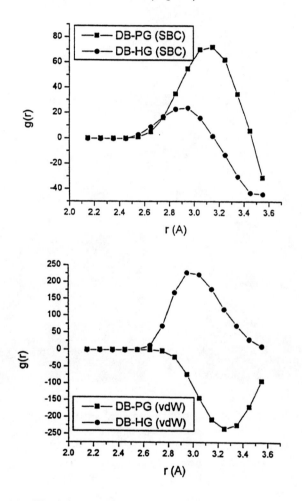

Figure 2. The pair correlation functions as calculated from the MD simulations using (a) the split- bond- charge model and (b) the van- der- Waals potential model. The excess from the average pair correlation function is shown for the pairs of sites on the surface of neighbouring molecules which reveals the dominant correlation at the nearest average contact separation of about 3 A. DB- double bond, PG- pentagon, and HG- hexagon.

We note that the diffuse scattering patterns in fullerene are qualitatively different at above and below 260 K. While the orientational correlations above 260 K are of short range and dynamical in nature, it appears there are small clusters percolating in three dimensions in a way different from the more ordered arrangement at below 260 K.

3. GEOPHYSICAL MINERALS

The major constituents [22-24] in the Earth's upper and lower mantle are the various phases of magnesium silicate with about 10% of magnesium replaced by iron in solid solution. The upper mantle, down to a depth of about 400 km, has largely olivin $(Mg_{2-x}Fe_x)SiO_4$ (which has the end members as forsterite Mg_2SiO_4 and fayalite Fe_2SiO_4) and pyroxene $(Mg_{2-x}Fe_x)Si_2O_6$ (with the magnesium end member known as enstatite). The lower mantle has principally the orthorhombic perovskite $MgSiO_3$. There are major seismic discontinuities at the depths of 400 Km, 660 Km, and 920 Km. etc., which may be due to phase changes or compositional changes. Besides the solid-solid phase transitions, the melting temperatures under pressure are also of interest since these data help in arriving at the upper bounds on the temperatures at various depths in the mantle.

In case of forsterite, the calculations based on quasiharmonic lattice dynamics [25] could reproduce the various thermodynamical properties including the equation of state and melting fairly well. These results are found to be in a good agreement with those from our recent MD simulations, which indicate that the effect of explicit anhamonicity in the forsterite structure is not very significant.

The phase diagram involving the various polymorphs of enstatite [26] has been investigated by comparison of the Gibb's free energy as obtained from the lattice statics and dynamics [14]. The various polymorphs are characterised by different stacking of the MgO_6 octahedra along with chains of SiO_4 tetrahedral units. The transition from orthoenstatite to protoenstatite at P = 0 is calculated at 860 K when the latter phase has a higher vibrational entropy at high temperatures. In experiments, the transition occurs at 1360 K [27]. MD simulation indicate the proto- ortho transition at 1050 K with a small hysteresis which occurs due to a finite heating/cooling rate (of the order of 1 K/ps). The small but significant difference between the MD and lattice dynamics may be due to the anharmonicity of phonons contributing to the vibrational entropy. In the MD simulation at P = 4 GPa, the ortho- phase remains stable on heating till melting, in agreement with experiment and the lattice dynamical result.

The MD simulations on orthoenstatite at high pressures have shown very interesting phase transitions. We have perhaps the first computer simulation of a pressure driven transition from this upper mantle phase to a lower mantle phase similar to the orthorhombic perovskite. The path followed is that in the first step the silicon atoms acquire a higher coordination from four to five oxygens and then in the next step silicon reaches the full octahedral 6-fold oxygen coordination together with the magnesiums increasing their oxygen coordination from 6 to 8. The MgO_8 units are still highly distorted due to the orthorhombic distortion of the perovskite phase. Both the steps involve a volume collapse of about 10% each. It is worth noting that in our simulation the transformation occurs as a displacive transition in contrast to a general belief that the enstatite $Mg_2Si_2O_6$ first decomposes into Mg_2SiO_4 plus SiO_2 and then recombines into $2MgSiO_3$ [28]. Our MD simulations may therefore help in understanding the transition process.

Simulations on the perovskite phase upto high pressure and temperatures [13] have provided further insight into the phase behaviour in the lower mantle [23,29]. The simulations indicate that the orthorhombic phase remains stable upto pressures beyond 100 GPa at low temperatures consistent with experiments. When the temperature is increased, say at 70 GPa, the orthorhombic distortion decreases and at about 5400 K we obtain a transition to a tetragonal phase. The tetragonal phase is found to show many interesting properties. The tetragonal distortion is very small, and the distortion fluctuates locally at a time scale a few ps. Thus the time- average symmetry may be observed as cubic. Our results thus suggest that the reported transition [30] from the orthorhombic to cubic perovskite at 70 GPa was essentially due to the laser heating under high pressure. The simulated phase transition is also accompanied by a significant diffusion of the oxygen atoms as in a solid ionic conductor. The results of the MD simulation [13] at 100 GPa as a function of temperature, as shown in Figs. 3 and 4, indicate the decrease in the orthorhombic distortion accompanied by the partial melting of the oxygen sublattices.

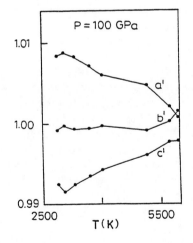

Figure 3. An MD simulation of the phase transition from the orthorhombic to tetragonal perovskite [13]. The lattice parameters are shown as a function of temperature at P= 100 GPa and are suitably normalised by the total volume such that any deviation from one indicates a noncubic character.

Further MD simulations at higher temperature show that the remaining sublattices melt at about 7000 K at 70 GPa at the time scale of 10 ps. This is somewhat large compared to the recent measurements of Zerr and Boehler [31] at 5000 K which itself was larger than previous estimates of about 3000 K. A simulated first- order transiton temperature is somewhat overestimated while heating due to the small time scale. However, as the simulated partial and full melting temperatures appeared to be too large, we studied the consequences of Fe/Mg disorder and other point- defects in the lattice. We found that the simulated transition temperatures were not much sensitive to the Fe/Mg disorder even with the 50:50 composition. A significant level of defects is not thermally activated in the simulation due to a rather small time scale. Therefore point- defects were explicitly introduced into the simulation cell. We introduced point-defects in the form of vacancies at about 1% of the sites randomly. This significantly lowered the melting temperatures by about 15% compared to that in the case of the perfect lattice.

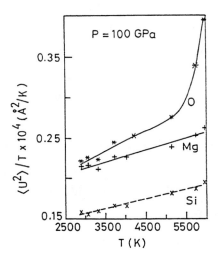

Figure 4. The variation of mean- square atomic displacements with temperature in the same MD simulations as in Fig. 3, indicating partial melting of the oxygen sublattices accompanying the orthorhombic to cubic phase transition in perovskite [13].

From a calculation of the vacancy formation energy, we estimate that the level of vacancy defects at equilibrium may be about 1% at 5000 K. The true level of defects in a real sample in experiments could be much higher depending on the sample history. It is therefore very likely that the lower melting temperatures and a broad range of values as observed in experiments may be accounted largely by the level of defects. We are thus able to qualitatively account for the melting transition from the simulations.

4. DISCUSSION

It is shown that a suitable combination of complementary scattering experiments, lattice dynamics and molecular dynamics simulation based on empirical potentials can provide valuable insight into the structures and thermodynamical properties of complex materials. It is important that the empirical potentials adequately describe the full energy range of the vibrational density of states which contribute to the free energy at the high temperature of interest. (Of course, in the case of molecular solids, the internal vibrations may be excluded if the distortions of the molecules can be ignored.) It is also important to ensure that the empirical potential is physically meaningful. It has been our endeavour that the potentials involve minimal number of adjustable parameters, describe a broad range of experimental data, and that the potentials be transferable among several materials involving the same atoms in similar environments as much as possible.

References

[1] Chaplot S.L., Indian J. Pure & Applied Physics **32**, 560 (1994).

[2] Rao K.R. and Chaplot S.L., in *Current Trends in Lattice Dynamics*, ed. Rao K.R. (Indian Physics Association, Bombay 1979) p. 589.

[3] Chaplot S.L., Current Sc. **55**, 949 (1986).

[4] Allen M.P. and Tildesley D.J., *Computer Simulation of Liquids* (Clarendon, Oxford, 1987).

[5] Car R. and Parrinello M., Phys. Rev. Letters **55**, 2471 (1985).

[6] Wentzcovitch R.M., Martins J.L. and Price G.D., Phys. Rev. Lett. **70**, 3947 (1993).

[7] Chaplot S.L., Phys. Rev. B **37**, 7435 (1988).

[8] Chaplot S.L., Phase Transitions **19**, 49 (1989).

[9] Chaplot S.L., Phys. Rev. B **42**, 2149 (1990).

[10] Chaplot S.L., Phys. Rev. B **45**, 4885 (1992).

[11] Rao K.R., Chaplot S.L., Choudhury N., Ghose S. and Price D.L., Science **236**, 64 (1987).

[12] Rao K.R., Chaplot S.L., Choudhury N., Ghose S., Hastings J.M., Corliss L.M. and Price D.L., Phys. Chem. Minerals **16**, 83 (1988)

[13] Chaplot S.L., Choudhury N. and Rao K.R., Intern. Conf. on Condensed Matter under High Pressure, Mumbai (1996).

[14] Choudhury N., Chaplot S.L. and Rao K.R., Intern. Conf. on Condensed Matter under High Pressure, Mumbai (1996).

[15] Chaplot S.L. and Pintschovius L., Fullerene Science and Technology **3**, 707 (1995).

[16] Pintschovius L. and Chaplot S.L., Z. Phys. B **98**, 527 (1995).

[17] Pintschovius L., Chaplot S.L., Roth G. and Heger G., Phys. Rev. Letters **75**, 2843 (1995).

[18] Chaplot S.L., Report B.A.R.C.-972 (1978).

[19] Axe J.D., Moss S.L. and Neumann P.A., in *Solid State Physics*, vol. **48**, eds. Ehrenreich H. and Spaepen F., (Acamedic Press, New York, 1994), p.149.

[20] Chaplot S.L., Pintschovius L., Haluska M. and Kuzmany H., Phys. Rev. B **51**, 17028 (1995).

[21] Chow P.C. *et al.*, Phys. Rev. Letters **75**, 2843 (1995).

[22] Jeanloz R. and Thompson A.B., Rev. Geophys. Space Phys. **21**, 51 (1983).

[23] Kawakatsu H. and Niu F., Nature **371**, 301 (1994).

[24] Angle R.J., Chopelas A. and Ross N.L., Nature **358**, 322 (1992).

[25] Choudhury N., Chaplot S.L. and Rao K.R., Phys. Chem. Minerals **16**, 599 (1989).

[26] Cameron M. and Papike J.J., Am. Mineral. **66**, 1 (1981).

[27] Yang H. and Ghose S., Phys. Chem. Minerals **22**, 300 (1995).

[28] Horuichi H. in *Dynamical Processes of Material Transport and Transformation in the Earth's Interior*, ed. Marumo F., (Terra Sci. Pub. Co., Tokyo, 1990)p. 189.

[29] Funamori N. and Yagi T., Geophys. Res. Lett. **20**, 387 (1993).

[30] Meade C., Mao H.K. and Hu J., Science **268**, 1743 (1995).

[31] Zerr A. and Boehler R., Science **262**, 553 (1993).

APPLICATIONS OF A NEW TIGHT-BINDING TOTAL ENERGY METHOD

D. A. Papaconstantopoulos,[1] M. J. Mehl,[1] and Brahim Akdim[2]

[1] *Complex Systems Theory Branch*
Naval Research Laboratory, Washington, D.C. 20375-5320, U. S. A.
[2] *Institute for Computational Sciences and Informatics*
George Mason University, Fairfax, VA 22030-4444, U. S. A.

1. INTRODUCTION

First-principles methods based on quantum mechanics via the density functional theory (DFT) [1] have been shown to be quite reliable in calculating such properties as the correct ground state of a crystal (except for certain magnetic systems), elastic constants, phonon frequencies, and even the Fermi surface of high-temperature cuprate superconductors.

Though DFT methods are very accurate, they are computationally slow, especially for problems containing a large number of transition metal atoms. For these systems, DFT requires solutions of secular equations of order $N = 100M$, where M is the number of atoms in the system. The computational time required to solve these equations scales as the cube of N. This is known as the "N-cubed" problem. "Order-N" methods, where the computational time scales linearly with N, have yet to be applied successfully to metals. Thus today's computers can only handle about 100 atoms per unit cell using current algorithms and technology when d-orbitals need to be included.

The study of real materials requires much larger unit cells. For example consider a dislocation in a metal. It is known that the interaction between two dislocations falls off as the cube of their separation. The simplest way to model any defect computationally is to periodically replicate the defect on the computer. This eliminates surface effects, and allows to use Bloch's theorem to block-diagonalize the Hamiltonian and speed up the calculation. Unfortunately, the periodicity also introduces an interaction between the original dislocation and its duplicates. The only way to eliminate this interaction is to use a large unit cell. Even for a simple, small dislocation, this requires a unit cell containing several thousand atoms. Because of the N-cubed problem, this computation is out of reach for DFT calculations with the current generation of computers.

To circumvent this difficulty approximate methods are used. The Embedded Atom Method (EAM), which is the most popular method for metals, is formally based on density functional theory but in practice uses a limited number of parameters to replicate first-principles and experimental results for such quantities as formation energies, elastic constants, and phonon frequencies. Because "atomistic" methods such as EAM are essentially pair potential calculations with a few modifications, they are very fast, and can handle systems involving tens of thousands of atoms on a supercomputer. These methods are also reasonably accurate when doing calculations on systems which are "close" to the structures included in the fit. However, these methods fail for many structures that are not fitted. Thus the EAM does not do well when dealing with surfaces or grain boundaries. In particular, it consistently underestimates the energy required to form a surface. In addition, atomistic methods do not explicitly include the quantum mechanical behavior of the electrons, as true DFT methods do.

The tight-binding method described in this paper explicitly includes quantum mechanics by mapping the results of first-principles calculations to an spd tight-binding basis. Our method is to take the results of the first-principles calculations on a few simple structures and choose our tight-binding parameters to replicate not only the energies of these structures, as is done in the EAM and related methods, but also to reproduce the band structure. As a result, the tight-binding method is applicable to wider range of systems than the atomistic methods. For example, by fitting only to face centered and body centered cubic structures of transition metals, we correctly predict:

- The ground state structure of the element, even when the ground state is hexagonal closed packed or α-Manganese structures.

- The elastic constants of the elements, to an accuracy almost as good as DFT calculations.

- Vacancy formation energies.

- Surface energies.

The tight-binding method still requires matrix diagonalization, and as such is subject to the N-cubed problem. The advantage of the method over DFT is that the tight-binding secular equation is of the order $N = 9M$, compared to $N = 100M$ for DFT. Since the computational time scales as the cube of N, this results in a speedup of three orders of magnitude over DFT calculations. As a result, the tight-binding method can handle systems including thousands of atoms and so can be used to study real materials, including those with defects, surfaces, grain boundaries and dislocations. Furthermore, because the method is based on quantum mechanics, any future algorithmic improvements found for DFT methods, including Car-Parinello methods and order-N algorithms, can be incorporated into the tight-binding method. Because of the reduction in the size of the secular equation from DFT to tight-binding, the tight-binding calculations will always be faster and able to handle larger systems than DFT.

Our method is unique in that it has been designed to handle any element, and is extendible to binary, ternary, and higher order alloys using the same techniques. In the next few years, as we develop tight-binding parameters for more systems, we expect our method to be widely used in computational materials physics. To encourage the use of this method, we have made the tight-binding parameters for twenty-nine elements available as a database on the World Wide Web, both through the AIP physics Auxillary Publication Service at (http://www.aip.org./epaps/E-PBRMD-54-4519-215-kb) and the Complex System Theory Branch web server at (http://cst-www.nrl.navy.mil/bind/) .

2. THEORY

We write a tight-binding (TB) Hamiltonian which, in general, is non-orthogonal,

$$(\tilde{H} - \epsilon\tilde{S})\Psi = 0 \tag{1}$$

and has nine orbitals per atom (s, p, and d). The matrix elements of this Hamiltonian contain the so-called Slater-Koster (SK) parameters [2, 3]which include on-site terms and hopping integrals. As an extension of the standard Slater-Koster approach we introduce [6−9] environment-dependent parameters as follows: we write the on-site TB parameters h_l in the form of a polynomial,

$$h_l(\rho) = \alpha_l + \beta_l\rho^{2/3} + \gamma_l\rho^{4/3} + \delta_l\rho^2 \tag{2}$$

where l is an angular momentum index, and ρ measures the density of neighboring atoms,

$$\rho = \sum_{j \neq i} q_j e^{-\lambda^2 R_{ij}}$$

where λ, α_l, β_l, γ_l, and δ_l are new parameters to be determined by fitting to first principles calculations, R_{ij} is the atoms positions and q_j is an effective charge for atom j. The hopping integrals as well as the overlap matrix integrals have the form

$$P_i(r) = (a_i + b_i r + c_i r^2)e^{-d_i^2 r} \tag{3}$$

where P_i represents the ten SK parameters ssσ, spσ, ppσ, ppπ, sdσ, pdσ, pdπ, ddσ, ddπ, and ddδ, r is the distance between atoms, and a_i, b_i, c_i and d_i are parameters to be determined by the fit. Both h_l and $P_i(r)$ are multiplied by a smooth cutoff function that restricts the calculation to about five shells of neighboring atoms.

A unique feature of our method is that we fit the above Hamiltonian simultaneously to the energy bands and the total energy of a given material. We write the total energy as follows:

$$E[n(r)] = \sum_{occ} \epsilon_n + F[n(r)] \tag{4}$$

where the first term represents a sum of the eigenvalues of the occupied states and F represents the remaining density functional total energy expression that depend on the charge density n(r).

D.A. Papaconstantopoulos, M.J. Mehl and Brahim Akdim

Here we introduce a quantity,

$$V_o(S, a) = \frac{F[n(r)]}{N_e} \tag{5}$$

where N_e is the number of valence electrons. Note that V_o depends on the structure S and the lattice parameter a. The substitution (5) allows us to write the total energy according to the following sum,

$$E[n(r)] = \sum_{occ}' \epsilon_n \tag{6}$$

where $\epsilon_n' = \epsilon_n + V_o$.

The result is that the sum of the shifted eigenvalues is equal to the total energy and therefore we avoid the use of a pair potential needed in other formalisms.

3. NUMERICAL PROCEDURE

For the case of monatomic materials, we first perform self-consistent Augmented Plane Wave (APW) calculations [10], typically for 5 different lattice constants in each of the fcc and bcc structures. The TB Hamiltonian described in the previous section is fitted to the resulting energy bands and total energies. The fitting of the energy bands is done using 89 and 55 k-points in the irreducible Brillouin zone for fcc and bcc respectively. We note that we fit at least six bands even if they are not all occupied and we block-diagonalize the Hamiltonian in order to avoid mismatching of different symmetries [3]. The total energy is fitted with a weight typically 300 times larger than that of the individual eigenvalues. We also restrict the behavior of the hopping integrals in Eq. 3 so that they are decreasing functions of distance and do not change sign. The number of parameters in Eqs. 2 and 3 are, for a non-orthogonal Hamiltonian, 97 and 306 for monatomic and diatomic materials respectively. For an orthogonal Hamiltonian, the number of parameters are 57 and 136 respectively. It is important to state our philosophy about the large number of parameters involved in the fitting. We remind the reader that we do not fit to experiment but to first-principles calculation and therefore these parameters should be viewed as intermediate quantities in changing the APW basis set to a tight-binding (i.e., linear combination of atomic orbitals) representation. In practice for most monatomic materials we have omitted the quadratic term in Eq. 3, thus reducing the number of parameters. Also, in many cases we exclude certain orbitals which would lead to further reduction of the number of parameters. For example, in C and Pb it is reasonable to omit the d-orbitals. For monatomic materials our fitting rms error is about 5 mRy for the bands and 0.5 mRy for the total energies. For diatomic materials our rms error for the bands is much larger ranging from 20 mRy to 50 mRy and for the total energy about 1 mRy.

4. RESULTS

In the bibliography the reader can find some of our recent papers[7–9] for a complete account of our results. In this paper we present some representative results as follows :

In Table 1 we give the relative energy per atom for 6 different structures in the third row of the transition metals. The fit was performed only for fcc and bcc structures to the APW results, and therefore the other structures are found as output of our TB scheme. Note that in every case we obtain the correct ground state, i.e. the elements Hf, Re and Os are predicted correctly to be hcp without having included this structure in the fit.

Table 1. Relative energies per atom of several structures for each of the metals examined by the tight-binding model discussed in the text. The energy of the experimental ground state structure is arbitrarily set to zero. All energies are calculated at the equilibrium volume found by the tight-binding fit, and are expressed in mRy. Below the common name of each phase is in its *Strukturbericht* designation.

Element	Structure					
	fcc	bcc	hcp	diamond	sc	αMn
	A1	A2	A3	A4	A_h	A12
Ba	0.9	0.0	0.1	85.3	23.6	2.8
Hf	1.2	7.1	0.0	380.2	113.	22.9
Ta	24.7	0.0	25.3	188.6	63.2	8.9
W	36.0	0.0	10.1	159.3	113.7	18.2
Re	13.4	27.7	0.0	28.6	55.6	0.2
Os	8.1	64.3	0.0	11.2	58.6	17.8
Ir	0.0	50.1	8.5	83.9	83.5	23.8
Pt	0.0	9.7	4.4	168.7	78.3	14.4
Au	0.0	1.4	0.6	96.1	24.1	8.7
Pb	0.0	3.0	1.4	81.4	55.5	

In Tables 2 and 3 we present our results for the elastic constants of the above metals and compare with experiment and first-principles calculations. The elastic constants are calculated by imposing an external strain on the crystal, relaxing any internal parameters (case of hcp crystals) to obtain the energy as a function of the strain [11]. These calculations are also an output of our TB approach, and especially for the hcp materials, they would be very costly to be performed from first-principles. For the cubic materials the elastic constants are consistent with the LAPW values and are usually to within 15% of experiment. This is the accepted standard of comparison between first-principles calculations and experiment. For the hcp materials we found experimental results only for Re and the agreement with our theory is very good.

Table 2. Elastic constants and bulk moduli for 5d cubic elements. Comparison is made between the results of our tight-binding parametrization (TB), first-principles full potential LAPW results (LAPW) [9], where available, and experiment [13] (Exp.). Calculations were performed at the experimental volume.

	C_{11}			C_{12}			C_{44}			B_0(GPa)		
	TB	LAPW	Exp	TB	LAPW	Exp	TB	LAPW	Exp	TB	LAPW	Exp
Ba	9			7			13			10		10
Ta	275	256	261	140	154	157	78	67	82	185	224	200
W	529	527	523	170	194	203	198	147	160	319	333	323
Ir	694	621	590	260	256	249	348	260	262	389	401	355
Pt	380	381	347	257	189	251	71	83	76	318	305	278
Au	184	200	189	154	173	159	43	33	42	187	205	173
Pb	55	54	46	39	41	39	53	20	14	44	45	43

Table 3. Elastic constants and bulk moduli for hexagonal close-packed elements. Comparison is made between the results of our tight-binding parametrization (TB) and experiment [13] (Exp). The tight-binding results include internal relaxation. Calculations were performed at the experimental volume, but at the c/a ratio which minimized the energy for that volume. Note that in a hexagonal crystal, $C_{66} = (C_{11} - C_{12})/2$.

	C_{11}		C_{12}		C_{13}		C_{33}		C_{44}		B_0(GPa)	
	TB	Exp	TB	Exp	TB	Exp	TB	Exp	TB	Exp	TB	Exp
Re	559	616	283	273	250	206	621	683	136	161	372	372
Os	754		272		274		831		232		441	418

Another application of our TB method is in the calculation of vacancy formation energies. These calculations are done with the atoms at the primitive lattice sites in the crystal (Fixed) and allowing relaxation around the vacancy (Relaxed). The results are summarized in Table 4. We have good agreement between theory and experiment, for Ta, Ir and Au while we have large discrepancies for W and Pt where the TB scheme overestimates the vacancy formation energy. It is worth noting that the experimental trend that the energies for Ta and W are larger than in the other metals is also found in our calculations. These calculations involve 108 atoms in a supercell. Such calculations, even on supercomputers, are very slow using first-principles methods.

Table 4. Tight-binding vacancy formation energies compared to experiment. Energies were computed using a 108 atom supercell. The experimental column shows a range of energies if several experiments have been tabulated. Otherwise the estimated error in the experiment is given.

Element	Vacancy Formation Energy (eV)		
	Tight-Binding		Experiment
	Fixed	Relaxed	
Ta	3.17	2.95	2.9 ± 0.4^a
W	6.86	6.43	4.6 ± 0.8^a
Ir	2.19	2.17	1.97^b
Pt	2.79	2.79	1.35 ± 0.09^a
Au	1.24	1.12	0.89 ± 0.04^a
Pb	1.73	1.71	

^{a}Ref. [4], ^{b}Ref. [5]

In Table 5 we present results of our TB methodology for surface energies comparing them to experiment, first-principles calculations, and atomistic models. The surface energy is expressed as the energy required to create a unit area of new surface, and is given by the following formula :

$$E_{surf} = \frac{1}{2A}(E_{slab} - NE_{bulk})$$

where A is the area occupied by one unit cell on the surface of the slab, E_{slab} is the total energy of the slab, N is the number of atoms in the unit cell, and E_{bulk} is the energy of one atom in the bulk at the lattice constant of the atoms in the interior of the slab [9]. These calculations were performed using a slab containing 25 atoms per unit cell. Our results for fcc metals conform to the relationship $E_{111} < E_{100} < E_{110}$, which shows that the close packed surfaces are most stable for fcc metals. The TB surface energies are uniformaly larger, and closer to experiment, than those obtained by the embedded atom method(EAM), which is known to underestimate surface energies in fcc metals. For the bcc metals our results are also good, for all unrelaxed surfaces we found $E_{110} < E_{100} < E_{111}$, and energies closer to experimental results than those obtained by the modified embedded atom method (MEAM).

We have also used our method to study grain boundaries, which play an important role in determining the properties of metals [17]. Since these structures are rather large, calculation of grain boundary formation energies cannot be easily performed using first-principles calculations. We have used the tight-binding method to determine the formation energy of two twin boundaries in Niobium, $\Sigma = 5[001](210)$ and $\Sigma = 5[001](310)$. These calculations were done by periodically replicating the grain boundary in the [2 1 0] and [3 1 0] directions respectively. The atoms were kept at

Table 5. Surface energies, calculated from the tight-binding theory (TB), by first-principles LDA method, by the embedded-atom method (EAM), or by modified embedded atom method (MEAM), compared to experiment. Energies are given in units of J/m^2.

Element	Orientation	TB	LDA	EAM	MEAM	Experiment
Ta	(111)	3.14			2.31	2.78[a]
	(110)	2.05		1.80[d]	2.17	
	(100)	3.00		1.99[d]	3.29	
W	(111)	6.75			2.25	2.99[b]
	(110)	4.30		2.60[d]	2.23	
	(100)	6.70	5.54[f]	2.81[d]	2.65	6.00[c]
Ir	(111)	2.59			2.84	3.00[a]
	(110)	3.19			3.06	
	(100)	2.95			2.91	
Pt	(111)	2.51		1.44[e]	1.66	2.49[a]
	(110)	2.97		1.75[e]	2.13	
	(100)	2.83		1.65[e]	2.17	
Au	(111)	1.48		0.79[e]	0.89	1.54[b]
	(110)	1.85		0.98[e]	1.12	
	(100)	1.69		0.92[e]	1.08	

[a]Ref. [14]
[b]Ref. [15]
[c]Ref. [16]
[d]Ref. [18]
[e]Ref. [19]
[f]Ref. [20]

their bulk positions except near the grain boundary, where the layer spacing was adjusted so that the nearest neighbor distance was the same as in the bulk. Thus our calculations yield an upper bound on the energy needed to form the boundaries. In both cases a supercell containing 40 atoms was large enough to eliminate the interaction between repeated twin boundaries. For the $\Sigma = 5[001](210)$ boundary we find a formation energy of 2.3 J/m^2, while for the $\Sigma = 5[001](310)$ boundary we find 2.0 J/m^2. This latter number should be compared to the embedded atom model calculation of Farkas [12], who found an energy of 1.7 J/m^2 for the unrelaxed cell. Since

EAM calculations often underestimate energies at surfaces, it is not surprising that this energy is somewhat lower than ours.

We have also applied our method to s-p metals and semiconductors and to calculate the phase diagrams of carbon. For carbon we fit the diamond, graphite and simple cubic structures and we were able to obtain as an output the band structure of K_3C_{60} in very good agreement with first-principles calculations.

We present here results of the simple s-p metal Pb. In this case we only fit four s-p bands and the number of our parameters is reduced to 41. In Fig.1 we present a comparison of the TB and APW energy bands at the lattice constant $a = 9.3a.u$. The fit has an rms deviation of 15 mRy for all four bands. Below the fermi level E_F the bands are indistinguishable. The only noticeable discrepancy is at the $W_{2'}$ state above E_F. As is shown in Fig.2 we have fitted perfectly the fcc and bcc structures and we obtain as an output of our method the hcp, L_{12} and simple cubic structures. We have not performed first-principles calculations for the output structures but our findings show the hcp between fcc and bcc which is what is expected. Also the other structures are found at much higher energies which is also a very reasonable result. We have calculated the elastic constants of Pb and included them in Table 2. The TB values for C_{11}, C_{12} and the bulk modulus B are in very good agreement with experiment but C_{44} is too large.

Pb–fcc a=9.3 a.u.

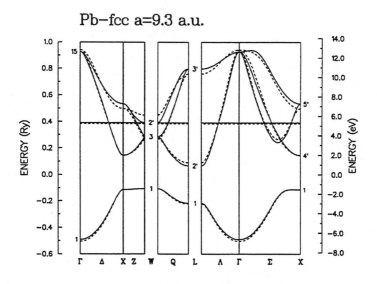

Figure 1. Comparison of APW (Solid lines) and TB (Dashed lines) energy bands for Pb.

At present we are extending this methodology to binary compounds. For the binary cases our results suggest that this approach will also give sufficient accuracy but with considerably more effort in the step of determining the TB parameters.

D.A. Papaconstantopoulos, M.J. Mehl and Brahim Akdim

In summary, we have presented selected results from a new tight-binding method that accurately predicts ground state properties of transition, noble metals, and simple metals.

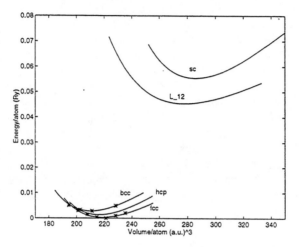

Figure 2. Energy as a function of volume/atom for Lead (Pb).

ACKNOWLEDGMENTS

This work was supported by the United States Office of Naval Research.

References

[1] P. Hohenberg and W. Kohn, Phys. Rev. **136**, B864 (1964).

[2] J. C. Slater and G. F. Koster, Phys. Rev. **94**, 1498 (1954).

[3] D. A. Papaconstantopoulos, *Handbook of Electronic Structure of Elemental Solids* (Plenum, New York, 1986).

[4] H. E. Shaefer, Phys. Stat. Sol. (a) **102**, 47 (1987).

[5] F. R. de Boer, R. Boom, W. C. M. Mattens, A. R. Miedema, and A. K. Niessen, *Cohesion in metals* Vol. 1 (North-Holland, Amsterdam, 1988).

[6] M.M. Sigalas and D. A. Papaconstantopoulos, Phys. Rev. B **49**, 1574 (1994).

[7] R. E. Cohen, M.J. Mehl, and D. A. Papaconstantopoulos, Phys. Rev. B **50**, 14694 (1994).

[8] M. J. Mehl and D.A. Papaconstantopoulos, Europhys. Lett. **31**, 537 (1995).

[9] M. J. Mehl and D. A. Papaconstantopoulos, Phys. Rev B54, 4519 (1996).

[10] L. F. Mattheiss, J. H. Wood, and A. C. Switendick, *Methods In Computational Physics*, Vol. 8, (Academic Press, Inc. NY, 1968).

[11] M. J. Mehl, B. M. Klein, and D. A. Papaconstantopoulos, in *Intermetallic Compounds*, Vol.I, J. H. Westbrook and R. L. Fleischer, eds. (John Wiley & Sons, Ltd, London, 1994).

[12] D. Farkas, *Private Communications.*

[13] G. Simmons and H. Wang, *Single Crystal Elastic Constants and Calculated Aggregate Properties: A Handbook*, 2nd Edition, (MIT Press, Cambridge, MA, 1971).

[14] M. I. Baskes, Phys. Rev. B **46**, 2727 (1992).

[15] W. R. Tyson and W. A. Miller, Surf. Sci. **62**, 267 (1977).

[16] E. Cordwell and D. Hull, Philos. Mag **19**, 951 (1969) .

[17] T. Takasugi, *Structure of Grain Boundaries in Intermetallic Compounds*, Vol.I, J. H. Westbrook and R. L. Fleischer, eds.(John Wiley & Sons,Ltd, London, 1994).

[18] A. M. Guellil and J. B. Adams, J. Mater. Res. **7**, 639 (1992).

[19] S. M. Foiles, M. I. Baskes, and M. S. Daw., Phys. Rev. B **33**, 7983 (1986).

[20] D. Singh, S. H. Wei, and H. Krakauer, Phys. Rev. Lett. **57**, 3292 (1986).

GROUND STATE PROPERTIES OF 3d AND 4d METALS: A STUDY VIA GGA MADE SIMPLE

K. Kokko[1] and M. P. Das[2]

[1]*Department of Physics, University of Turku*
FIN-20014 Turku, Finland

[2]*Department of Theoretical Physics, IAS, RSPhysSE*
The Australian National University, Canberra, ACT 0200, Australia

ABSTRACT

Generalized gradient approximation (GGA) for the exchange correlation energy has become recently available in a simple form to be used in the place of local density approximation (LDA). We have applied this simplified GGA in a self-consistent LMTO method to study the equilibrium volume and bulk moduli of 3d and 4d transition metals. We have obtained systematic improvements of the results in comparison with those in LDA.

1. INTRODUCTION

Density functional theory (DFT) is considered as a standard model for low energy physics describing atoms, molecules and solids [1], [2]. In DFT the exact functional for the exchange correlation energy (E_{xc}) is unknown, so one has to rely on trial functionals satisfying necessary constraints. In this respect the so called local density or local spin density approximation (LDA/LSDA) has been considered as a major achievement in the electronic structure theory of materials. Generally the LDA works well, but for certain properties and if results with chemical accuracy are needed the theory has to go beyond the local density level.

The last decade witnessed several interesting developments in the exchange correlation functional [3] and in 1990's a number of attempts have been made to incorporate the inhomogeneity effects in the energy functional [4]. One way to include the inhomogeneity effects into the exchange-correlation energy is to use so called generalized gradient approximations (GGA) [5]

$$E_{xc}^{GGA}[\rho] = \int d^3 r f(\rho, \nabla \rho), \qquad (1)$$

where ρ is the electronic density. In their recent work Perdew *et al.* [6] have obtained a simple form of the GGA in which all parameters (other than those in LSDA) are fundamental constants. This functional is easy to implement for any electronic structure calculations. In view of the availability of the new simplified GGA functional [6] it is our purpose to present some ground state properties of all the 3d and 4d transition metals to show what difference it makes to the already known results.

2. CALCULATIONS AND RESULTS

The calculations were performed using the scalar relativistic self-consistent-field LMTO method [7], [8] with the atomic sphere approximation (ASA) and the combined correction terms. The valence states were expanded with spherical harmonics up to $l_{\max} = 3$. The core states were taken from atomic calculations and treated with the frozen-core approximation in the band structure program. The $\bar{k}$-space integrations were done with 916 and 819 $\bar{k}$points in the irreducible wedge of the fcc and bcc Brillouin zones, respectively.

All the calculations were performed using the non-spin-polarized scheme thus referring to the paramagnetic ground state. We used the observed crystal structures for the fcc and bcc metals, but the elements with more complicated crystal structures were treated as fcc. For the exchange-correlation potential we used two different approximations: the LDA presented by Perdew and Wang [9] and the recent simplified version of the GGA by Perdew *et al.* [6].

Our LDA calculations predict correctly the general trends of r_{WS}^0 and B within the 3d and 4d transition metal series, except in the case of the magnetic 3d metals (Fig. 1). The obtained discrepancy between the calculated and experimental results for the 3d metals from Cr to Ni is expected because the calculations refer to the paramagnetic state which is not the true ground state of these materials. Largest discrepancies occur in the middle of the magnetic series (Mn, Fe and Co), the r_{WS}^0 being 5–6 % too small and the B for Mn being three times too large. Both r_{WS}^0 and B show a parabolic behavior as a function of the number of d electrons (n_d). The relation between r_{WS}^0 and B, namely smaller r_{WS}^0 implies larger B, was also obtained. For the 3d series the minimum of the r_{WS}^0 curve is at $n_d = 7$ and the maximum of the B curve is $n_d = 5$, the corresponding values for the 4d series are $n_d = 7$ and 6. In the case of the 4d metals we can compare this with the experiments where the minimum of r_{WS}^0 and the maximum of B are both at $n_d = 6$.

In comparing our calculations with the experimental data in a more quantitative level (Fig. 2), we exclude the cases of Mn, Fe and Co where the magnetization plays the most important role. For the 3d metals the LDA gives 2–3 % too small r_{WS}^0 and 30–60 % too large B compared to the experimental data. Apart from the slight decrease when going from Sc to Ti the discrepancy in r_{WS}^0 increases with the increasing filling of the d band (increasing n_d). For the 4d metals the discrepancy in r_{WS}^0 is generally smaller than for the 3d metals. Also in this case the calculated results are smaller than the experimental ones: the deviation is 1.5 % for Y, then decreasing to the level of 0.5 % for Zr, and staying at that level up to Tc, after which it starts to increase again and reaches the largest value of 3 %for Ag. The calculated bulk modulus of the 4d metals is 20–60 % larger than the experimental one. Like in the case of r_{WS}^0 the smallest relative discrepancy is found in the middle of the series.

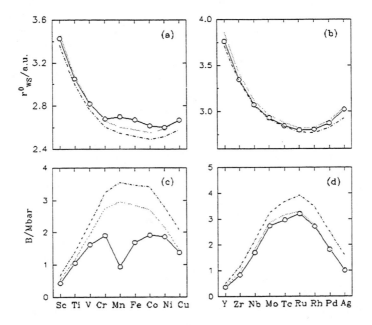

Figure 1. Equilibrium Wigner-Seitz radius (r_{WS}^0) and bulk modulus (B) of 3d and 4d transition metals. Experimental data [10], [8] is shown by open circles, dash-dotted and dotted curves refer to LDA and GGA results, respectively.

Generally the GGA improves our results. Bulk modulus is improved in all cases but for the r_{WS}^0 in the beginning of the 4d series (from Y to Tc) the 0.5–1.5 % underestimation of the LDA changes to the 1–2.5 % overestimation of the GGA. However, taking into account that the ASA slightly overestimates r_{WS}^0 [11] the simplified GGA is expected to improve the r_{WS}^0 for all 4d transition metals provided the more accurate FP treatment will be used. It is interesting to note that the GGA correction to the LDA results has a general trend within both the 3d and 4d transition metals. The relative correction due to the GGA is smallest at the middle of the series being typically of 1.5–3.5 times smaller than at the beginning and at the end of the series for both r_{WS}^0 and B. This can be explained in the following way. The lattice expansion and the filling of the d band by the antibonding states both increase charge density inhomogeneities which leads to larger gradient corrections. When going from the beginning of the transition metal series towards the end of the series the volume decreases (Fig. 1) leading to the gradual decrease of the gradient corrections. However, at the end of the series the filling of the antibonding states increases the inhomogeneities of the electronic density leading to the reentrant increase of the gradient corrections at the end of the series. The same effect can be seen also in the FP-LMTO results for 4d transition metals [11]. Due to the nonlinearity of the GGA corrections as a function of n_d the parabolic shape of the r_{WS}^0 and B curves changes slightly. This shows up, for instance, in the change of the minimum position of the r_{WS}^0 curve of the 4d metals from Rh to Ru which is also the experimental minimum point. The changes due to the GGA compared to the LDA results are the following. For the 3d metals the r_{WS}^0 is increased by 2–3.5 % and the B is decreased by 20–50 %. For the 4d metals the corresponding changes are 1.5–4 % for r_{WS}^0 and 15–55 % for B.

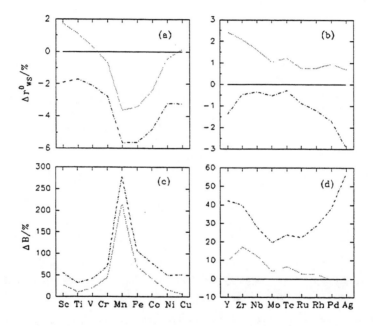

Figure 2. The differences between the calculated and experimental values of r_{WS}^0 and B (in percents). Dash-dotted and dashed curves refer to LDA and GGA results, respectively.

There has been several theoretical investigations for cohesive properties of transition metals using various calculational methods and different approximations. Here, however, it is possible to refer only to a few of them. The Korringa-Kohn-Rostoker muffin-tin calculations of Moruzzi *et al.* [12] have shown that sophisticated LDA calculations are able to produce the observed general trends of the experimental r_{WS}^0 and B of transition metals. The LMTO-ASA calculations of Andersen *et al.* [13] give r_{WS}^0 which are a few percent smaller than the experimental ones and the discrepancy is somewhat larger than that found in the work of Moruzzi *et al.* [12]. Using the results of Ozoliņš and Körling [11] one can obtain the change of r_{WS}^0 and B due to the GGA compared to the corresponding LDA results. The average change for the 3d and 4d transition metals is 2 % (2 %) for r_{WS}^0 and -11 %(-14 %) for B, the first numbers refer to the ASA calculations, the FP values are in parenthesis. This can be compared with our results for the simplified GGA which is 2 % for r_{WS}^0 and -22 % for B.

In conclusion it has been shown that the new simplified GGA systematically improves the calculated equilibrium volumes and bulk moduli of the 3d and 4d transition metals compared to the LDA results. Gradient corrections are larger both at the beginning and at the end of the transition metal series. This phenomenon is attributed to the bonding properties of the d states of the transition metals.

ACKNOWLEDGMENTS

We thank John Perdew for sending us their paper (Ref. 6) before publication and for providing the computer code of the simplified GGA and Timo Korhonen for discussions and comments concerning the present work. K.K. acknowledges the Department of Theoretical Physics, The Australian National University for hospitality and computational facilities and Suomalainen Tiedeakatemia for the financial support during his stay in Canberra.

References

[1] S. Lundqvist and N. H. March (eds.), *Theory of the Inhomogeneous Electron Gas*, (Plenum Press, New York, 1983).

[2] R. M. Dreizler and E. K. U. Gross, *Density Functional Theory. An Approach to the Quantum Many-Body Problem*, (Springer-Verlag, Berlin Heidelberg, 1990).

[3] D. J. W. Geldart and M. Rasolt, in *The Single-Particle Density in Physics and Chemistry*, N. H. March and B. M. Deb (eds.), (Academic Press, London, 1987).

[4] J. P. Perdew *et al.*, Phys. Rev. B **46**, 6671 (1992).

[5] E. K. U. Gross and R. M. Dreizler (eds.), *Density Functional Theory*, NATO ASI Series B: Physics Vol. 337, (Plenum Press, New York, 1995).

[6] J. P. Perdew, K. Burke, and M. Ernzerhof, Phys. Rev. Lett. **77**, 3865 (1996).

[7] O. K. Andersen, Phys. Rev. B **12**, 3060 (1975).

[8] H. L. Skriver, *The LMTO Method. Muffin-Tin Orbitals and Electronic Structure*, (Springer-Verlag, Berlin, 1984).

[9] J. P. Perdew and Y. Wang, Phys. Rev. B **45**, 13244 (1992).

[10] C. Kittel, *Introduction to Solid State Physics*, (John Wiley & Sons, Inc., New York, 1976).

[11] V. Ozoliņš and M. Körling, Phys. Rev. B **48**, 18304 (1993).

[12] V. L. Moruzzi, J. F. Janak, and A. R. Williams, *Calculated Electronic Properties of Metals*, (Pergamon Press Inc., New York, 1978).

[13] O. K. Andersen, O. Jepsen, and D. Glötzel in *Proceedings of the International School of Physics "Enrico Fermi" Course LXXXIX*, F. Bassani, F. Fumi, and M. Tosi (eds.), (North-Holand, Amsterdam, 1985).

COEXISTENCE OF SPIN DENSITY WAVE AND STRUCTURAL TRANSITIONS WITH THE SUPERCONDUCTING TRANSITION

S. N. Behera,[1] Priyatama Deo,[2] S. K. Ghatak,[3] Haranath Ghosh,[1]
Manidipa Mitra,[1] and Srikantha Sil[1]

[1] *Institute of Physics, Bhubaneswar 751005, Orissa, India*
[2] *Physics Department, Ravenshaw College, Cuttack 753003, Orissa, India*
[3] *Department of Physics and Meteorology, Indian Institute of Technology
Kharagpur 721302, West Bengal, India*

ABSTRACT

A microscopic mean field theory is developed for the coexistence of three different long range orders in a system, i. e. the band Jahn-Teller distortion of the lattice, the spin density wave state, and superconductivity. The interplay between the spin density wave and the superconducting order parameters is discussed in detail when the latter has symmetries other than the s-wave. The effect of electron correlation on the coexistence of band Jahn-Teller and superconducting transitions has been emphasized.

1. INTRODUCTION

The appearance of a long range order when a material undergoes a phase transition on varying the external parameters such as temperature and pressure is rather well understood. Quite often some systems exhibit more than one such transition with two or more long range orders coexisting with each other. In these circumstances, it is of interest to ask how the different order parameters interplay with each other. The interplay of magnetism and superconductivity (SC) in some materials had been the subject matter of extensive investigation, since the discovery of the later phenomenon. It is well known that not only a magnetic field but even the presence of a small percentage of magnetic impurities in a superconducting material can destroy superconductivity. Therefore, for all practical purposes the coexistence of SC with ferromagnetic (FM) long range order can be ruled out. But its existence with antiferromagnetic (AFM) long range order has been realized because of the rather large coherence lengths of the conventional superconductors. Similarly, the interplay of a structural phase transition (SPT) with SC has also been a subject of much research. As a system approaches the structural transition temperature, it is well known that one or more of the phonons will soften and act as a precursor to the transition. In the case of the low temperature phonon mediated SC, the strength of the attractive interaction between a pair of electrons gets enhanced when the phonon frequency softens, thereby increasing the superconducting transition temperature (T_C). On the other hand a strong electron-phonon interaction may drive a lattice instability, thereby invalidating the simple minded BCS-theory. Therefore, it is expected that both long range magnetic and structural transitions will have profound effect on superconductivity. In this paper we shall present the results of the interplay between AFM order, SPT and SC derived from microscopic mean field theoretic calculations [1].

The SPT under consideration will be an electronically driven band Jahn-Teller (BJT) distortion. The BJT transition arises from the competition between the lowering of the electronic energy due to the lifting of the orbital degeneracy of the bands at the Fermi level, and the increase in elastic energy due to the appearance of the distortion (strain) in the lattice [2]. The system will optimize the distortion so as to maximize the gain in energy, which is the case when the split lower band is maximally occupied and the upper band is minimally occupied. Thus the difference in the occupation probabilities of the two bands will couple to the lattice strain which defines the BJT-Hamiltonian. The BJT transition is favored if the Fermi level (FL) lies in a peak in the density of states (DOS).

The AFM long range order being considered, also owes its origin to itinerant electrons in the system and is in the form of a spin density wave (SDW). Generally, the SDW state results from a Fermi surface (FS) instability due to the existence of nested pieces of FS separated by a wave vector $\vec{Q}$. While the nesting amounts to an electron hole symmetry such that the quasiparticle energy satisfies the requirement $\epsilon_{k\pm Q} = -\epsilon_k$, the wave vector $\vec{Q}$ determines the super-periodicity due to the formation of the SDW, which may or may not be commensurate with the lattice [3]. The interaction responsible for any form of magnetism is usually the Coulomb repulsion, hence a good starting point for the description of the SDW state is the Hubbard model [4]. The interatomic Coulomb repulsion (U) term in the Hamiltonian is usually treated in the mean field approximation allowing for a spin polarization in the z-direction or spin flip fluctuations on the same site which give rise to the longitudinal and transverse SDW-states.

Finally, it is well known that SC arises due to an effective attractive interaction between two electrons, which is responsible for the formation of two-electron bound states known as Cooper pairs. For conventional low temperature superconductors, this interaction arises from phonon mediation as envisaged in the BCS theory. However, for the families of the high temperature superconductors (HTSC) the mechanism for SC is not yet understood [5]. Hence, it will be assumed that SC arises in the system due to some Boson mediated pairing mechanism, without specifying any interaction responsible for the attractive interaction. Furthermore, in the context of the BCS theory, usually the isotropic s-wave singlet pairing is considered, so that the superconducting order parameter (Δ) does not have an explicit dependence on the direction of the wave vector $(\vec{k})$. But if the pairing is mediated by say the quanta of AFM spin fluctuations the pairing order parameter is known to possess the d-wave symmetry. Therefore, in the present paper while discussing the coexistence of SDW and SC, the order parameter for the later will be assigned different pairing symmetries, to demonstrate how the interplay of the two long range orders differ from one pairing symmetry to the other [6].

The problem of interplay between the SDW state and SC has been formulated by many authors [1,7] earlier, specifically for the s-wave symmetry of the later order parameter. Similarly, the interplay between the BJT distortion and SC has also been considered [8]. But there exist systems like the cuprate HTSCs where all the three phenomena can coexist. Hence in Sec. 2 we formulate the problem for the simultaneous coexistence of all the three together. Section 3 is devoted to the discussion of results for the coexistence of the SDW and SC as well as the BJT-distortion and SC. In either of these discussions the role of the coulomb correlation will be emphasized keeping in mind the fact that the HTSC cuprates are strongly correlated systems. Finally we conclude in section 4 by pointing out the qualitative agreement between the predictions of the present mean field calculations and the existing experimental data in some systems where the coexistence has been observed.

2. MEAN FIELD FORMULATION OF THE THEORY FOR THE COEXISTENCE OF SDW STATE, BJT - DISTORTION AND SC

As pointed out in the introduction, while for the BJT - transition it is necessary that the Fermi level of the system lies in a degenerate band with the electron densities interacting with strain, the occurrence of the SDW - state requires one to take into account the Coulomb interaction and assume the presence of nesting in the FS. Of course SC arises due to a pairing interaction. Therefore, the simplest starting point is a model Hamiltonian given by

$$H = H_0 + H_{Coul} + H_{BJT} + H_P \qquad (1)$$

where H_0 the tight binding Hamiltonian for a doubly degenerate band is given by

$$H_0 = \sum_{i\sigma,\alpha} t_0 n_{i\alpha\sigma} + \sum_{i\neq j,\alpha,\sigma} t_{ij}^{\alpha} C_{i\alpha\sigma}^{\dagger} C_{i\alpha\sigma} \qquad (2)$$

t_0 being the energy of the α-th ($\alpha=1,2$) orbital. $n_{i\alpha\sigma} = C_{i\alpha\sigma}^{\dagger} C_{i\alpha\sigma}$ is the number operator for the electrons and t_{ij}^{α} is the hopping integral for the α-th orbital, with all the other symbols having their usual meaning. The most important contributions for the Hamiltonian describing the Coulomb interaction $H_{coul.}$ are the intrasite intraorbital Hubbard like term (U) and the intrasite inter-orbital term (U'); the later arising because of the degeneracy of the band, i.e.

$$H_{Coul} = \frac{1}{2}U\sum_{i\alpha\sigma} n_{i\alpha\sigma}n_{i\alpha-\sigma} + U'\sum_{i,\alpha\neq\beta,\sigma\sigma'} n_{i\alpha\sigma}n_{i\beta\sigma'} \qquad . \tag{3}$$

The BJT interaction which involves the coupling of the strain to the difference in occupation of the two orbitals is given by

$$H_{BJT} = -Ge\sum_{i\sigma}\left[n_{i1\sigma} - n_{i2\sigma}\right] \tag{4}$$

where G is the strength of the interaction and e is the lattice strain. Finally, the Hamiltonian for the pairing interaction as usual is given by

$$H_P = -V\sum_{k\alpha} C^\dagger_{k\alpha\uparrow}C^\dagger_{-k\alpha\downarrow}C_{-k\alpha\downarrow}C_{k\alpha\uparrow} \tag{5}$$

where V is the strength of the interaction arising from the Boson exchange mechanism. In eqn. (5) only intra-band pairing is taken into account. To treat the problem in the mean field approximation eqn. (5) is reduced to the standard BCS form; while the Hartree-Fock approximation is used for the second term in eqn. (3). The Hubbard term (1st term) in eqn. (3) in the generalized Hartree-Fock approximation separates into charge and spin parts, the later involving the SDW state as follows:

$$H^1_{Coul} = \frac{1}{2}U\sum_{i,\alpha}\langle n_{i\alpha}\rangle n_{i\alpha} + \sum_{i\alpha}\vec{\sigma}_{i\alpha}\vec{B}_{i\alpha} \tag{6}$$

where $n_{i\alpha} = \sum_\sigma n_{i\alpha\sigma}$ the total number of electrons with up and down spins on site i, the magnetic moment on site i being

$$\vec{\sigma}_{i\alpha} = \left(C^\dagger_{i\alpha\uparrow} \ C^\dagger_{i\alpha\downarrow}\right) \vec{\tau} \begin{pmatrix} C_{i\alpha\uparrow} \\ C_{i\alpha\downarrow} \end{pmatrix}$$

$\vec{\tau}$ being the Pauli matrices, with the electron operators written in the Nambu representation and $\vec{B}_i^\alpha$ is the average magnetic field at site i produced by the other electrons, which when written in terms of the electron operators has the form

$$B_{i\alpha z} = -US_{i\alpha z} = -\frac{1}{2} U \left[\langle n_{i\uparrow}^{\alpha} \rangle - \langle n_{i\downarrow}^{\alpha} \rangle \right] \qquad (7)$$

$$B_{i\alpha\pm} = -US_{i\alpha\pm} = -U \begin{bmatrix} \langle C_{i\alpha\uparrow}^{\dagger} C_{i\alpha\downarrow} \rangle \\ \langle C_{i\alpha\downarrow}^{\dagger} C_{i\alpha\uparrow} \rangle \end{bmatrix} \qquad (8)$$

In fact, eqns. (7) and (8) represent respectively the longitudinal and transverse SDW order parameters. Therefore, the total mean field Hamiltonian written in the momentum representation becomes

$$H = \sum_{k\alpha\sigma} [\epsilon_{\alpha}(k) - \mu] n_{k\alpha\sigma} + \Delta_{SDW} \sum_{k\alpha} \left(C_{k+Q,\alpha\uparrow}^{\dagger} C_{k\alpha\downarrow} + h.c. \right) + \Delta_{SC} \sum_{k\alpha} \left(C_{k\alpha\uparrow}^{\dagger} C_{-k\alpha\downarrow}^{\dagger} + h.c. \right) \qquad (9)$$

where the SDW and SC order parameters are defined as

$$\Delta_{SDW} = -U \sum_{k\alpha} \langle C_{k+Q\alpha\uparrow}^{\dagger} C_{k\alpha\downarrow} \rangle \qquad (10)$$

$$\Delta_{SC} = -V \sum_{k\alpha} \langle C_{k\alpha\uparrow}^{\dagger} C_{-k\alpha\downarrow}^{\dagger} \rangle \qquad (11)$$

and the renormalized electron band energies are given by

$$\epsilon_{\alpha}(k) = t_0^{\alpha} + \epsilon(k) \mp Ge \qquad (12)$$

with $\epsilon(k)$ being the tight binding band dispersion and

$$t_0^{\alpha} = t_0 + \frac{1}{2} U \langle n_{k\alpha} \rangle + U' \langle n_{k\beta} \rangle \qquad (13)$$

t_0^{α} being the energy of the renormalized αth atomic orbital and the band degeneracy is removed by the Jahn-Teller interaction which split the two bands by an amount $2Ge$. In eqn. (10) only the transverse SDW state is considered for simplicity. It is easy to see from eqn. (13) that the energy of the first orbital not only depends on its own occupation probability but also on the occupation probability of orbital 2 and vice versa. The total Hamiltonian as given by eqn. (9) is a sum total of two independent single particle Hamiltonians for the two orbitals; and hence can be solved to determine the order parameters for the SDW-state, SC and the strain. The equation for the order parameters are as given below,

(i) for the SC order parameter (Δ_{SC})

$$1 = -V\sum_k \frac{1}{(E_{k+}^2 - E_{k-}^2)}\left(\frac{\tanh\left(\frac{\beta E_{k-}}{2}\right)}{E_{k-}}\left[E_{k-}^2 - \epsilon(k)^2 - (Ge)^2 - \Delta_{SC}^2 + 4\Delta_{SDW}^2\right] \right.$$

$$\left. - \frac{\tanh\left(\frac{\beta E_{k+}}{2}\right)}{E_{k+}}\left[E_{k+}^2 - \epsilon(k)^2 - (Ge)^2 - \Delta_{SC}^2 + 4\Delta_{SDW}^2\right] \right) \qquad (14)$$

(ii) for the strain (e)

$$1 = 2G\sum_k \frac{1}{(E_{k+}^2 - E_{k-}^2)}\left(\frac{\tanh\left(\frac{\beta E_{k-}}{2}\right)}{E_{k-}}\left[E_{k-}^2 + \epsilon_1(k)\epsilon_2(k) - \Delta_{SC}^2 + 4\Delta_{SDW}^2\right] \right.$$

$$\left. - \frac{\tanh\left(\frac{\beta E_{k+}}{2}\right)}{E_{k+}}\left[E_{k+}^2 + \epsilon_1(k)\epsilon_2(k) - \Delta_{SC}^2 + 4\Delta_{SDW}^2\right] \right) \qquad (15)$$

and (iii) the SDW order parameter (Δ_{SDW})

$$1 = -2U\sum_k \frac{1}{(E_{k+}^2 - E_{k-}^2)}\left(\frac{\tanh\left(\frac{\beta E_{k-}}{2}\right)}{E_{k-}}\left[E_{k-}^2 - \epsilon_1(k)\epsilon_2(k) + \Delta_{SC}^2 - 4\Delta_{SDW}^2\right] \right.$$

$$\left. - \frac{\tanh\left(\frac{\beta E_{k+}}{2}\right)}{E_{k+}}\left[E_{k+}^2 - \epsilon_1(k)\epsilon_2(k) + \Delta_{SC}^2 - 4\Delta_{SDW}^2\right] \right) \qquad (16)$$

where the quasi-particle energies in the coexistent phase are given by

$$E^2_{k\pm} = \left[\epsilon(k)^2+(Ge)^2+\Delta^2_{SC}+4\Delta^2_{SDW}\right]\pm2\left[\epsilon(k)^2(Ge)^2+4\Delta^2_{SDW}\Delta^2_{SC}+4\Delta^2_{SDW}(Ge)^2\right]^{\frac{1}{2}} \quad (17)$$

If we take $t_0^\alpha = 0$ then, because of the nesting property of the tight binding bands, the Jahn-Teller split bands acquire the following symmetry property,

$$\epsilon_{k\pm Q,1} = -\epsilon_{k,2} \quad and \quad \epsilon_{k\pm Q,2} = -\epsilon_{k,1} \quad . \quad (18)$$

Equations (14-16) have to be solved self consistently in order to determine the temperature dependence of the various order parameters. Indeed, the study can even be generalized to the case of different band fillings by adding yet another equation for the chemical potential, to this set of three equations for the order parameters. Even though eqns. (14-16) describe the coexistence of all the three order parameters together, it will be simpler and even be of interest to analyze the results for the coexistence of two of these long-range orders at a time. These results will be presented in the next section.

3. RESULTS AND DISCUSSION

In what follows, we shall first consider the coexistence of SDW and SC and then discuss the results for the coexistence of BJT-distortion and SC. While considering the interplay of the former two order parameter, the formalism for SC will be generalized to take into account order parameter symmetries other than the s-wave which is implied in the formulation of sec. 2. Similarly, while considering the coexistence of BJT-distortion and SC, the correlation will be treated in the Slave Boson formalism to take into account the fact that the system exhibiting the above phenomena may be a strongly correlated one.

3.1 Interplay of SDW State and SC

Since the coexistence of only the SDW state and SC is being considered, it suffices to take into account only a non-degenerate band, and the equations for the order parameters in this case can be recovered from eqns. (14) and (15) by setting Ge = o. In this special case the quasi-particle energies given by eqn. (17) reduces to

$$\tilde{E}_{k\pm} = \left[\epsilon(k)^2 + (\Delta_{SC}\pm2\Delta_{SDW})^2\right]^{\frac{1}{2}} = \left[\epsilon(k)^2+\Delta^2_\pm\right]^{\frac{1}{2}} \quad (19)$$

indicating that the effective order parameter shows some kind of interference between the SDW and SC states [7]. The equations for the order parameters reduce to

$$\Delta_{SC} = \frac{V}{2}\sum_{k}\left[\frac{\Delta_{+}\tanh\dfrac{\beta\tilde{E}_{k+}}{2}}{\tilde{E}_{k+}} + \frac{\Delta_{-}\tanh\dfrac{\beta\tilde{E}_{k-}}{2}}{\tilde{E}_{k-}}\right], \qquad (20)$$

for the SC order parameter, and

$$\Delta_{SDW} = \frac{U}{2}\sum_{k}\left[\frac{\Delta_{+}\tanh\dfrac{\beta\tilde{E}_{k+}}{2}}{\tilde{E}_{k+}} - \frac{\Delta_{-}\tanh\dfrac{\beta\tilde{E}_{k-}}{2}}{\tilde{E}_{k-}}\right], \qquad (21)$$

for the SDW order parameter. Equation (20) for the SC order parameter corresponds to s-wave pairing in which case: both the interaction strength V as well as the order parameter Δ_{SC} are constants independent of the wave vector $\vec{k}$. Equations (20) and (21) has been solved numerically earlier [1] within the Bilbro-McMillan approximation [9] for the density of states at the FS. It was demonstrated that the interplay between the two order parameters can exhibit very interesting phenomena like the reentrant and coexistent behavior of the SDW and SC states. On the other hand, for symmetries of the order parameter other than the s-wave there will be an explicit momentum dependence of these quantities, which requires the generalization of eqn. (20) to

$$\Delta_{SC}(k) = \frac{1}{2}\sum_{k'} V(k,k')\left[\frac{\Delta_{+}(k')\tanh\dfrac{\beta\tilde{E}_{k'+}}{2}}{\tilde{E}_{k'+}} + \frac{\Delta_{-}(k')\tanh\dfrac{\beta\tilde{E}_{k'-}}{2}}{\tilde{E}_{k'-}}\right] \qquad (22)$$

and accordingly the effective order parameters $\Delta_{\pm}$ acquire a k-dependence and become $\Delta_{\pm}(k)$ in eqn. (21) too. On assuming a separable form for the k-dependence of the interaction strength, i.e. $V(k,k') = v(k) v(k')$; it is evident from eqn. (22) that the symmetry of the order parameter is the same as that of the strength of the interaction $v(k)$. Depending upon whether $v(k) = v$, a constant, or $v(k) = v(\cos k_x + \cos k_y)$, i. e. same symmetry as the lattice in the case of a square lattice or $v(k) = v(\cos k_x - \cos k_y)$ the symmetry of the SC order parameter is either s-wave or extended s-wave or d-wave respectively. In these cases the eqns. (21) and (22) can be solved self-consistently together with an equation which determines the band filling i.e. the deviation from half filling, given by

$$n = 1 - \frac{1}{2}\sum_{k}[\epsilon(k)-\mu]\left[\frac{\tanh\dfrac{\beta\tilde{E}_{k+}}{2}}{\tilde{E}_{k+}} + \frac{\tanh\dfrac{\beta\tilde{E}_{k-}}{2}}{\tilde{E}_{k-}}\right] = 1-x \qquad (23)$$

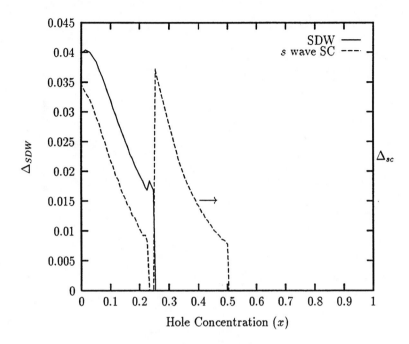

Figure 1. The magnitudes of the order parameters Δ_{SDW} and Δ_{SC} for the spin density wave and s-wave superconducting states respectively at 5 K plotted as a function of the dopant (hole) concentration (x). The values of the strengths of interactions are chosen to be U = 0.125 and V = 0.1875 in units of the band width. There is coexistence of the order parameters for x ≤ 0.25.

The result of the self consistent evaluation of eqns. (21) - (23) for the s-wave SC is shown in Fig. 1, for a particular choice of the parameters V (=0.1875) and U (=0.125) in terms of the width of the tight binding band for a square lattice. The plot shows the magnitudes of the SDW and SC order parameters at a temperature of 5 K as a function of the hole concentration (x). At close to half filling of the band, the FS being a square, there is perfect nesting hence the SDW state turns out to be very stable. But since the strength of the pairing interaction is assumed to be much larger than that of the Coulomb repulsion (V), SC also appears for low dopant concentrations. The magnitudes of both decrease with increasing x, and SC disappears first around x ≈ 0.24, and then the SDW state vanishes at x = 0.25. But with the vanishing of the SDW state the SC-order parameter increases suddenly showing a reentrant kind of behavior. Superconductivity

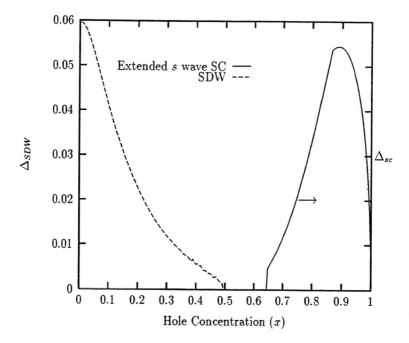

Figure 2. Same as in Fig. 1, but for the coexistence of SDW and extended s-wave SC. The regions of x for the two order parameters are segregated.

persists in the range $0.25 \leq x \leq 0.5$, without any SDW state. For > 0.5 neither SC nor the SDW-state exist. Thus, there is the SC state at relatively large dopant concentrations and the coexistence of SDW and s-wave SC for low dopant concentrations. In contrast to this behavior, the SDW-state and extended s-wave SC never coexist within the same sample with a given dopant concentration as shown in Fig. 2. While the SDW-state exists for $0 < x \leq 0.5$, the extended s-wave SC exists for $0.65 \leq x \leq 1$. Finally, the behavior of this phase diagram for d-wave SC and the SDW state is shown in Fig. 3. As usual, the SDW state is the stable state for dopant concentrations (x) close to zero. But SC appears in a narrow range of x ($0.25 < x \leq 0.4$) completely suppressing the SDW state. For $0.4 < x \leq 0.5$ the SDW-state reappears but without SC. Thus, in this case there is no coexistence of the SDW state and d-wave SC, similar to the case of extended s-wave SC, with the difference that while d-wave SC appears in a range of x close to the low dopant concentration, the extended s-wave SC appears for high dopant concentrations. One can also look at mixed symmetry and the temperature dependence of the order parameters [7].

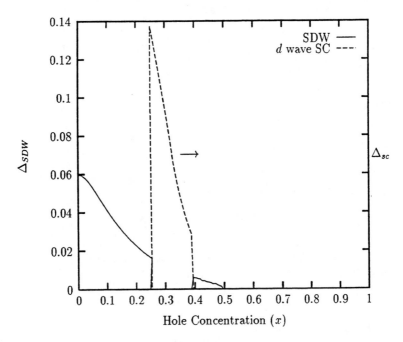

Figure 3. Same as in Fig. 1, but for the coexistence of SDW and d-wave SC. The two long range orders occur in the same region of x, but when one appears the other is suppressed.

3.2. Interplay Between BJT-distortion and SC

The consideration of BJT-distortion necessitates the presence of the doubly degenerate band and the equations for the strain, and the SC order parameter can be deduced from eqns. (14) and (15) by setting $\Delta_{SDW} = 0$. In this case, the quasi-particle energy as obtained from eqn. (17) takes the form

$$\eta_{k\pm} = \left[(\epsilon(k)\pm Ge)^2 + \Delta_{SC}^2\right]^{\frac{1}{2}} \equiv \left[\epsilon_{1,2}^2 + \Delta_{SC}^2\right]^{\frac{1}{2}} \tag{24}$$

where $\epsilon_{1,2}(k) = (\epsilon(k) \pm Ge)$ are the two Jahn-Teller split bands. The equations for the order parameters and the band filling in this case are given by

$$1 = \frac{V}{2} \sum_k \left[\frac{\tanh \frac{\beta \eta_{k+}}{2}}{\eta_{k+}} + \frac{\tanh \frac{\beta \eta_{k-}}{2}}{\eta_{k-}} \right] \qquad (25)$$

for the SC with s-wave pairing symmetry,

$$(Ge) = G \sum_k \left[\frac{\epsilon_2(k)\tanh \frac{\beta \eta_{k-}}{2}}{\eta_{k-}} - \frac{\epsilon_1(k)\tanh \frac{\beta \eta_{k+}}{2}}{\eta_{k+}} \right] \qquad (26)$$

for the lattice strain parameter (Ge), and finally an equation for the dopant concentration (x) which is just the generalization of eqn. (23) for the case of the two Jahn-Teller split bands given by

$$x = \frac{1}{2} \sum_k \left[\frac{\epsilon_1(k)}{\eta_{k+}} \tanh \frac{\beta \eta_{k+}}{2} + \frac{\epsilon_2(k)}{\eta_{k-}} \tanh \frac{\beta \eta_{k-}}{2} \right] \qquad (27)$$

Equations (25) - (27) can be solved self consistently to determine the interplay between SC and BJT-transition [2,8]. However, in these equations the effect of electronic correlation has not been accounted for. In order to incorporate this effect the first term in eqn. (3), namely, the intra site intratomic Coulomb repulsion (U) can be treated by the Slave Boson method [10]. In the saddle point approximation this amounts to a renormalization of the electron energy band $\epsilon(k)$ appearing in eqns. (24) - (27) to be replaced by $\bar{\epsilon}(k) = \bar{q}\epsilon(k)$; where the renormalization factor $\bar{q} = (1-u^2)$ with $u = U/U_C$ which drives the metal-insulator transition at $U = U_C$.

The effect of correlation on the coexistence of the two order parameters e and Δ has been studied in detail [2, 11]. As u increases the bands get narrowed, resulting in an enhancement of the density of states of the undistorted system which favors the growth of strain. Consequently, the degeneracy of the bands is lifted and the density of states of the distorted lattice reduces at the Fermi level, which in turn reduces the SC transition temperature (T_c) as well as the magnitude of the order parameter. Besides at Tc when the SC appears in the system the growth of strain is arrested [11]. The strain being proportional to the difference in the occupation probabilities of the two Jahn-Teller split bands, the effect of intra and inter atomic correlations U and U' on these occupations is shown in Fig. 4. The occupations of the two bands, therefore, is also a measure of the orthorhombicity of the distorted lattice; which grows with increasing U and U' as is evident from the figure. Furthermore, with increasing orthorhombicity the SC gets suppressed, as can be inferred from the shifting of the point of arrest of the growth of the strain to lower temperatures.

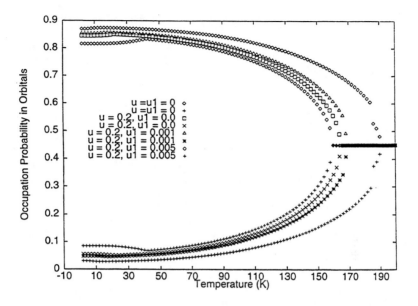

Figure 4. The electron occupation probabilities, of the lower (upper curves) and upper (lower curves) Jahn-Teller-split bands which is proportional to the orthorhombicity of the lattice, are plotted as a function of temperature, for different values of the correlations u and u'. The arresting and lowering (increasing) of the occupation of lower (upper) band at T_c = 40 K for the uncorrelated (u = 0, u' = 0) case and subsequent suppression of T_c with increasing values of correlation (u and u') is noteworthy.

4. CONCLUSION

In this paper, the interplay between three different long range orders, i. e. the SPT in the form of a BJT-distortion, the AFM in the form of a SDW-state and SC has been studied within a mean field formalism. The coexistence of two of these at a time, namely, the SDW-state and SC has been examined in detail when the SC order parameter has the symmetry of either the s-wave or the extended s-wave or the d-wave. Similarly, the interplay between the BJT-distortion of the lattice and the SC has also been considered. The effect of electron correlation on the coexistence of these two order parameters has been emphasized. There are many physical systems where the coexistence of two or more of these order parameters has been observed. For example, the SDW-state and SC are known to coexist in heavy fermion systems and organic superconductors, where as possible candidates for BJT-transition and SC are the high temperature cuprate and fulleride superconductors. The cuprates could also be the candidates for the coexistence of all the three phenomena together. Many of the features of the coexistence predicted in the present model calculations indeed show qualitative agreement with the experimental observations for the cuprates, which has been pointed out in [8].

ACKNOWLEDGMENT

One of us (S.N.B.) would like to acknowledge the Indo-French and Indo-U.S. collaboration projects for partial travel support. The Department of Physics, Virginia Commonwealth University in Richmond provided the financial support for a visit during which the paper was written. He would like to acknowledge the excellent hospitality provided by Professor Bijan K. Rao during this visit.

REFERENCES

[1] S.N. Behera, Priyatama Deo, and Haranath Ghosh in *Proceedings of the XX International Workshop on Condensed Matter Theories* held at the University of Poona, India during Dec. 9-14 (1996) (to be published by Nova Science Publishers N.Y. 1997).

[2] S.N. Behera, Haranath Ghosh, Manidipa Mitra, and S.K. Ghatak in *Proceedings of the International Workshop on High Temperature Superconductivity: Ten years after its discovery* held at Jaipur, India during Dec. 16-21, 1996 (to be published by Narosa Publishers, New Delhi, 1997).

[3] G. Grüner, *Density Waves in Solids*, (Addison Wesley, N.Y., 1994).

[4] J. Hubbard, Proc. Roy. Soc. A **276**, 238 (1963).

[5] S.N. Behera, P. Entel, and D. Sa in *Local Order in Condensed Matter Physics*, Eds. S.D. Mahanti and P. Jena (Nova Science Publishers Inc., N.Y. 1995).

[6] Haranath Ghosh, S. Sil, and S.N. Behera, unpublished calculations (1997).

[7] S. Bhattacharya and S.N. Behera Physica C **167**, 112 (1990) Phase Transitions **19**, 15 (1989) and the papers referred to therein.

[8] Haranath Ghosh, S.N. Behera, S.K. Ghatak, and D.K. Ray, Physica C **274**, 107 (1997).

[9] G. Bilbro and W.L. McMillan, Phys. Rev. B **14**, 1887 (1976).

[10] G. Kotliar and A.E. Ruckenstein, Phys. Rev. Lett. 57, 1362 (1986); M. Lavagna, Phys. Rev. B 41, 142 (1990); P. Entel, S.N. Behera, J. Zielinski, and E. Kaufmann Int. J. Mod. Phys. B 5, 271 (1991).

[11] Haranath Ghosh, Manidipa Mitra, S.N. Behera, and S.K. Ghatak, to be published elsewhere. (1997).

AN APPROACH TO NONEXTENSIVE, NONEQUILIBRIUM SYSTEMS: OLD REALIZATIONS AND A NEW DREAM ?

A. K. Rajagopal
Naval Research Laboratory
Washington D. C. 20375 - 5320, U. S. A.

ABSTRACT

There are many novel materials being fabricated (or are being thought of) especially in cluster or nanometric forms. In some of the already known such systems, one has observed persistence of long-time memory and fractal space-time behavior, which may be due to the range of interactions being comparable to the system size. Under these circumstances, the system may be nonextensive and in nonequilibrium situation. We outline here a theoretical approach to understand such systems based on the Lindblad evolution of the density matrix for nonequilibrium state combined with the statistical mechanics of nonextensive system founded on Tsallis entropy. Some examples are cited as illustrations of our scheme.

1. INTRODUCTION

As I was preparing this talk a few weeks back, Professor Bijan Rao faxed me the Scientific Program of this Meeting. I found to my pleasant surprise that the first talk by Professor Morrel Cohen and the last talk by me form a complementary pair! Immediately I called Professor Rao and did not reach him but I redirected my call to Professor Jena, who was as excited about this coincidence as I was and suggested a change in the Title of my Talk, to reflect this complementarity! For me the old realizations of using the known techniques of statistical mechanics and quantum theory in trying to understand phenomena involving nanosystems need change; and the new dream is to present a theory which generalizes the old theories to understand the physics of these materials! I will outline such a theory and try to justify these two statements in the rest of the Talk. This being the last presentation, I will keep it short, because "Brevity is the soul of wit".

Any construction of theory is empty if it is not founded on experimental observations, especially in the study of the physics of materials. The need for new approaches to understand the novel materials is compelling due to the observation of fractally structured space-time dependence of physical properties and persistence of long-time memory (anomalous diffusion [13] in some surface growths, for instance) on the one hand, and the possibility of structural phase transformations in nanometer size crystals (reported by Alivisatos [1]), with implications for preparation of high energy density metastable materials, on the other. Also, nanometric device structures that are being thought of or even being made, are operating at pico and atto second time scales, requiring new paradigms than the classical ones (equivalent circuits with L, C, R) used so far. Many of these properties may be due to the range of interactions being comparable to the system size. Under these circumstances, the system is nonextensive and can be in nonequilibrium conditions. In my opinion, the best approach to these questions is via the study of solutions of the appropriate dynamical equations for the density matrix of the system, given suitable initial conditions. Let me remind you that in the usual conventional situation, one starts with the Liouville-von Neumann (LvN) equation for the density matrix with the initial condition that the system be prepared in some thermodynamic equilibrium (Gibbs) state. The thermodynamic equilibrium itself is specified by the maximum entropy principle subject to known constraints of given physical quantities such as energy, etc. In the study of macroscopic systems, we employ a non responsive heat bath, von Neumann entropy valid for extensive systems, and unitary time evolution of the density matrix. In the next Section, we suggest the use of Tsallis entropy [11] instead, for the nonextensive system. In the third Section, we advocate the use of Lindblad [6, 7] time evolution of the density matrix for dealing with the "metastable" or nonequilibrium aspect, which may be thought of as a manifestation of more dynamic role of the heat bath in producing this nonunitary time evolution. In the final Section, we summarize the paper with some remarks on the actual calculation of system properties in such a scheme.

2. TSALLIS ENTROPY FOR NONEXTENSIVE SYSTEMS

We begin with the introduction of the system density matrix as a hermitian, traceclass (with $\text{Tr}\rho = 1$), positive operator, ρ. In Tables I, II, III we display the definitions and properties of the entropy functionals associated with the systems that are extensive (von Neumann) and nonextensive (Tsallis). The problem of statistical mechanics of nonextensive systems arises whenever we have long range interactions as in gravitational physics or equally, whenever we have the range of interactions exceeds or of the same order as, the system size. The well-known properties of the density matrix are that it describes a pure state if $\rho^2 = \rho$ and a mixture if $\rho^2 < \rho$. From Table I, we note that the nonextensive property is indexed by a parameter q in the definition of the Tsallis entropy, which depends on the range of interactions in the system under consideration. It is a real number. When q=1, we recover the von Neumann entropy valid for extensive systems. In both of these definitions, the pure state is characterized by the corresponding entropies being zero. q is not known a priori. Tsallis makes another change in the usual formulation of the mean value of a physical quantity, so that the known mathematical properties of the entropy such as Legendre transformation, variational principle (e.g., maximum entropy) are retained with the least amount of changes in the interpretation of the formalism. We will see this philosophy of changing the old and tested framework to adopt to the new situations again when we introduce the dynamics of nonequilibrium quantum statistical mechanics in the next section.

Table I. Definitions of entropy

Extensive - von Neumann	Nonextensive - Tsallis [11]
$S = -k_B Tr \rho \ln \rho$	$S_q = \dfrac{k}{(q-1)} Tr(\rho - \rho^q)$, q a real number
$\langle A \rangle = Tr A \rho = mean\ value\ of\ A$	$\langle A \rangle_q = Tr A \rho^q = q\text{-}mean\ value\ of\ A$

Table II shows how close the Tsallis formalism for nonextensive statistical mechanics is to the known framework for extensive systems. In Table III, the important dissimilar properties of the extensive and nonextensive features appearing in the two entropies are displayed. The parameter q in the Tsallis entropy plays an important role as a marker of the nonextensivity and can be related to the range of interactions present in the system under consideration.

3. DYNAMICS OF NONEQUILIBRIUM SYSTEMS - LINDBLAD EQUATION

The Lindblad equation is extensively used in quantum optics (see [8] for references) and in studies of open nanostructures [8], and in discussions of dissipation in quantum systems [8], where this equation is derived from ordinary quantum mechanics for a subsystem interacting with an environment. From considerations of generators of quantum dynamical semigroups, Lindblad [7], and independently and almost simultaneously, from considerations of completely positive dynamical semigroups of N-level systems, Gorini, Kassakowski, and Sudarshan [6] derived what is now commonly known as the Lindblad equation. The Lindblad equation for the density matrix, ρ, under only the very general requirements of linearity, locality in time, and conserving probability (traceclass nature of ρ), thus going beyond the original Markov dynamics, was derived almost a decade after the appearance of [6, 7] by Banks *et al.* [2]. This theory includes the nonequilibrium Markov dynamics recently used by Eyink [5]. A variational approach for this equation has recently been put forward both for the Markov dynamics (Fokker-Planck equation) (Eyink [5]) and for Lindblad dynamics (Rajagopal [10]). In this connection we may also mention that the nonequilibrium Green function methods of Keldysh and Schwinger which were

A.K. Rajagopal

Table II. Similar properties of the entropies

Extensive - von Neumann	Nonextensive - Tsallis [11]
Positively: $S \geq 0$, equality for pure states.	Positively: $S_q \geq 0$, equality for pure states.
The microcanonical ensemble has equiprobability.	The microcanonical ensemble has equiprobability.
S is concave. **Implies maximum entropy principle.**	S_q is concave for $q > 0$, and convex for $q < 0$. **Implies maximum entropy principle for $q > 0$ and minimum entropy principle for $q < 0$.**
For canonical ensemble $\rho = \exp(-\beta H)/Z_1$, $Z_1 = Tr \exp(-\beta H)$, The partition function.	For canonical ensemble, $\rho = Z_q^{-1}[1 - \beta(1-q)H]^{(1/1-q)}$, $Z_q = Tr[1 - \beta(1-q)H]^{(1/1-q)}$, the q - partition function. Note that there is a natural cutoff for $q < 1$.
The Legendre transform structure of free energy defined by $F = -\beta^{-1}\ln Z_1 = U - TS$, $U = Tr H\rho$; $U = \dfrac{\partial}{\partial \beta}(\beta F)$, or $S = -\dfrac{\partial F}{\partial T}$, and $\dfrac{\partial S}{\partial U} = \dfrac{1}{T}$.	The Legendre transform structure of free energy defined by $F_q = -\beta^{-1}\left(\dfrac{Z_q^{1-q}-1}{1-q}\right) = Q_q - TS_q$, $U_q = Tr H\rho^q$; $U_q = \dfrac{\partial}{\partial \beta}(\beta F_q)$,or $S_q = -\dfrac{\partial F_q}{\partial T}$, and $\dfrac{\partial S_q}{\partial U_q} = \dfrac{1}{T}$.
Causality (Kramers-Kronig), time-reversal symmetry, and Onsager reciprocity obeyed by the Kubo dynamic linear response function.	Causality (Kramers-Kronig), time-reversal symmetry, and Onsager reciprocity obeyed by the q-Kubo dynamic linear response function.[9]

Table III. Dissimilar properties of the entropies

Extensive - von Neumann	Nonextensive - Tsallis [11]
Additivity property of S: If $\rho(A \cup B) = \rho(A) \otimes \rho(B)$, then $S(A \cup B) = S(A) + S(B)$.	Additivity property of S_q: If $\rho(A \cup B) = \rho(A) \otimes \rho(B)$, then $S_q(A \cup B) = S_q(A) + S_q(B)$ $+ (1-q)S_q(A)S_q(B)$. Super additive for $q < 1$ (entropy of whole is greater than the sum of its parts) and sub additive for $q > 1$.
Additivity property of mean values: $\langle O \rangle(A \cup B) = \langle O \rangle(A) + <O>(B)$.	Additivity property of mean values: $\langle O \rangle_q(A \cup B) = \langle O \rangle_q(A) + \langle O \rangle_q(B)$ $+ (1-q)[\langle O \rangle_q(A)S_q(B) + \langle O \rangle_q(B)S_q(A)]$.
The Kubo dynamic linear response function obeys the fluctuation-dissipation theorem.	The q-Kubo dynamic linear response function obeys (a different form of) the fluctuation-dissipation theorem. [9]

based on unitary time evolution of LvN framework need to be modified by using the time-dependent density matrix solutions of the equations discussed in this paper. In these works, the importance of forward and backward time evolution was explicitly taken into account to incorporate the time ordering into the framework. In Table IV we exhibit the similarities and differences in the von Neumann and Lindblad evolutions.

It should be remarked that the arbitrariness in the choice of the Q-operators and their associated coefficients have their parallels in the choice of the coupling of the system to the environment (e.g., linear) and its subsequent approximate treatment in obtaining the kernels. Thus we have here a more general framework for exploring a wider class of models for studying dissipation. The environment coupling model is not applicable in systems with long range interactions where we cannot separate out the subsystem, for example, in nanometric systems where the size of the system is smaller than the range of correlations, or as in gravitational systems.

A.K. Rajagopal

Table IV. Comparison of von Neumann and Lindblad evolutions

Unitary Evolution - von Neumann	Nonunitary Evolution - Lindblad [2,7]
$$\frac{d\rho}{dt}=-\frac{i}{\hbar}[H,\rho]\ .$$ ρ is hermitian. The initial condition is usually specified to suit thephysical situation under study. For example, one may choose to maximize entropy (von Neumann or Tsallis) given some physical constraints such as energy.	$$\frac{d\rho}{dt}=-\frac{i}{\hbar}[H,\rho]$$ $$-\frac{1}{2\hbar}\sum_i A^{(i)}\left(Q_i^\dagger Q_i\rho+\rho Q_i^\dagger Q_i-2Q_i\rho Q_i^\dagger\right)\ .$$ The operators Q_i are arbitrary and represent the decay processes. ρ is hermitian iff $A^{(i)}$ are real. The initial condition is usually specified to suit the physical situation under study. For example, one may choose to maximize entropy (von Neumann or Tsallis) given some physical constraints such as energy.
ρ is traceclass. $\mathrm{Tr}\rho(t)=\mathrm{Tr}\rho(0)=1$.	ρ is traceclass. $\mathrm{Tr}\rho(t)=\mathrm{Tr}\rho(0)=1$.
ρ is a positive operator; its eigenvalues are positive.	ρ is a positive operator; its eigenvalues are positive, iff $A^{(i)}$ are positive.
von Neumann entropy $S=-\mathrm{Tr}\rho\ln\rho$ is independent of time and so no possibility of transition from pure to mixed state and vice versa.	von Neumann entropy $S=-\mathrm{Tr}\rho\ln\rho$ depends on time; $\frac{ds(t)}{dt}\geq0$ provided $Q_i^\dagger=Q_i$. Thus possibility of transition from pure to mixed state and vice versa.
Tsallis entropy $S_q=\frac{k}{(q-1)}Tr(\rho-\rho^q)$ is independent of time and so no possibility of transition from pure to mixed state and vice versa.	Tsallis entropy $S_q=\frac{k}{(q-1)}Tr(\rho-\rho^q)$ depends on time; provided, $\frac{dS_q(t)}{dt}\geq0$ for $q\geq1$, or $q<1$, and so possibility of transition from pure to mixed state and vice versa. (AKR,1996,unpublished)
Action principle-variational approach [4]	Action principle-variational approach [10]

4. POSSIBLE APPLICATIONS AND CONCLUDING REMARKS

Let me first make a few general remarks concerning the two topics presented here. First of all, the Tsallis entropy is a generalization of the von Neumann entropy relevant to a much larger class of systems and is rapidly becoming a useful way of examining a large variety of physical and mathematical problems which could not be handled in a coherent theoretical way. For a list of some of these, one may see ReL[3, 9]. The Lindblad equation incorporates dynamics that takes mixed states to pure states and vice versa thus generalizing the von Neumann equation for unitary dynamical evolution. It is used in quantum optics more so than in any other field because one has developed some sense of the choice of the relevant Q operator appropriate to the optics problems, which usually involve a Gaussian fonn of the density matrix in structure. The general method of solving this equation is not known and the variational approach to it has only been given very recently [10]. This equation provides a way of approaching nonequilibrium problems within the quantum mechanical principles. What is needed is a way to understand the types of Q operators to be chosen for the given problem, their physical meaning, etc. The combination of the two proposed here goes one more step in suggesting the Tsallis entropic method be used as providing the initial condition for the solution of the Lindblad equation, so that we have a framework to examine the nonequilibrium and nonextensive systems that are fast becoming objects of interesting experimental studies [1]. Even in the context of Fokker-Planck equations [5], this suggestion ought to be considered.

The types of problems that may be examined in the next century concern with nanosize, very fast (femto- and atto-second) time scale devices. The usual Golden Rule treatments of time- dependent problems become of doubtful significance because non adiabatic processes will play the central role. Moreover, very short time dynamics involving out of equilibrium phenomena will require a framework of the type suggested here. This proposed work is for the near future and I must say, only formal framework is put forward so far. In [10], a Gaussian model of nonisoentropic dynamics is outlined, which is a generalization of isoentropic model for early universe [4] as well as early-time behavior of a model system! From the nature of the framework I have outlined here, it should be clear that this suggestion is interdisciplinary in its structure, ranging from field theory and nuclear physics on one extreme to condensed matter at the other end. This then is the dream!

ACKNOWLEDGMENTS

This work is supported in part by the Office of Naval Research. It is a pleasure to thank Professor Tsallis for reading the draft of this article and making valuable suggestions. It should be mentioned that he updates a compendium of all the works on the Tsallis entropy from time to time, which he has always supplied me generously. I had the pleasure of meeting Dr. Rameswar Bhargava, who informed me of his exciting observations of luminecence from doped nanocrystafline materials which seem to require explanation along the lines outlined here. I thank him for sharing his unpublished work as well as discussions.

References

[1] P. Alivisatos, *Size dependence of structural metastability in semiconductor nanocrystals*, Seminar given at Naval Research Laboratory, October 7, 1996. See also R. N. Bhargava, J. Luminescence **70**, 85 (1996) on new effects observed in Doped Nanocrystalline Materials, possibly due to phenomena addressed in the text.

[2] T. Banks, L. Susskind, and M. E. Peskin, Nucl. Phys. B **244**, 125 (1984).

[3] B. M. Boghosian, Phys. Rev. E **53**, 4754 (1996). (Relaxation in 2-D turbulence using Tsallis statistics)

[4] O. Eboli, R. Jackiw, and So-Young Pi, Phys. Rev. D **37**, 3557 (1988). See also R. Jackiw, *Diverse Topics in Theoretical and Mathematical Physics* (World Scientific, Singapore, 1995); p.283.

[5] G. L. Eyink, J. Stat. Phys. **83**, 955 (1996); Phys. Rev. E **54**, 3419 (1996); F. J. Alexander and G. L. Eyink, Phys. Rev. Lett. **78**, 1 (1997). (Variational approach to (Fokker-Planck type) nonequilibrium equations)

[6] V. Gorini, A. Kassakowski, and E. C. G. Sudarshan, J. Math. Phys. **17**, 821 (1976).

[7] G. Lindblad, Commun. Math. Phys. **48**, 119 (1976).

[8] G. Mahler and V. A. Weberruss, *Quantum Networks* (Springer, New York, 1995).

[9] A. K. Rajagopal, Phys. Rev. Lett. **76**, 3469 (1996). (Dynamic linear response theory)

[10] A. K. Rajagopal, Phys. Lett A **228**, 66 (1997); and paper in preparation (Action principle for Lindblad evolution)

[11] C. Tsallis, J. Stat. Phys. **52**, 479 (1988); E. M. F. Curado and C. Tsallis, J. Phys. A **24**, L69 (1991); **24**, 3187 (1991); **25**, 1019 (1992); C. Tsallis, Phys. Lett. A **206**, 389 (1995). (Theoretical foundations)

[12] C. Tsallis, S. V. F. Levy, A. M. C. Souza, and R. Maynard, Phys. Rev. Lett. **75**, 3589 (1995). (Statistical foundation of ubiquity of Lévy distributions in nature)

[13] C. Tsallis and D. J. Bukman, Phys. Rev. E **54**, R2197 (1996). (Anomalous diffusion)

PARTICIPANTS

S. C. Agrawal
Physics Department
IIT, Kanpur
Kanpur 16, U. P.
India
sca@iitk.ernet.in

J. R. Anderson
Dept. of Physics
University of Maryland
College Park, MD 20742
U.S.A.
ja26@umail.umd.edu

J. Barth
Inst. de Physique Experimentale
EPFL PHb-Ecublens
CH-1015 Lausanne
Switzerland
johannes.barth@ipe.dp.epfl.ch

C. Basu Choudhury
S.N.Bose National Centre for Basic Sciences
Block JD, Sector III, Salt Lake
Calcutta 91
India
chhanda@bose.ernet.in

Inder P. Batra
PHP 802, IBM Almaden Research Centre
650 Harry Rd.
San Jose, CA 95120
U.S.A.
ipbatra@almaden.ibm.com

S. N. Behera
Institute of Physics
Bhubaneswar 751 005
India
snb@iop.ren.nic.in

R. N. Bhargava
P.O. Box. 820
Briarcliff Manor
N.Y. 10510
U.S.A.
amib2@aol.com

P. P. Biswas
S.N.Bose National Centre for Basic Sciences
Block JD, Sector III, Salt Lake
Calcutta 91
India
ppb@bose.ernet.in

C. Bréchignac
Laboratoire Aimé Cotton
CNRS - Campus d'Orsay
91405 Orsay Cedex
France

J. Q. Broughton
Naval Research Laboratory, Code 6690
4555 Overlook Avenue SW
Washington, DC 20375-5320
U.S.A.
broughto@dave.nrl.navy.mil

M. Broyer
Univ. Claude Bernard-Lyon I
Lab. de Spect. Ionique et Mol.
43 Blvd. du 11 November 1918
69622 Villeurbanne Cedex
France
broyer@cognac.univ-lyon1.fr

S. L. Chaplot
Solid State Physics Division
Bhabha Atomic Research Centre
Bombay 400 085
India
chaplot@magnum.barc.ernet.in

Morel H. Cohen
Exxon Research & Engg.
Clinton Township, Route 22
E. Annandale, NJ 08801
U.S.A.
makane@erenj.com

Amal K. Das
Institute of Physics
Bhubaneswar 751 005
Orissa
India
amal@iop.ren.nic.in

G. P. Das
Bhabha Atomic Research Centre
Mumbal 400 085
India
gpd@magnum.barc.ernet.in

M. P. Das
Dept. of Physics
Australian National University
Canberra ACT 0200
Australia
mpd105@phy.anu.edu.au

N. Das
S.N.Bose National Centre for Basic Sciences
Block JD, Sector III, Salt Lake
Calcutta 91
India
nitya@bose.ernet.in

Sashikala Das
Ohio
USA
sdas@desire.wright.edu

T. P. Das
Dept. of Physics
State University of New York
Albany, NY 12222
U.S.A.
tpd56@csc.albany.edu

P. H. Dederichs
Institut für Festkörperforschung
Forschungzszentrum Jülich
D-52425 Jülich
Germany
I.gerken@kfa-juelich.de

B. N. Dev
Institute of Physics
Bhubaneswar 751 005
Orissa
India
dev@iop.ren.nic.in

Soma Dey
Institute of Physics
Bhubaneswar 751 005
Orissa, India
soma@iop.ren.nic.in

R. P. Dutta
S.N.Bose National Centre for Basic Sciences
Block JD, Sector III, Salt Lake
Calcutta 91, India
radhika@bose.ernet.in

M. H. Engineer
Bose Institute
A.P.C. Road
Calcutta 9
India
meher@boseinst.ernet.in

Peter Entel
Dept. of Theoretical Physics
Gerhard-Mercater University, Duisberg
47048 Duisburg, Germany
entel@thp.Uni-Duisburg.de

S. Ghatak
Physics Department
IIT, Kharagpur
Kharagpur, W. Bengal
India
skghatak@phy.iitkgp.ernet.in

Haranath Ghosh
Institute of Physics
Bhubaneswar 751 005
Orissa
India
hng@iop.ren.nic.in

Sanjit Ghosh
Institute of Physics
Bhubaneswar 751 005
Orissa, India
sanjit@iop.ren.nic.in

S. Ghosh
S.N.Bose National Centre for Basic Sciences
Block JD, Sector III, Salt Lake
Calcutta 91, India
subhra@bose.ernet.in

Bikash C. Gupta
Institute of Physics
Bhubaneswar 751 005
Orissa
India
bikash@iop.ren.nic.in

L. C. Gupta
Tata Institute of Fundamental Research
Mumbai 400 005
India
lcgupta@tifrvax.tifr.res.in

G. C. Hadjipanayis
Dept. of Physics and Astronomy
University of Delaware
Newark, DE 19716-2570
U.S.A.
hadji@strauss.udel.edu

W. Hergert
Fachnberich Physik
Martin Luther Universitat
Fr-Bach Plaz 6
0-06099 Halle, Germany
hergret@physik.uni-halle.de

K. P. Jain
Dept. of Physics
Indian Institute of Technology
Delhi
India
kpjain@physics.iitd.ernet.in

J. Jellinek
Chemistry Division
Argonne National Laboratory
Argonne, IL 60439 , U.S.A.
jellinek@anlchm.chm.anl.gov

P. Jena
Physics Department
Virginia Commonwealth University
Richmond VA 23284-2000
U.S.A.
jena@gems.vcu.edu

T. Katsuyama
Central Research Laboratory
Hitachi Ltd.
Kokubunji, Tokyo 185
Japan
katsuyama@crl.hitachi.co.jp

E. Kaxiras
Dept. of Physics
Harvard University
Lyman Lab., 324A
Cambridge, MA 02138, U.S.A.
kaxiras@cmtek.harvard.edu

A. P. Kulshreshtha
Dept. of Science and Technology
Government of India
New Delhi 110 016, India
apk@alpha.nic.in

Vijay Kumar
Materials Science Division
IGCAR
Kalpakkam 603 102
India
kumar@igcar.ernet.in

Goutam Kuri
Institute of Physics
Bhubaneswar 751 005
Orissa, India
kuri@iop.ren.nic.in

S. D. Mahanti
Physics Department
Michigan State University
East Lansing, MI 48824
U.S.A.
mahanti@pa.msu.edu

M. J. Manninen
Physics Department
University of Jyvaskyla
P.O.Box 35 SF-40351 Jyvaskyla
Finland
manninen@jyflas.jyu.fi

T. P. Martin
Max Planck Inst. FKF
Helsenberg - Str. 1
Stuttgart 70569
Germany
martin@vaxff3.mpi-stuttgart.mpg.de

Albert Masson
ENSCP - II
13 Rue Ernet Cresson
Paris 75014
France

J. W. Mintmire
Naval Research Laboratory, Code 6119
4555 Overlook Avenue SW
Washington, DC 20375-5320
U.S.A.
mintmire@alchemy.nrl.navy.mil

S. G. Mishra
Institute of Physics
Bhubaneswar 751 005
Orissa
India
mishra@iop.ren.nic.in

Manidipa Mitra
Institute of Physics
Bhubaneswar 751 005
Orissa, India
manidipa@iop.ren.nic.in

A. Mookerjee
S.N.Bose National Centre for Basic Sciences
Block JD, Sector III, Salt Lake
Calcutta 91
India
abhijit@bose.ernet.in

R. N. Mukherjee
Institute of Physics
Bhubaneswar 751 005
Orissa
India
rabi@iop.ren.nic.in

S. Mukherjee
Saha Institute of Nuclear Physics
Calcutta 64
India
sugata@hp2.saha.ernet.in

David Nagel
Naval Research Laboratory, Code 6600
4555 Overlook Avenue SW
Washington, DC 20375-5320
U.S.A.
nagel@dave.nrl.navy.mil

K. K. Nanda
Institute of Physics
Bhubaneswar 751 005
Orissa
India
nanda@iop.ren.nic.in

P. Nayak
Dept. of Physics
Sambalpur University
Orissa
India

H. C. Padhi
Institute of Physics
Bhubaneswar 751 005
Orissa, India
hcp@iop.ren.nic.in

A. K. Pal
Dept. of Materials Science
IACS, Jadavpur
Calcutta 32
India
msakp@iacs.ernet.in

D. Papaconstantopoulos
Naval Research Laboratory, Code 4693
4555 Overlook Avenue SW
Washington, DC 20375-5320
U.S.A.
papacon@dave.nrl.navy.mil

G. Pari
S.N.Bose National Centre for Basic Sciences
Block JD, Sector III, Salt Lake
Calcutta 91
India
pari@bose.ernet.in

Biresh Patel
Institute of Physics
Bhubaneswar 751 005
Orissa, India
patel@iop.ren.nic.in

K.G.Prasad
Tata Institute of Fundamental Research
Bombay 400 005
India
kgp@tifrvax.tifr.res.in

B. Prevel
Univ. Claude Bernard-Lyon I
Dept. Physique des MatériAux
43 Blvd. du 11 November 1918
69622 Villeurbanne Cedex
France
prevel@dpm.univ-lyon1.fr

S. B. Raj
Institute of Physics
Bhubaneswar 751 005
Orissa, India
raj@iop.ren.nic.in

A. K. Rajagopal
Naval Research Laboratory, Code 6860
4555 Overlook Avenue SW
Washington, DC 20375-5320
U.S.A.
rajagopal@estd.nrl.navy.mil

B. K. Rao
Physics Department
Virginia Commonwealth University
Richmond VA 23284-2000
U.S.A.
brao@gems.vcu.edu

K. V. Rao
Dept. of Condensed Matter Physics
Royal Institute of Technology
Stockholm
Sweden S10044
rao@cmp.kth.se

B. B. Rath
Associate Director
Naval Research Laboratory, Code 6000
4555 Overlook Avenue SW
Washington, DC 20375-5320, U.S.A.
rath@anvil.nrl.navy.mil

G. V. Raviprasad
Institute of Physics
Bhubaneswar 751 005
Orissa, India
ravi@iop.ren.nic.in

D. Sahoo
Institute of Physics
Bhubaneswar 751 005
Orissa
India
dsahu@iop.ren.nic.in

S. N. Sahu
Institute of Physics
Bhubaneswar 751 005
Orissa
India
sahu@iop.ren.nic.in

T. Sahu
Dept of Physics
Berhampur University
Orissa
India
ts@bpuruniv.ernet.in

Umesh A. Salian
Institute of Physics
Bhubaneswar 751 005
Orissa
India
umesh@iop.ren.nic.in

R. R. dos Santos
Inst. de Fisica
Univ. Fed. Fluminense
Campus de Praia Ver, etha
AV Lotpraneastr 24210-340
Niteroi RJ
Brazil
rrds@if.uff.br

B. Sanyal
S.N.Bose National Centre for Basic Sciences
Block JD, Sector III, Salt Lake
Calcutta 91
India
biplab@bose.ernet.in

S. N. Sarangi
Institute of Physics
Bhubaneswar 751 005
Orissa
India
sarangi@iop.ren.nic.in

Subir Sarkar
School of Physical Sciences
Jawaharlal Nehru University
Delhi 110 067
India
sarkar@jnuniv.ernet.in

K. Sekar
Institute of Physics
Bhubaneswar 751 005
Orissa
India
sekar@iop.ren.nic.in

S. Sil
Institute of Physics
Bhubaneswar 751 005
Orissa
India
kanta@iop.ren.nic.in

R. K. Soni
Physics Department
IIT, Delhi
Delhi 16
India
ravisoni@physics.iitd.ernet.in

G. M. Stocks
171 Whippoorwill Drive
Oak Ridge TN 37830
U.S.A.
gms@gmsis.ms.ornl.gov

K. R. Subbaswamy
Department of Physics
University of Kentucky
Lexington, KY 40511-8433
U.S.A.
swamy@pa.uky.edu

B. Sundravel
Institute of Physics
Bhubaneswar 751 005
Orissa, India
sundar@iop.ren.nic.in

R. Superfine
Dept. of Physics
University of North Carolina
CB 3255, Phillips Hall
Chapel Hill, NC 27599-3255, U.S.A.
rsuper@physics.unc.edu

S. Tripathi
Tata Institute of Fundamental Research
Bombay 400 005
India
tripathi@tifrvax.tifr.res.in

D. N. Tripathy
Institute of Physics
Bhubaneswar 751 005
Orissa
India
dnt@iop.ren.nic.in

G. S. Tripathy
Dept. of Physics
Berhampur University
Orissa
India

Shikha Verma
Institute of Physics
Bhubaneswar 751 005
Orissa
India
shikha@iop.ren.nic.in

A. F. Voter
T12, B268, Los Almos National Laboratory
P.O.Box 1663
Los Almos, NM 87545
U.S.A.
afv@gold.lanl.gov

G. Wright
Naval Research Laboratory
4555 Overlook Avenue SW
Washington, DC 20375-5320
U.S.A.
gwright@smap.org

AUTHOR INDEX

Subject Index

Subject Index